GEODETIC NETWORK ANALYSIS AND OPTIMAL DESIGN: CONCEPTS AND APPLICATIONS

by

Shanlong Kuang, Ph.D.

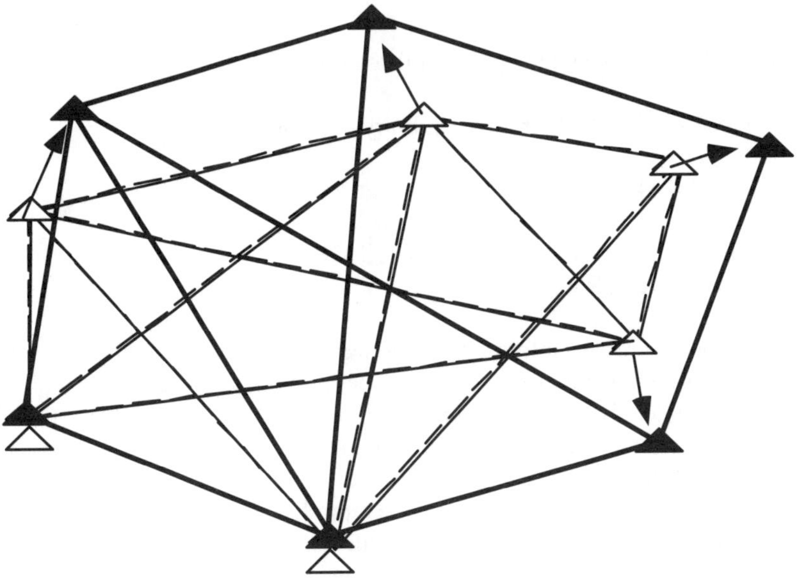

Ann Arbor Press, Inc.
Chelsea, Michigan

Library of Congress Cataloging-in-Publication Data

Catalog record is available from the Library of Congress.

ISBN 1-57504-044-1

ANN ARBOR PRESS, INC.
121 South Main Street, Chelsea, Michigan 48118

Printed in the United States of America
1 2 3 4 5 6 7 8 9 0

About the Author

Dr. Shanlong Kuang holds B.Sc.E. and M.Sc.E degrees in surveying engineering from Wuhan Technical University of Surveying and Mapping in P.R. China, and a Ph.D. from the University of New Brunswick in Canada. His current major areas of expertise include geodetic network analysis and optimization; precision engineering surveys; deformation monitoring and analysis; GPS applications; optimal estimation; and applied geodesy for engineering and geoscience projects, etc. He is the author of numerous technical research articles on geodetic engineering and surveying published in quality geodetic journals worldwide. In April 1994 he was awarded a postdoctoral research fellowship by the Natural Sciences and Engineering Research Council (NSERC) of Canada in recognition of his academic achievements. In June 1995 obtained an honorable mention award of The Institution of Surveyors, Australia, R.D. Steele Prize for his article, "On Optimal Design of Leveling Networks" published in the journal *The Australia Surveyor.* During 1992 to 1993 he was a lead geodesist on the Superconducting Super Collider project in Dallas, Texas, designing and implementing control surveys to guide the driving of a 54 mile underground tunnel of oval shape, and for the setting out and alignment of accelerators of high energy physics. Dr. Kuang worked as Director of Geodetic Surveys at ASC American Surveying Consultants, P.C., in Illinois, 1994–1996. He is currently President of GPS Surveying Systems, Inc., in Chicago, Illinois.

About the Author

Foreword

This book presents a systematic approach to its topic: *geodetic network analysis* and *optimal design*, which are two of the most important processes in establishing a geodetic network. Conceptually, network optimal design deals with how to develop a most economic survey campaign that will guarantee to meet the preset project criteria prior to entering the field, thus preventing the survey project from failing, while network analysis is concerned with how to process and analyze the collected survey data after the network has been measured, in order to produce reliable network results. Although these activities have been in practice for centuries, it is only in the last two decades or so that significant progress in both theoretical development and practical applications has been made. Numerous books in geodesy and surveying have maybe a chapter or two on the subjects, but so far, to the best knowledge of the author, none is devoted solely to them. This book is designed to make amends for that situation. It covers the optimal design and analysis of both geodetic positioning networks and networks established for deformation monitoring purposes.

The author has seen an urgent need for a systematic discussion of the concepts and techniques of network analysis and optimal design to provide an up-to-date knowledge of the field and promote their practical applications in a way that is acceptable to surveying professionals from a variety of backgrounds. The reason is that although some of the concepts and methods presented have been under development for more than two decades, and some have just been accomplished recently, they are still not very popular and are mainly comprehensive to the small majority of the surveying profession; that is, the researchers or experts in the academic environment. This may partially be due to some sophisticated theoretical background involved. Since the majority of the surveying profession, i.e., the practicing surveyors/engineers, usually have not much or limited interest in a profound understanding of the entire theory behind the proposed techniques, what they need are simple examples and a good understanding of the physical meaning of the concepts and procedures involved. Furthermore, with the development of advanced surveying instruments and methodologies; e.g., an automated surveying system such as a total station and the NAVSTAR Global Positioning System (GPS) technique, the traditional demands on a surveyor being able to make observations quickly and accurately have been diminished, and the focus has been placed on his ability to analyze the quality of the data, and process them to produce reliable results. That is, emphasis of the surveying profession has been transferred from observation to survey design, data processing, and analysis in order to produce reliable results. Sophisticated network design and analysis techniques that might have been considered a "luxury" for routine survey practice and were mainly limited to experts are now becoming an indispensable tool for the design and computation of an efficient control survey. The cost of employing experts to analyze the data is high. Even with the availability of commercial computer software packages for network design and adjustment, which may greatly facilitate these activities on a surveyor's part, a good understanding of the terminology, concepts, and procedures involved is, however, instrumental in properly interpreting the results and therefore successfully designing and/or analyzing a survey campaign.

Development of network optimal design and analysis techniques requires a sound knowledge of mathematical calculus, matrix algebra, mathematical statistics, and geodesy. This book concentrates on explaining the concepts and the step-by-step procedures involved. Theoretical discussions and complex derivations of formulae, whenever they do not add to the understanding of the concepts, have been ignored. Continuous reference is, however, made throughout the text to the literature for readers who want to study the theoretical backgrounds. This book should prove to be a valuable tool for readers from a variety of backgrounds in the surveying profession, including practicing surveyors/engineers, university undergraduate and graduate students, software developers of surveying, etc. It is also a starting point for researchers looking for new developments.

Shanlong Kuang, Ph.D.

Introduction

"Geodetic positioning" is the determination of the coordinates of one, two, or a group of three or more points on land, at sea, or in space with respect to a predefined coordinate system, and it is accomplished by making measurements that link the unknown point(s) to points with known coordinate values that could be either terrestrial points or extraterrestrial objects, such as stars or satellites, or both. Depending on the number of unknown points under study, geodetic positioning is classified into *point positioning*, *relative positioning*, and *network positioning,* which deal with determining the coordinates of one point, the relative location of one point with respect to another, and the relative locations among a group of three or more points, respectively (cf. Vanicek and Krakiwsky, 1986). A geodetic network is therefore defined as being any geometric configuration of three or more terrestrial survey points that are connected either by geodetic measurements made among themselves, such as horizontal directions, angles, azimuths, spatial distances, etc., or by astronomical or space techniques; for instance, the NAVSTAR Global Positioning System (GPS), or both. Point positioning and relative positioning may be considered special cases of network positioning.

With the development of science and technology, a wide variety of disciplines have prompted the need for a network of appropriately distributed points (geodetic control) of known horizontal and/or vertical coordinates. As discussed in the works of Vanicek and Krakiwsky (1986), among others, the main areas where geodetic control networks are needed include mapping, boundary demarcation (i.e., positioning and staking out of international, provincial, and local boundaries), urban management, engineering projects (e.g., to lay out the various components under construction in predesigned locations), geographic information system (GIS), hydrography, environmental management, ecology, earthquake-hazard assessment, aerial photography, space research, astronomy, geophysics, etc. A common feature for all these applications is that a framework of points with known horizontal and/ or vertical coordinates with respect to a predefined coordinate system(s) is required to provide position reference. A geodetic network is also an invaluable tool for deformation monitoring. Repeated geodetic measurements on a network of points have been used to monitor and/or predict crustal movements, slope creep studies, ground subsidence, and the deformation of man-made structures such as water power dams, underground tunnels, bridges, high buildings, etc.

Depending on their extent, geodetic networks may be categorized as being of a local, regional, continental (national), or global scale. Here we concentrate on networks of local or regional scales. In general, the establishment of a geodetic network involves the following three stages:

1. Network design
2. Execution (i.e., the field campaign), and
3. Network analysis.

First, *network design* answers the essential questions of where the network points should be placed (i.e., to develop a network configuration) and how the network should be measured (i.e., to develop an observing plan) in order to achieve the required network quality, which is usually set by the network user (i.e., the client), with a reasonably low cost. It is carried out prior to entering the field to ensure that the resulting network will meet the project criteria. Second, the *execution* process turns a designed network into reality. It deals with both the monumentation of the proposed network stations and the actual field measurement techniques. Third, after a network has been measured, *network analysis* deals with the processing and analysis of the collected geodetic data and subsequently reporting the network results together with their quality to the agency that requests the control.

This book deals with the first and the last process, described above, in establishing a geodetic network; that is, network design and network analysis, which are contained in Part I and Part II of this book, respectively. For discussion on monumentation and geodetic measurement techniques, the reader is referred to Kahmen and Faig (1988), Cooper (1982), and Richardus (1984), among others. Logically, network design should be discussed first. However, since network design requires some input of the network analysis techniques, the latter comes first instead. The step-by-step procedures of network analysis are outlined below:

- Accuracy analysis of observations
- Observation data pre-processing
- Pre-adjustment data screening
- Least squares network adjustment
- Post-adjustment data screening
- Quality analysis of the results, and finally
- Reporting network results and their quality to the user, i.e., the agency requesting the control.

The above procedures, except for the final project reporting, are treated in Chapters 1 to 6, respectively, and Chapter 7 deals with the analysis of a geodetic network established for deformation monitoring purposes. First, accuracy analysis of observations (Chapter 1) is to determine the appropriate variance of a geodetic observable (i.e., a proposed measurement) or an actual measurement according to the instrument and the observational procedures used to obtain the measurement. Good knowledge of the achievable accuracy of an observable is very important at the stage of network design to evaluate the possible effects of the instrument and/or the environment on the measuring process and subsequently on the network results. Whereas, after a survey campaign has been completed, good knowledge of the actually achieved accuracy of the survey is crucial for a proper statistical assessment of the results. Second, observation data pre-processing (Chapter 2) deals with correcting the raw observations for any possible meteorological and/or instrumental effects, and further, depending on the coordinate system one has chosen for coordinate computation, the observations may have to be reduced to a three-dimensional geodetic reference frame, or onto the surface of a reference ellipsoid, or to a conformal mapping plane to account for the effects of the variations of the earth's gravity field and/or the geometry of the earth. A least squares network adjustment (Chapter 4) is required for an overdetermined network in which the total number of independent observations is greater than that of independent unknowns. Since any overdetermined network has more than one unique solution for the coordinates of unknown points, in order to obtain the best solution based on all of the available information, an additional optimum criterion must be imposed, and the most frequently used criterion in geodetic science is the least squares criterion that proposes to minimize the weighted sum of squares of inconsistencies among observations. Due to the high sensitivity of the least squares estimation method to gross errors and/or systematic effects existing in the observation data, however, both the pre-adjustment and post-adjustment data screening techniques (Chapters 3 and 5) have been developed and they must be applied to detect their existence and eliminate them from the data in order to produce reliable results. The former is applied prior to the network adjustment to detect gross errors of large magnitude mainly based on checking condition closures, whereas the latter concentrates on detecting and eliminating gross errors of marginal magnitude after a least squares network adjustment based on statistical testing of the estimated observational residuals. Finally, after all the statistical tests are passed, the system is considered to be free from significant gross errors, and the network results are considered reliable. The quality of the network results (Chapter 6) is then evaluated and presented together with the network results to the network user; that is, the agency that requests the control.

The concepts and procedures of both geodetic and deformation *network optimal design* or *network optimization* are discussed in Chapters 8 to 11. Network design is the first step towards establishing a geodetic network. The underlying reason for this is that in order to prevent the whole operation from failing, the surveyors/engineers in charge should know about the result of their work according to the preset objectives before any measurement campaign is started. As is well known, at the design stage of a geodetic network, the fundamental problem that a surveying engineer faces is how to decide on its configuration, i.e., the point location, how to choose the types of geodetic observations, and how to measure the network. The purpose of *network optimal design* or *network optimization* is therefore to design an optimum network configuration and an optimum observing plan in the sense that they will satisfy the preset network quality requirements at a minimum cost (cf. Grafarend, 1974; Cross, 1985; Schmitt, 1985; Schaffrin, 1985b; Kuang, 1991, etc.). Of interest to surveying engineers is that it enables surveyors to avoid unnecessary observations, and therefore may result in saving considerable time and effort in the field while adhering to the required network quality. Furthermore, in regard to the various aspects of network analysis described above, an optimized surveying scheme will help in identifying and eliminating gross errors in observations as well as minimizing the effects of undetectable residual gross errors on unknown parameters, leading to an enhanced network reliability.

Mathematically, optimization of geodetic networks means minimizing or maximizing an objective function that represents the "quality of the network." For geodetic positioning networks, three general criteria are used to evaluate this quality; that is, precision, reliability, and economy. In the case of deformation monitoring networks one more quality measure has to be added; that is, the sensitivity. Procedures of network optimization include:

- Setup of network quality criteria
- Mathematical modeling of design problems, and
- Solution of the mathematical models.

A review of the network design problems and the development of their solution methods is first given in Chapter 8. The mathematical setup of the various network quality criteria, i.e., the objective functions, which will be used in the mathematical modeling of network optimization problems is then discussed in Chapter 9. Chapters 10 and 11 deal with the formulation and solution of mathematical models for the optimal design of geodetic positioning and deformation monitoring networks, respectively. Finally, some optimization examples are given in Chapter 12 that demonstrate in detail how the optimization methodology is applied in practice.

Contents

PART I: NETWORK ANALYSIS

Part I

Network Analysis

1 Accuracy Analysis of Observations

Accuracy and precision are statistical terms. Conceptually, accuracy refers to the degree of closeness of an estimate to its true value, while precision refers to the degree of closeness of observations to their means. In the surveying community, however, these two terms are often mixed, and the term "accuracy" is prevailingly used as a substitute for precision. This is true when observations are free of blunders and systematic biases. Quantitatively, precision is expressed by variance, which is defined as the mean of squares of the differences between measurements of a random variable and its true value.

This chapter discusses the determination of the precision of an observable (i.e., a proposed observation) or an observation according to the instrument(s) and observational procedures that have been proposed for use or have been used to obtain the measurement. In this context, usually two different procedures are distinguished; i.e., the *a priori* accuracy analysis and the *a posterior* accuracy analysis. The a priori accuracy analysis is made before any measurement is made by analyzing the contributions of all the possible error sources involved in the survey, while the a posterior accuracy analysis is performed after a redundant set of measurements have been made to evaluate the actually achieved accuracy of the survey based on only the observational values and the geometry of the surveyed network. In either case, it is assumed that the measurement is subject to the effects of only random errors; i.e., any gross errors and systematic effects are assumed to have been removed by either observing or mathematical procedures or a combination of the two. In this case, the word "accuracy" can be used as a substitute for "precision." These two terms are, therefore, used in our context without distinction.

The a priori knowledge of the accuracy of the proposed observations is needed at the network design stage in order to study how the network and the final results are influenced by both the instruments and the environmental conditions, while the purpose of the a posterior accuracy analysis is to provide reliable information about the variances and covariances (if applicable) of all the different types of observations obtained in a survey campaign in order to perform a successful estimation of the unknown parameters and the network quality analysis. This chapter starts with a review of the observation errors in surveying projects in §1.1, and the techniques for the a priori and a posterior accuracy analysis are discussed in §1.2 and §1.3, respectively.

1.1 An Overview of Observation Errors in Surveying Projects

Observation errors are inherently associated with any measurement activity. Their causes may be classified into internal and external factors. The former refers to the instrumental factors and the observer's human limitations, and the latter accounts for the unpredictable and uncontrollable effects of the environment in which the measuring process takes place. First of all, no surveying instrument can ever be made perfectly in construction. Examples include: any linear or circular scales are graduated evenly only to a certain extent due to physical limitations, and the optical center of an EDM (Electromagnetic Distance Measuring device) can hardly be made to coincide with the survey mark. On the other hand, lack of proper adjustment of any survey equipment according to the manufacturer's instructions before their use for data acquisition will certainly introduce errors in the measurements. The environmental factors that may affect a measuring process can be numerous, the most popular of which is the atmospheric refraction, which is caused by changes in temperature, humidity, and atmospheric pressure in the environment where a measuring process takes place. Atmospheric refraction causes the bending of light rays, and it, therefore, will affect any optical and/or electro-optical measuring process. Finally, personal errors are inevitably introduced by the observer due to the limitations of humans' vision and hearing abilities whenever they are involved in a measuring process.

Conventionally, observation errors are classified into random errors, gross errors, and systematic errors. At first, ***random errors*** are errors the occurrence of which does not follow a deterministic pattern. In mathematical statistics, they are considered as stochastic variables, and despite their irregular behavior, the study of random observational errors in any well-conducted measuring process or experiment has indicated that random observational errors follow the following empirical rules:

1. A random observational error will not exceed a certain amount.
2. Positive and negative random observational errors may occur at the same frequency.
3. Errors that are small in magnitude are more likely to occur than those that are large in magnitude.
4. The mean of random errors tends to zero as the sample size tends to infinite.

In mathematical statistics, the above observations form the basis for inferring that random observational errors follow the normal or Gaussian distribution. This fundamental result was in fact obtained by Gauss (1809), and has been elaborated on in practically every statistics textbook dealing with observational errors. The frequency with which a true error Δ occurs is usually described by the so-called histogram that, for Gaussian distribution, takes the shape as shown by Figure 1.1, in which the abscissa represents the magnitude of random errors that are sorted into groups of error Δ, and the height of the rectangles plotted for each group is proportional to the number of errors within the respective group. The smoothed step function of the rectangles of the histogram represents the theoretical error distribution curve that has been derived by Gauss as follows:

$$\phi(\Delta) = \frac{\sigma}{\sqrt{\pi}} e^{-\sigma^2 \Delta^2} \qquad (1.1)$$

where $\phi(\Delta)$ represents the relative frequency, e the exponential function, and σ a constant related to the measuring accuracy, respectively.

The effects of random observational errors on the measurements can be rigorously accounted for by adopting a proper statistical estimation method; for instance, the "least squares method," in the data post-processing stage, as discussed later in Chapter 4.

Gross errors or blunders are usually considered as errors of large magnitude, and are mainly caused by an observer's mistakes; failure of equipment; misinterpretation; etc. For instance, the wrong reading and/or wrong recording of a measurement, or the wrong sighting of a target will certainly produce wrong results. The effects of gross errors are of local nature, and since gross errors are not random variables, they cannot be rigorously accounted for by statistical methods. Due to these reasons, in any standard survey practice, proper observational procedures have been designed to allow for the blunders to be detectable and thus rejected either in the field measuring process or in the office data processing stage as completely as possible. For instance, repeating the measurement of a distance, an angle, or an azimuth several times allows for the computation of a mean value, and any single observation that significantly differs from its mean value indicates the existence of a blunder. Occupying a station and/or sighting a target several times and making redundant observations by closing a geometrical and/or algebraic figure are also good means for detecting any potential blunders. Nevertheless, it should be noted that despite the design precautions, some blunders, especially those of moderate or small magnitude as compared with the measurement precision, may still remain. Their detection and rejection should be carried out following the statistical testing theory and other techniques as discussed later in Chapters 3 and 5.

Unlike random errors and blunders, ***systematic errors*** or ***systematic effects*** possess a certain pattern in their signs and magnitudes. Their existence would introduce global variations and therefore they may be modeled by a mathematical function that may take the form of a "Constant" (i.e., the value and sign remain the same all through the measuring process), a linear function, or a nonlinear function, depending on the nature of the systematic errors under study. A good example for this is that the "Zero Error," "Scale Error," and "Cyclic Error" of an EDM instru-

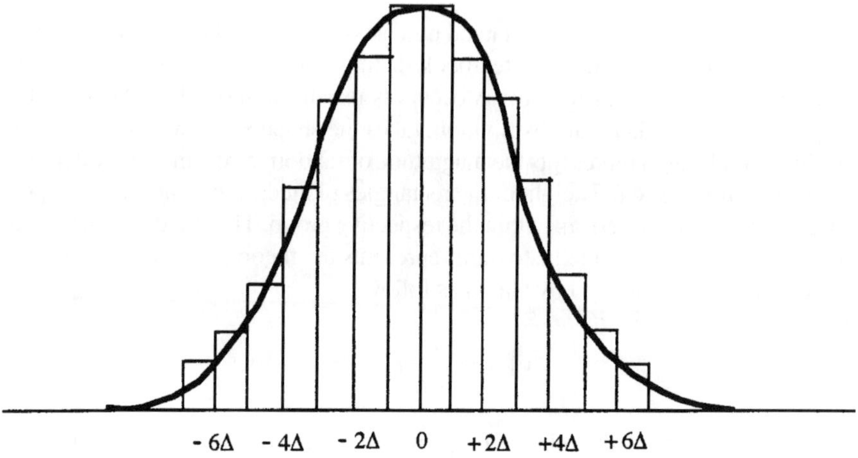

Figure 1.1. Histogram of observations

ment is a constant, a linear function, and a Sine function, respectively (cf. Chapter 2). Theoretically, systematic errors indicate an inconsistency between the observations and the functional model.

Despite the fact of the modelability of the effects of systematic errors, it is not common in practice to extend the classical observational models to include parameters for systematic errors, although some models have been proposed; for instance, in photogrammetry for bundle adjustment extending the collinearity equations by including additional parameters that model the lens distortions (Moniwa, 1977). The reason for this is that in most cases the factors causing systematic errors are not fully known, and therefore it is very difficult, if not impossible, to find an analytical function that could adequately represent the effects of all systematic errors for a certain measuring process under study. Therefore, according to the causes of systematic errors, in order to minimize their effects in a measuring process, it is common practice to, first of all, perform proper instrument calibration before its use for data acquisition. Based on the performance of a specific measuring instrument, some systematic effects may be determined a priori by designing some certain measuring and computation procedures and then applied to the raw observations after data acquisition (for instance, the aforementioned "Zero Error," "Scale Error," and "Cyclic Error" for an EDM instrument). The effects of systematic errors due to environmental factors may also be accounted for by formulating corrections to the raw observations, and they may also be eliminated or greatly reduced by designing a proper observational procedure. Good knowledge of the measuring process, its functional model, and the environment is certainly crucial to its success. For instance, different observational procedures have been proposed for the measurement of horizontal directions, geodetic leveling, and traversing in a tunnel, etc., in order to eliminate or minimize the effects of atmospheric refraction on these geodetic measuring processes, although these effects may be determined a priori according to some well-defined propagation models of electromagnetic waves to apply corrections to the raw observations. In designing observational procedures, it should be noted that systematic errors may not be detected by having repetitive measurements since they may affect the repeated observations in much the same way. One should also keep in mind that systematic errors may by no means be eliminated completely. Residual systematic effects may be treated as randomized errors, and their existence may also be analyzed in the data post-processing stage by statistical method.

Finally, it should be noted that there are cases where clear distinction between local systematic errors and gross errors of small magnitude cannot be made. Both kinds of errors have the same effect on the observations, and that is also applicable to the large random errors and very small gross errors.

1.2 The a Priori Accuracy Analysis of Observations

Before performing any actual measurements, surveyors may usually acquire information about the expected precision of the proposed observations from the precision of the instruments as claimed by the manufacturers who usually determine their accuracy through experimental measurements. It should be noted, however, that the nominal accuracy of an instrument claimed by the manufacturer represents in general an average situation and may significantly differ from the actual situation under which observations are made.

The a priori accuracy estimation of a type of observable can be done using the following procedures:

1. Understand the principle/procedure of the data acquisition.
2. Identify all the possible error sources, both random and systematic error sources.
3. Analyze the effects of each individual random and systematic error on the raw data. Major parts of the effects of systematic errors will have to be removed by applying corrections to the raw data, and the residual systematic effects, which are caused by the uncertainty in the determination of the systematic errors, are considered randomized errors.
4. Compute the total effect of all the random errors and residual systematic errors on the observable by the law of random error propagation. For instance, if we assume that the effects of all the different types of errors on a type of observable under study are statistically independent, the variance of the observable would be the sum of the squares of each individual effect.

There are a variety of types of observables that can be used for the purpose of geodetic network positioning and land surveying. These observables may be divided into two categories; i.e., the conventional terrestrial geodetic observables, such as azimuth, spatial distances, horizontal directions and angles, zenith distances, etc., and space observables, such as Global Positioning System (GPS) measurements. Furthermore, each type of observable may be made using different instruments. This section is not intended to analyze every type of observable and every type of instrument. Instead, we confine ourselves to the error analysis and accuracy estimates of some basic conventional terrestrial geodetic observables such as EDM distances, horizontal directions and angles, and zenith distances made by a theodolite, azimuths by a gyro-theodolite, and geodetic leveling. The main purpose here is to show the reader how to approach the a priori accuracy analysis of geodetic observables.

It should be noted that the a priori accuracy analysis here is confined only to the accuracy of raw observations. In actual network computations, raw observations have to be reduced to a certain reference frame as discussed later in Chapter 2; for instance, the local geodetic reference frame, a mapping plane, or the geoid (for height networks), by applying gravimetrical and/or geometrical corrections. Therefore, any inaccuracies resulting from these corrections must be accounted for. This can also be done by applying the law of error propagation through the reduction formula as discussed in Chapter 2.

1.2.1 Angular Measurements

This section discusses the accuracy analysis of measuring horizontal directions/angles and zenith distances with a theodolite and measuring azimuths with a gyro-theodolite. We assume that no mechanical and/or optical system failure exists; that is, the theodolite is in correct adjustment, or that any misalignment or other errors can be eliminated by suitable observation procedures. In this case, internal errors related to the instrument and/or observer are identified as pointing, reading, and instrument leveling errors, and the major external errors related to the environment are centering error and atmospheric refraction that causes the ray of light between two stations to be curved. Atmospheric refraction is further divided into lateral and vertical refraction. The former affects horizontal directions/angles when the lines of sight pass close to objects that are significantly different in temperature than the surrounding air, such as in the tunnel, while the latter contaminates the measurement of zenith distances when the vertical temperature gradient along the line of sight exists. In the following, the effect of each individual error source on an observable is analyzed first, and the variance of the observable is then calculated according to the law of random error propagation.

Effects of Pointing Error, Reading Error, and Instrument Leveling Error

The ***pointing error*** σ_p is defined as the error in pointing a target. This error arises mainly due to the limited optical resolution of the instrument in use, the limited human vision of the observer, and the variations of the atmospheric conditions. The existence of pointing error prevents us from repeating the same observation exactly. The magnitude of pointing error is directly related to the telescope magnification of the instrument. Under average visibility and thermal turbulence conditions with a well-designed target, the pointing error for a single pointing at distances larger than a few hundred meters is estimated to be (Chrzanowski, 1977):

$$\sigma_p = 30''/M \text{ to} \sim \sigma_p = 60''/M \tag{1.2}$$

where M is the telescope magnification of the instrument.

Under poor visibility or large thermal turbulence the expected pointing error would certainly be larger. The contribution of personal error to the pointing error has been neglected in the above formula, and it should be considered if the pointing capability of different instrument observers varies considerably when evaluating the pointing error of different observers.

The ***reading error*** σ_r refers to the observer's inability to repeat the same reading. For optomechanic theodolites, this error is mainly a function of the least count or smallest angular division of the theodolite. Graduation errors in either the horizontal circle or the micrometer scale would certainly contribute to the reading error. But, in practice, observation procedures are designed to minimize the graduation errors; i.e., to take the mean of many evenly spaced "zeros" between 0 and 180° for the horizontal circle, and using the full range of the micrometer scale for measurement of an individual set of directions (Nickerson, 1978). For theodolites using coincidence micrometers, the reading error is expected to be (typically for d=1" or 0.5" instruments)

$$\sigma_r = 2.5\, d'' \tag{1.3}$$

where d″ is the least count in seconds of the theodolite (typically for d=1" or 0.5" instruments).

The ***instrument leveling error*** is caused by the insensitivity of the spirit levels used to level the instrument. The sensitivity of spirit levels is measured by their bubble value, which is the angular value necessary to displace the bubble through one of the divisions marked on the top of the spirit level. For a spirit level system, the accuracy for leveling the instrument is estimated to be (Chrzanowski, 1977; Cooper 1971)

$$\sigma_v = 0.2\, v'' \tag{1.4}$$

where v″ is the bubble value of the spirit level in seconds, while for a split level system, which is centered by a coincidence reading system and used by many manufacturers on the vertical circle bubble, the leveling accuracy could be 10 times better; i.e.,

$$\sigma_v = 0.02\, v'' \tag{1.5}$$

Many modern theodolites have automatic compensators for the vertical circle. The effect of the instrument leveling error on a measured horizontal direction can be expressed by

$$\sigma_L = \sigma_v \cot Z \qquad (1.6)$$

where Z is the zenith distances to the target.

Effects of Centering Error

Centering error refers to the inability to exactly align the center of the instrument and/or the target with the survey mark. The magnitude of centering error depends on the method and equipment used. Under good conditions (i.e., no wind, equipment in good condition), the expected centering errors for different types of centering equipment are listed below (Chrzanowski, 1977):

Method of Centering	Expected Error
String plumb bob	1 mm/m
Optical plummet	0.5 mm/m
Plumbing rods	0.5 mm/m
Forced or self-centering	0.1 mm

These values are, of course, only approximate, and are also dependent on the particular equipment and the conditions under which they are used. Self centering refers to the method frequently used in traversing; i.e., leaving the tribrachs attached to the tripods and exchanging only the instrument and the targets. Assuming the general case in which both the instrument and the targets have a centering error, the influence of the centering errors on a measured horizontal angle is derived as follows (Chrzanowski, 1977):

$$\sigma_\beta = (\rho'')\sqrt{\frac{\sigma_{c1}^2}{s_1^2} + \frac{\sigma_{c2}^2}{s_2^2} + \frac{\sigma_{c3}^2}{s_1^2 s_2^2}(s_1^2 + s_2^2 - 2 s_1 s_2 \cos\beta)} \qquad (1.7)$$

where σ_{c_1} and σ_{c_2} are the centering errors of the targets, and σ_{c_3} the centering error of the theodolite, respectively; β is the angle measured; s_1 and s_2 are the distances to the targets; and $\rho'' = 206265''$.

In the case that the centering errors of the targets and instruments are about the same, and the distances are about equal, the above equation reduces to

$$\sigma_\beta = \frac{2\rho''\sigma_c}{s}\sqrt{\left(1 - \frac{\cos\beta}{2}\right)} \qquad (1.8)$$

For horizontal directions it becomes

$$\sigma_d = \sqrt{2}\,\frac{\rho''\sigma_c}{s} \qquad (1.9)$$

where both the instrument and the target are assumed to have centering error of σ_c. Obviously, the effect of horizontal centering errors on zenith distances may be neglected.

Effects of Atmospheric Refraction

When radiation propagates through the infinite "layers" of the atmosphere, it undergoes some change in both speed and direction due to refractivity (i.e., the refractive index), which is caused by the heterogeneous densities of the air. Therefore, in the atmosphere, an optical path is not a straight line due to atmospheric refraction, and the curvature of the refracted optical path may differ considerably from one point to another along the line of sight and from one instant of time to another (see Figure 1.2).

The curvature of the optical path or other electromagnetic waves is derived as follows (cf. Rüeger, 1980, etc.):

$$\frac{1}{\rho} = -\frac{1}{n}\frac{\partial n}{\partial y} \tag{1.10}$$

where $\partial n/\partial y$ is the gradient of the refractive index. The refractive index of the air can be expressed as

$$n = 1 + (n_g - 1)\frac{273.16\,p}{(273.16 + t)1013.25} - \frac{11.20 \times 10^{-6}}{(273.16 + t)}e \tag{1.11}$$

where n is the group refractive index valid for atmospheric conditions described by t, p, and e, which represent the "dry bulb" temperature of air (in $^{\circ}$c), the atmospheric pressure (in mb), and the partial water vapor pressure (in mb), respectively, and n_g is the group refractive index calculated at 0°c, 1013.25 mb, and 0.03% of CO_2.

For visible and the near infrared spectrum, the group refractive index n_g is calculated using the following formula (cf. Rüeger, 1980):

$$(n_g - 1) \times 10^6 = 287.604 + 3\left(\frac{1.6288}{\lambda^2}\right) + 5\left(\frac{0.0136}{\lambda^4}\right) \tag{1.12}$$

with λ being the effective wave length, in micrometers, of light in a vacuum.

According to Equation (1.11), differentiating n with respect to y, one obtains the gradient of the refractive index as follows:

$$\frac{\partial g}{\partial y} = (n_g - 1)\left[\frac{273.16}{(273.16+t)1013.25}\frac{\partial p}{\partial y} - \frac{11.20\times 10^{-6}}{(273.16+t)}\frac{\partial e}{\partial y} + \right.$$
$$\left. + \frac{1}{(273.16+t)^2}\left(11.20\times 10^{-6}e - \frac{273.16p}{1013.25}\right)\frac{\partial t}{\partial y}\right] \tag{1.13}$$

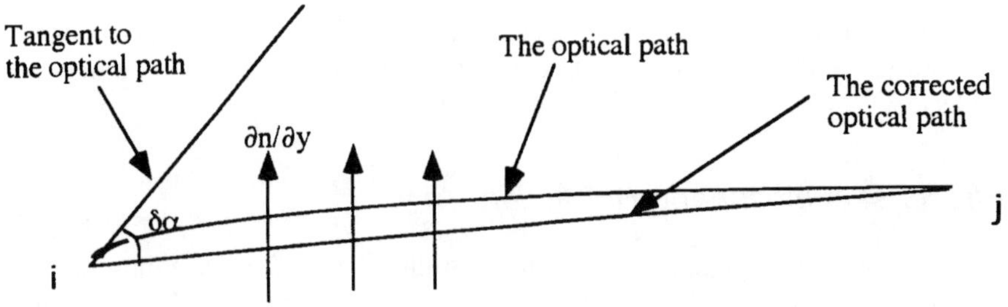

Figure 1.2. A refracted optical path

In practice, the influence of both the air pressure gradient and the partial water vapor pressure on angular measurements may be neglected, and that leads to the following simplified form of the above equation:

$$\frac{\partial n}{\partial y} = -(n_g - 1)\left(\frac{273.16p}{(273.16+t)^2 1013.25}\right)\frac{\partial t}{\partial y} \tag{1.14}$$

Substituting the above formulae (1.14) into Equation (1.10) and replacing n by its close approximation 1.0, one obtains the curvature of an optical path as follows:

$$\frac{1}{\rho} = (n_g - 1)\frac{273.16p}{(273.16+t)^2\, 1013.25}\frac{\partial t}{\partial y} \tag{1.15}$$

where the temperature gradient must be determined, either horizontally for lateral refraction or vertically for vertical refraction, in a direction perpendicular to the optical path or line of sight. In order to trace the optical path during the measurements, the horizontal or vertical gradients of temperature and barometric pressure should be measured at a number of points along the optical path in conjunction with the survey.

The atmospheric refraction is often expressed in terms of a coefficient of refraction k that is defined as follows:

$$k = \frac{R}{\rho} \tag{1.16}$$

where R is the mean radius of the earth, and ρ is the curvature of the optical path as defined by Equation (1.15). Under average conditions, the value of k is about 0.13 for visible spectrum, and 0.25 for microwaves. Otherwise, the values of k may range from –12 to +1 or even more in measurements over glacier surfaces, hot sand, etc., particularly in the first 100 m above the ground (Chrzanowski, 1977).

Theoretically, the effects of atmospheric refraction on the angular measurements are of systematic nature, and therefore corrections must be applied to the raw data before they can be used. These corrections are often referred to as meteorological corrections. In practice, to evaluate the effects of atmospheric refraction on angular measurements, the refracted optical path is usually accepted as being a circular curve with the curvature calculated for a mean gradient of n. In this case, the effects of atmospheric refraction on a measured horizontal direction and a zenith distance are, respectively, as follows:

$$\delta d = \frac{k_h s}{2R} \tag{1.17}$$

$$\delta Z = \frac{k_v s}{2R} \tag{1.18}$$

where s is the length of the line of sight, and k_h and k_v represent the horizontal and vertical coefficient of refraction, respectively.

Since an observed horizontal angle at a station can be expressed as the difference of two directions originating from the same station (see Figure 1.3), the effect of lateral refraction on a horizontal angle is therefore

$$\delta\beta = \frac{1}{2R}\left(k_h^{ik} s_{ik} - k_h^{ij} s_{ij}\right) \tag{1.19}$$

where k_h^{ij}, s_{ij} and k_h^{ik}, s_{ik} represent the length and the lateral coefficient of refraction of the lines of sight along i – j and i – k, respectively.

The random error associated with atmospheric refraction is due to the inaccuracy in the determination of the lateral and vertical refraction that is calculated from the measurements of horizontal and vertical temperature gradient, respectively. The temperature gradient is the most difficult item to determine. It is a function of many things including density of the air, temperature, soil characteristics under the sight line, the height of the sight line above the ground, wind speed, etc. Therefore, even if corrections for atmospheric refraction are made, the residual effects of refraction, due to the inaccuracy in determining the temperature gradient, on a horizontal direction, a horizontal angle, and a zenith distance can be estimated respectively as follows:

$$\sigma_d = \frac{s}{2R}\sigma_{k_h} \tag{1.20}$$

$$\sigma_\beta = \sqrt{2}\,\frac{s}{2R}\sigma_{k_h} \tag{1.21}$$

$$\sigma_z = \frac{s}{2R}\sigma_{k_v} \tag{1.22}$$

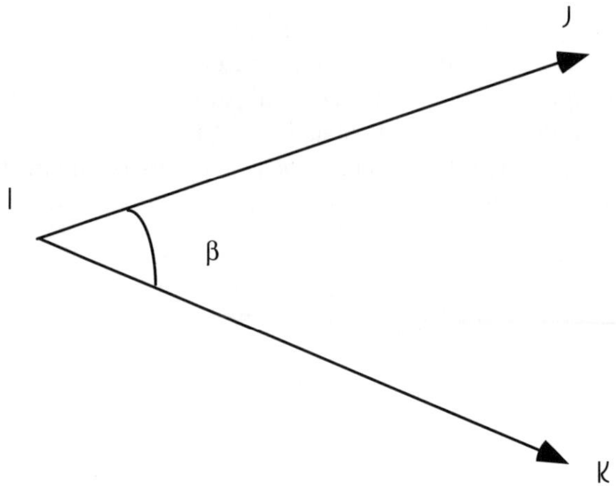

Figure 1.3. A horizontal angle

where σ_{k_h} and σ_{k_v} are the standard error in determining the coefficient of lateral and vertical refraction, respectively, and they are related to the standard error of the temperature gradient; i.e., $\sigma_{\partial t/\partial y}$, through (1.15) and (1.16) as follows

$$\sigma_k = (n_g - 1)\frac{273.16\ p\ R}{(273.16 + t)^2\ 1013.25}\sigma_{\partial t/\partial y} \tag{1.23}$$

For horizontal directions and angles, the only way to determine the horizontal temperature gradient is to observe the temperature along the line of sight, and since this is not usually feasible in ordinary survey practice, the only recourse is to avoid situations where the line of sight passes close to a temperature anomaly (Nickerson, 1978). If corrections for atmospheric refraction are totally neglected, an angular measurement would error by the refraction angle as determined by Equations (1.17) to (1.19).

For the measurement of zenith distances, the vertical refraction can also be determined by simultaneous reciprocal zenith distances. Assuming that the refraction angle ε will be the same at both ends of the line sight ij (see Figure 1.4), the refraction angle is calculated by

$$\varepsilon = \frac{180 - (Z_{ij} + Z_{ji})}{2} \tag{1.24}$$

where Z_{ij} and Z_{ji} are the zenith distances measured simultaneously at stations i and j, respectively.

The above approach implies that the directions of the verticals at i and j are parallel in the plane of the line of observation between i and j. This is expected for lines that are not too long (e.g., < 10 km) and are not in a gravity disturbed area.

Other alternatives to handle vertical refraction include modeling the vertical refraction angle into a network adjustment (mainly in three-dimensional space) including measured zenith angles to determine it analytically (cf. Vincenty, 1973).

Accuracy Estimates

The a priori accuracy of a proposed observation is calculated as the expected total effect of the various internal and external errors. Assuming that the effects of different error sources on the angular measurements are statistically independent, the total effect, i.e., the variance of an observable, can be calculated, according to the law of random error propagation, as the sum of squares of each individual effect. Summarizing the above developments, formulae for the accuracy estimation of angular measurements are given below.

Horizontal Directions

The a priori variance of a horizontal direction measurement obtained from a single sighting is estimated as follows:

$$\sigma_{d_i}^2 = \sigma_p^2 + \sigma_r^2 + \sigma_v^2 \cot^2 Z + (\rho'')^2 \frac{2\sigma_c^2}{s2} + \left(\frac{s}{2R}\right)^2 \sigma_{k_h}^2 \qquad (1.25)$$

where σ_p, σ_r, σ_v, σ_c, and σ_{k_h} represent the pointing error, reading error, instrument leveling error, instrument and target centering error, and the error in determining the coefficient of lateral refraction, respectively; s and Z are the length and zenith angle of the line of sight, respectively; k_h is the coefficient of lateral refraction, and finally R is the mean radius of the earth.

Conventionally, in order to minimize both the internal and external effects, horizontal directions are measured in multiple sets and the zero setting is changed between "sets" of directions with two pointings and readings (i.e., face left and face right) of the same direction within each set. The mean direction $\bar{d}$ calculated from all the sets of observations is used as final product, i.e.,

$$\bar{d} = \frac{1}{n_s} \sum_{i=i}^{n_s} \frac{(d_{Li} + d_{Ri} - 180°)}{2} \qquad (1.26)$$

where n_s is the total number of sets measured, and d_{Li} and d_{Ri} are the face left and face right readings within each set, respectively. If no releveling and recentering of the instrument happen between sets, the variance of the mean direction is calculated according to error propagation as follows

$$\sigma_{\bar{d}}^2 = \frac{\sigma_p^2 + \sigma_r^2}{2n_s} + \sigma_v^2 \cot^2 Z + (\rho'')^2 \frac{2\sigma_c^2}{s^2} + \left(\frac{s}{2R}\right)^2 \sigma_{k_h}^2 \qquad (1.27)$$

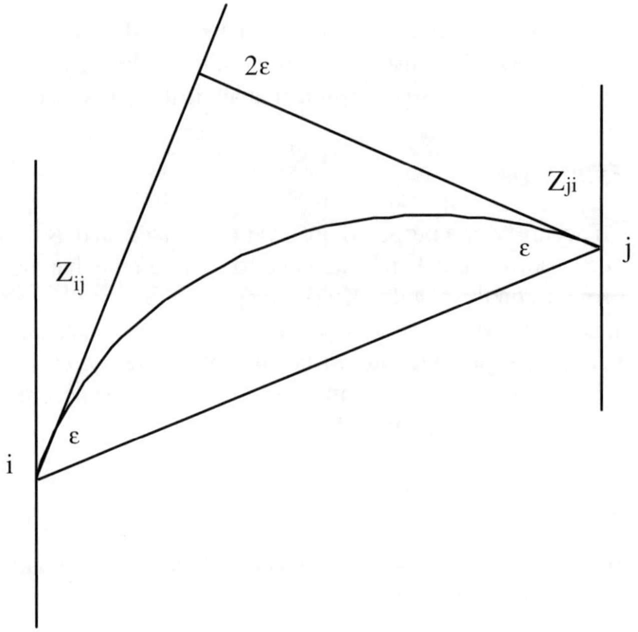

Figure 1.4. Reciprocal zenith distance measurement

If, however, the instrument is releveled and recentered between sets, the accuracy of the mean direction will be

$$\sigma_d^2 = \frac{\dfrac{\sigma_p^2 + \sigma_r^2}{2} + \sigma_v^2 \cot^2 Z + (\rho'')^2 \dfrac{2\sigma_c^2}{s^2}}{n_s} + \left(\frac{s}{2R}\right)^2 \sigma_{k_h}^2 \tag{1.28}$$

Horizontal Angles

As discussed before, horizontal angles are essentially the difference of two directions. Thus the variance of a horizontal angle derived from two horizontal directions that are obtained from a single sighting is

$$\sigma_{\beta_i}^2 = 2\left(\sigma_p^2 + \sigma_r^2 + \sigma_v^2 \cot^2 Z\right) + \\ + (\rho'')^2 \frac{4\sigma_c^2}{s^2}\left(1 - \frac{\cos\beta}{2}\right) + 2\left(\frac{s}{2R}\right)^2 \sigma_{k_h}^2 \tag{1.29}$$

Similarly, if a mean horizontal angle $\bar{\beta}$ is derived from means of horizontal directions measured in n_s sets, the variance is

$$\sigma_{\bar{\beta}}^2 = \frac{\sigma_p^2 + \sigma_r^2}{n_s} + 2\sigma_v^2 \cot^2 Z + \\ + (\rho'')^2 \frac{4\sigma_c^2}{s^2}\left(1 - \frac{\cos\beta}{2}\right) + 2\left(\frac{s}{2R}\right)^2 \sigma_{k_h}^2 \tag{1.30}$$

where no releveling and no recentering have been assumed between sets. Otherwise, if the instrument is releveled and recentered between sets, the variance will be

$$\sigma_{\bar{\beta}}^2 = \frac{\sigma_p^2 + \sigma_r^2 + 2\sigma_v^2 \cot^2 Z + (\rho'')^2 \frac{4\sigma_c^2}{s^2}\left(1 - \frac{\cos\beta}{2}\right)}{n_s} + \\ + 2\left(\frac{s}{2R}\right)^2 \sigma_{k_h}^2 \tag{1.31}$$

Zenith Distances

In principle, a zenith distance is the difference between the direction of the vertical axis of the theodolite and the direction defined by the optical axis of the telescope pointed at the target, and the latter is the actual quantity being measured by the vertical circle of a theodolite. Similar to the treatment of horizontal directions, the accuracy of a zenith distance obtained from a single sighting can be estimated by

$$\sigma_{z_i}^2 = \sigma_p^2 + \sigma_r^2 + \sigma_v^2 + \left(\frac{s}{2R}\right)^2 \sigma_{k_v}^2 \tag{1.32}$$

where σ_v accounts for the leveling error of the vertical circle index, and σ_{k_v} is the error in determining the coefficient of vertical refraction. σ_p, σ_r, and s mean the same as before. If a zenith distance is measured n_s sets, the variance for the mean zenith distance Z is

$$\sigma_z^2 = \frac{\sigma_p^2 + \sigma_r^2 + \sigma_v^2}{2n_s} + \left(\frac{s}{2R}\right)^2 \sigma_{k_v}^2 \tag{1.33}$$

where the vertical circle index is assumed to be releveled for each observation, and that is usually the case in practice.

Gyro-Azimuths

The accuracy of azimuth observation depends on the method used. In the case of azimuth determination using a gyro-theodolite, the azimuth is computed by

$$A = M - N + E \tag{1.34}$$

where A represents the measured astronomical azimuth, M the horizontal circle reading for the reference mark, N the horizontal circle reading for north as determined by the gyro, and E the calibration value. Assuming no correlation between M, N, and E, the variance of A is written as

$$\sigma_A^2 = \sigma_M^2 + \sigma_N^2 + \sigma_E^2 \tag{1.35}$$

where σ_M^2 can be calculated using the same formula as used for direction measurement, while σ_N^2 and σ_E^2 represent the error in the determination of the north by a gyro instrument. Main factors include mislevelment of the gyro, drift effects, changes in band torque equilibrium position, changes in angular momentum of the gyro, changes in the angle between the optical axis of the theodolite and axis of the gyro reading system, and errors in the given latitude. Due to these various causes, it is difficult to accurately evaluate all these error sources *a priori.* Thus it is suggested that one evaluate the accuracy of a gyro azimuth *a posterior* by computing the sample variance from a set of redundant observations as follows (cf. §1.3.2):

$$\sigma_A^2 = \frac{1}{n-1}\sum_{i=1}^{n}\left(A_i - \bar{A}\right)^2 \tag{1.36}$$

where n is the number of observed gyro-azimuths, A_i an individual azimuth, and $\bar{A}$ the mean of the set of observed gyro-azimuths. For the accuracy analysis of azimuth determination by astronomical method, the interested reader is referred to Nickerson (1978).

1.2.2 EDM Distances

Spatial distances may be measured by different methods. Examples include mechanical methods (e.g., steel tape or Invar tape), optical methods (e.g., stadia tachometry, subtense bar), and EDM (Electromagnetic Distance Meter), etc. EDM is now the most frequently used method in surveying. There are also many ways to measure distances by electronic means. Typical examples are the pulse method, phase difference method, Doppler method, and interferometry (Rüeger, 1980). However, all common EDM instruments used in surveying are based on the phase difference method. EDM instruments are classified according to the type of carrier wave used. Those using light or IR waves are classified as electro-optical instruments, and those based on radio waves are generally called microwave instruments. Short-range instruments with infrared light sources are used for all types of surveys, while microwave instruments are used to measure long distances. In this context, we confine ourselves to the discussion of eletro-optical instruments only.

In the electro-optical instruments, the phase difference measurement is implemented as follows: the measuring signal that is modulated on the carrier wave in the emitter travels to the reflector and back to the EDM instrument, where it is picked up by the receiver. In the receiver, the phases of the outgoing and the incoming signals are compared and the phase lag $\Delta\Phi$ is measured. The equation for distance using phase measurement is as follows:

$$s = \frac{1}{2}\lambda\left(m + \frac{\Delta\Phi}{2\pi}\right) \tag{1.37}$$

where s is the measured distance, $\Delta\Phi$ the measured phase difference between transmitted and reflected wave in radians, λ the modulation wavelength being used, and finally m is the integer number of wavelength in twice the distance that is resolved by introducing more than one wavelength in the EDM instrument.

As with angular measurements, the errors affecting a distance measurement can also be divided into internal and external components. For electro-optical instruments, the internal errors have been identified as zero error, cyclic error, and phase measurement error, which are mainly due to imperfections of the instruments in use, while external errors are mainly due to atmospheric refraction.

Effects of Zero Error, Cyclic Error, and Phase Measurement Error

At first, ***zero error*** results from the inaccurate knowledge of the difference between the electrical center of the instrument and the point used to center the instrument over the station. For instruments using light waves, this value is usually small, but for microwave devices, the value can be quite significant. The effect of zero error on a measured distance is a constant. The zero error of an EDM instrument is usually determined by the manufacturer. It can also be determined by the user through measuring distances on calibrated baselines as discussed in Rüeger (1980), among others. Assuming a zero error of z_0, a measured distance must be corrected by applying the following correction:

$$\delta s_z = -z_0$$

The ***cyclic error*** of an electro-optical instrument is caused primarily by electric cross-talk within the instrument. The error can be modeled over the defined unit length of the instrument (i.e., the period of the cyclic error) by the use of a sine curve, as given by the following equation (cf. Rüeger, 1980):

$$s_i = s_1 + x_{ref.} - A \sin\left[\frac{2\pi}{U}(s_i + B)\right] \tag{1.38}$$

where U is the unit wavelength, A the amplitude of the cyclic error correction, B the phase of the cyclic error correction, s_1 the distance from the instrument to the first reflector position, s_i a measured distance to the present reflector position, and finally, $x_{ref.}$ is a known distance between the present reflector position and the first reflector position. With a set of measured distances that vary over one unit length of the instrument, the unknown parameters A and B are solved by the method of least squares. The cyclic error correction to a measured distance is given by:

$$\delta s_c = A \sin\left[\frac{2\pi}{U}(s + B)\right] \tag{1.39}$$

Finally, the error in phase difference measurement is due to limited resolution of the measurement technique used. This is an error of random nature. The accuracy of determining the phase difference $\Delta\Phi$ depends on the method used. The digital method of phase detection, which is used in the modern infrared equipment, gives a resolution of 0.001 to 0.0003 of a cycle. By differentiating Equation (1.37) with respect to $\Delta\Phi$, one obtains the effect of a random error in phase difference measurement on the measured distance as follows:

$$\sigma_s = \frac{\lambda}{4\pi}\sigma_{\Delta\phi} \tag{1.40}$$

Effects of Atmospheric Refraction

Due to atmospheric refraction, the velocity of electromagnetic waves in the atmosphere is not equal to the ideal value that would be obtained in a vacuum. Additionally, the paths followed by electromagnetic waves in the atmosphere are not straight lines. Atmospheric refraction introduces errors in the wavelength because the actual refractive index (n) at the time the measurement is taken differs from the reference refractive index (n_{ref}) fixed by the manufacturer for the instrument. This error, if not corrected, will create a systematic scale error in the measured distances. Since the actual wavelength of the light wave in space is expressed by

$$\lambda = \frac{\lambda_0}{n} = \frac{c_0}{f}\frac{1}{n} \tag{1.41}$$

where c_0 is the speed of light in a vacuum, and f the modulation frequency being used. The reference wavelength $\lambda_{ref.}$ calculated under the reference refractive index $n_{ref.}$ specified by the manufacturer is:

$$\lambda_{ref.} = \frac{c_0}{f}\frac{1}{n_{ref.}} \tag{1.42}$$

The error in the wavelength is therefore

$$\delta\lambda = \lambda - \lambda_{ref.} = \frac{c_0}{f}\left(\frac{1}{n} - \frac{1}{n_{ref.}}\right) = \lambda_{ref.}\frac{n_{ref.} - n}{n} \tag{1.43}$$

The effect of the error in the wavelength on the measured distance is therefore

$$\delta s_\lambda = \frac{1}{2}\left(m + \frac{\Delta\Phi}{2\pi}\right)\delta\lambda = s\frac{n_{ref.} - n}{n} \tag{1.44}$$

This correction is also called the first velocity correction. Another systematic error introduced by atmospheric refraction is called the second velocity correction that accounts for the non-uniformity of the curvature of the wave path due to the heterogeneous refractive index along the wave path. This correction is negligible for electro-optical instruments, but can be significant for microwave instruments. For more details on this subject, the reader is referred to Rüeger (1980).

Accuracy Estimates

The effects of the above discussed zero error, cyclic error, and atmospheric refraction are all systematic errors. Corrections must be applied to the measured distances before they can be used. The random errors associated with these systematic effects are due to the inability to determine these systematic effects exactly. Assuming that the effects of different error sources are statistically independent, the variance of a measured distance can be calculated by:

$$\sigma_s^2 = \left(\frac{\lambda}{4\pi}\right)^2 \sigma_{\Delta\Phi}^2 + \sigma_z^2 + \sigma_c^2 + \frac{\sigma_n^2}{n^2}s^2 \tag{1.45}$$

where $\sigma_{\Delta\Phi}$, σ_z, σ_c, and σ_n are the standard error in determining the phase difference $\Delta\Phi$, the zero error z_0, the cyclic error, and the refractive index n, respectively.

According to Equation (1.11), the variance in determining the refractive index can be calculated by:

$$\sigma_n^2 = (n_g - 1)^2\left\{\left[\frac{273.16}{(273.16 + t)1013.25}\right]^2 \sigma_p^2 + \left[\frac{273.16p}{(273.16 + t)^2 1013.25}\right]^2 \sigma_t^2\right\} \tag{1.46}$$

where σ_p and σ_t are the standard error in determining the air pressure and temperature, respectively. The effects of water vapor pressure are negligible for electro-optical instruments.

If phase measurement is made m_1 times and meteorological readings m_2 times, the variance for the mean distance will be

$$\sigma_s^2 = \frac{1}{m_1}\left(\frac{\lambda}{4\pi}\right)^2 \sigma_{\Delta\Phi}^2 + \sigma_z^2 + \sigma_c^2 + \frac{1}{m_2}\frac{\sigma_n^2}{n^2} s^2 \tag{1.47}$$

The above formula explains why the accuracy of EDM distances is usually expressed in terms of a constant and a distance-dependent error as follows:

$$\sigma_s^2 = a^2 + b^2 s^2 \tag{1.48}$$

where the constant part a accounts for the effects of zero error, cyclic error, and phase measurement error, while the distance-dependent parameter b describes mainly the effect of atmospheric refraction.

1.2.3 Geodetic Leveling

Geodetic leveling is used to determine the height differences between accessible terrain points with respect to a defined datum. Internal errors come from both the level instrument and the level rods. As with angular measurements, major internal errors associated with the level instrument are ***pointing error***, ***reading error***, ***instrument leveling error***, and ***collimation error***, and those associated with the leveling rod are ***rod scale error*** and ***rod index error***. The rod scale error refers to a wrong scale of the rod, and the rod index error represents a constant offset of the zero mark on each rod from the base of the plate. There are a variety of external error sources associated with geodetic leveling, and those of major concern are ***vertical atmospheric refraction***, instrument and turning point ***sinking*** and ***rebound***, and ***earth curvature***.

Effects of Internal Errors

First, the pointing error in geodetic leveling is due to the limited optical resolution of the telescope of the level instrument. Similar to angular measurements, under good visibility and atmospheric conditions, the effect of pointing error on a level measurement is estimated by

$$\sigma_p = \frac{1}{\rho''}\frac{30''}{M}s \sim \frac{1}{\rho''}\frac{60''}{M}s \tag{1.49}$$

where s is the length of the line of sight, M the magnification of the telescope of the level instrument, and $\rho'' = 206265''$.

Second, unlike angular measurements, the reading error in geodetic leveling mainly comes from the effect of the non-verticality of the level rod as shown in Figure 1.5.

A leveling rod is usually equipped with a box level that serves to keep the rod vertical. The limited sensitivity of the level bubble prevents the rod from being perfectly vertical. From Figure 1.5, the effect of the non-verticality of the level rod on a level measurement can be estimated by

$$\begin{aligned} \sigma_r &= l\left(1-\cos\frac{v_r''}{\rho}\right) \\ &\approx \frac{l}{2}\left(\frac{v_r''}{\rho}\right)^2 \end{aligned} \tag{1.50}$$

where l is the height of sighting on the rod, and v_r'' is the sensitivity of the level bubble on the rod in arc-seconds. For rods used in geodetic leveling, the sensitivity can amount to 10'.

As with angular measurements, the leveling of a level instrument is limited by the sensitivity of the level bubble. Misleveling of the instrument will cause the line of sight to deviate from the horizontal axis (see Figure 1.6). The effect of the instrument leveling error on a level measurement can be approximated by

$$\sigma_L = s\frac{\sigma_v}{\rho''} \tag{1.51}$$

where σ_v can be evaluated using formula (1.4).

Collimation error of a level instrument represents a systematic deviation of the line of sight from the horizontal plane as defined by the gravity vector at the instrument. The effect of collimation error on the leveling data is the same as instrument leveling error. However, since collimation error is a systematic error, its effect can be removed by instrument calibration and by balancing the lengths of the backsights and foresights. The effects of rod scale error and rod index errors are also of a systematic nature. The rod scale error can be removed or reduced to a negligible amount by calibration before the beginning of the leveling campaign, and the effect of rod index error can be totally removed by using an even number of setups, or one setup with the same rod used for both the backsight and foresight.

Effects of External Errors

At first, the effect of vertical atmospheric refraction on a measured height difference can be expressed by

$$\delta h_r = \frac{k_{vF} s_F^2 - k_{vB} s_B^2}{2R} \tag{1.52}$$

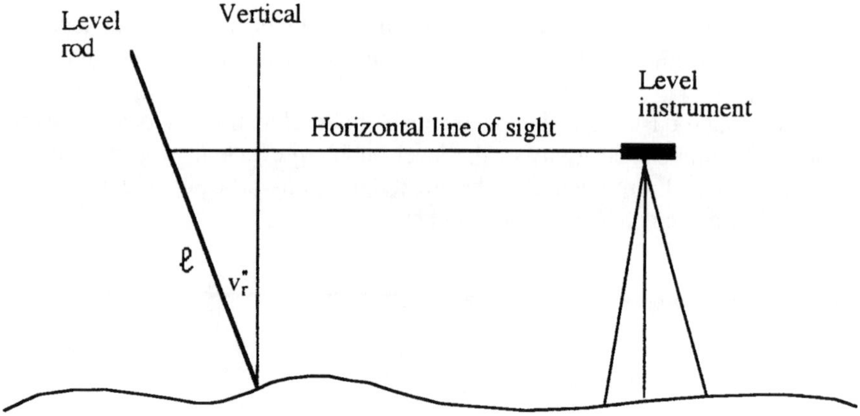

Figure 1.5. Effect of the non-verticality of the level rod

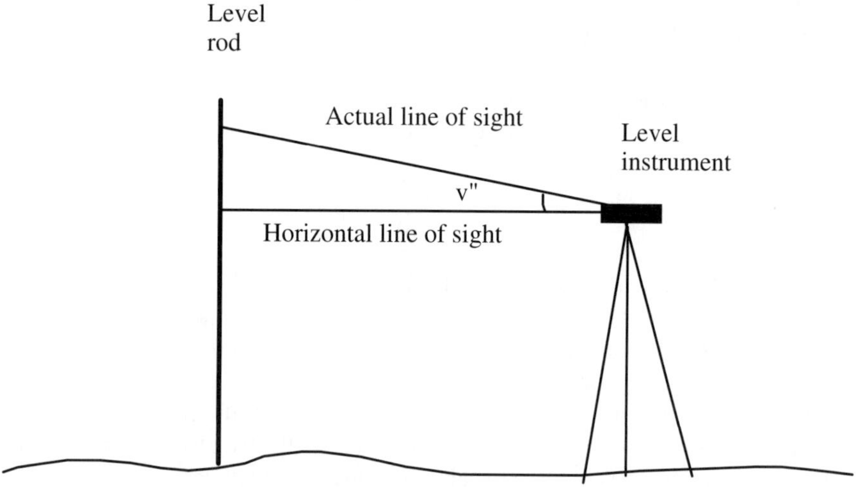

Figure 1.6. Effect of the instrument leveling error

where k_{v_F}, s_F and k_{v_B}, s_B are the coefficient of vertical refraction and sight length for the foresights and backsights, respectively. This systematic error can be offset by applying corrections to the raw data, and the residual error due to the inaccuracy in determining the coefficient of vertical refraction could be negligible due to the short sight lengths involved in geodetic leveling and by balancing the foresights and backsights during the survey.

Sinking is caused by the weight of the instrument or turning point when set up on the ground, and rebound is caused by the response to the instrument or turning point when set up on spongy material. These errors can be minimized by following a proper observation procedure and setting up the instrument and level rods on stable locations.

Finally, the effect of earth curvature on a leveled height difference is

$$\delta h_e = \frac{s_F^2 - s_B^2}{2R} \tag{1.53}$$

where s_F and s_B are the foresights and backsights, respectively, and R is the radius of the earth. This effect is considered since the level instrument and the rods are set up with respect to the direction of the gravity, which is normal to a curved equipotential surface, while the line of sight of a geodetic level describes a plane tangent to this surface. This effect is removed by balancing the backsights and foresights.

Accuracy Estimates

The accuracy of a leveled height difference depends on the field procedure followed. When using the double simultaneous observing procedure with double scale (high and low scales) rods (cf. FGCS, 1984), four readings are taken at each setup, and the height difference at each setup is calculated by

$$\Delta h_i = \frac{(B_{Li} - F_{Li}) + (B_{Hi} - F_{Hi})}{2} \tag{1.54}$$

where B_{Li}, F_{Li}, and B_{Hi}, F_{Hi} are the backsight and foresight readings on the low and high scales, respectively. The variance of Δh_i is then determined by the law of error propagation as follows

$$\begin{aligned} \sigma_{\Delta h_i}^2 &= \frac{4\left(\sigma_p^2 + \sigma_r^2 + \sigma_L^2\right)}{4} \\ &= \sigma_p^2 + \sigma_r^2 + \sigma_L^2 \end{aligned} \tag{1.55}$$

where σ_p, σ_r, and σ_L have been defined in Equations (1.49) to (1.51).

Finally, the variance of a leveled section is estimated by

$$\sigma_{section}^2 = m \cdot \sigma_{\Delta h_i}^2 \tag{1.56}$$

where m is the total number of setups for the leveled section. The effects of certain residual systematic errors may increase the above values that have to be considered for each specific case.

1.3 The a Posterior Accuracy Analysis of Observations

After a survey campaign has been completed, the a posterior accuracy analysis of the survey is needed to confirm the a priori error budget and to analyze the actually achieved accuracy of the survey. This knowledge is very important at the network computation stage in order to assign appropriate variances and covariances to observations for the estimation of the unknown parameters under study and for the proper statistical assessment of the results. This section discusses a few of the most frequently used methods for the a posterior accuracy estimation of observations. Unlike the a priori accuracy analysis, the a posterior accuracy analysis is based on only the actual observational values and the network geometry.

1.3.1 Estimation of Variance from True Errors

Much effort has been made to estimate the accuracy of geodetic measurements based on closing errors in condition equations. According to the error theory, if a measured value and the true value of an observable are denoted by l_i and μ, respectively, the true error in l_i is defined by

$$\Delta_i = l_i - \mu \tag{1.57}$$

If a series of observations have been made with equal accuracy, and assume that the observations are statistically independent, the variance of an observation can be estimated by

$$\sigma_l^2 = \sum_{i=1}^{n} \frac{\Delta_i^2}{n} \tag{1.58}$$

where n is the number of true errors. A typical example for this is to estimate the variance of angular measurements from misclosures of triangles using Ferrero's formula. Since the theoretical sum of a triangle is 180 degrees (neglecting the effects of the earth's curvature), due to errors of measurement the actual sum of an observed angle differs from the theoretical value by

$$w_i = (\beta_{i1} + \beta_{i2} + \beta_{i3}) - 180° \tag{1.59}$$

where β_{i1}, β_{i2}, and β_{i3} are the observed values of the three angles of a triangle respectively, and w_i represents the true error of the measured triangle. It is also called the misclosure of the triangle. If independent angles are measured with equal accuracy, the variance of a measured angle is calculated by

$$\sigma_\beta^2 = \sum_{i=1}^{n} \frac{w_i^2}{3n} \tag{1.60}$$

with n being the total number of independent triangles measured. This is also known as Ferrero's formula.

Another example is to estimate the accuracy of geodetic leveling from loop closures. Since the theoretical sum of height differences, after being subject to appropriate gravimetrical corrections, along a loop is zero, the true error associated with a level loop is then

$$\Omega_i = \sum_{j=1}^{m_i} \Delta h_j - 0 \tag{1.61}$$

where Δh_j is the height difference for level section j and m_i the total number of level sections for the loop i, respectively. Assuming that all the loop closures are statistically independent, the variance of a measured height difference in unit square root of section length can be estimated by

$$\sigma^2_{\Delta h} = \sum_{i=1}^{n} \frac{\left(\Omega_i / \sqrt{s_i}\right)^2}{n} \tag{1.62}$$

where s_i is the length of loop i, usually in kilometers or miles, and Ω_i the misclosure of loop i in meters or feet, respectively. n is the total number of loops.

Theoretically, the above formulae (1.60) and (1.62) represent the situation where all the measured triangle closures and level loop closures in use are statistically independent. It is usually difficult in practice for this assumption to hold true, depending on the observing procedures used. Nevertheless, these formulae may give a good estimate of the overall accuracy of the measurements.

1.3.2 Estimation of Variance from Residuals

Unfortunately, in most of the surveying applications, the true value μ of an observable is unknown, and it is therefore approximated by its estimate $\hat{\mu}$. The true error of a measurement l_i is then approximated by

$$v_i = \hat{\mu} - l_i \tag{1.63}$$

where v_i is called the residual of observation l_i. If more than one observation is made with equal accuracy, the variance of the observation can be estimated by

$$\sigma^2_l = \sum_{i=1}^{n} \frac{v_i^2}{r} \tag{1.64}$$

where r is the degrees of freedom, and n the total number of observations.

An application of the above formula (1.64) is to estimate the variance of a set of repeated observations made for the same observable; that is,

$$\sigma^2_l = \sum_{i=1}^{n} \frac{\left(l_i - \bar{l}\right)^2}{n-1} \tag{1.65}$$

where $\bar{l}$ represents the arithmetic mean of the observations calculated by

$$\bar{l} = \frac{1}{n}\sum_{i=1}^{n} l_i \tag{1.66}$$

1.3.3 The Method of Minimum Norm Quadratic Unbiased Estimation (MINQUE)

A geodetic network usually involves different types of observables, and each type of observable may be made using different instruments of unequal accuracy. The above formulae (1.58) and (1.65) apply only to the case of homogeneous observations. In the case where a network involves different types of observables; e.g., angles, distances, etc., and the same type of observable is made with different instruments of heterogeneous accuracy, these formulae are apparently insufficient. In this case, the method of Minimum Norm Quadratic Unbiased Estimation (MINQUE) is recommended. It has been developed for the estimation of variance-covariance components of observations, and it is applicable to mixed types of observations with heterogeneous accuracy. The theoretical background of MINQUE has been elaborated in a series of papers of Rao (1970; 1971; 1973; 1979; etc.). Applications of MINQUE in the survey community can be found, among others, in Grafarend et al. (1980), Welsch (1981), Schaffrin (1983), Chen et al. (1990), and Kuang (1993d). Basic equations of the MINQUE method are given below.

Estimation of the Variance-Covariance Components

The MINQUE method starts with the following Gauss-Markov model (cf. Chapter 4):

$$E\{l\} = \mathbf{A}\,\mathbf{x} \tag{1.67}$$

$$D\{l\} = \sum_{1}^{k} \theta_i\,\mathbf{T}_i = \mathrm{C}_l \tag{1.68}$$

where $E\{\ \}$ and $D\{\ \}$ represent the statistical expectation and dispersion operators, respectively. l and $\mathbf{x}$ are the vector of observations and unknown parameters, respectively; $\mathbf{A}$ is the design matrix; θ_1, θ_2, ..., and θ_k are called the variance-covariance components to be estimated; and $\mathbf{T}_1$, $\mathbf{T}_2$, ..., and $\mathbf{T}_k$ are the corresponding coefficient matrices.

Denote

$$\mathbf{T} = \sum_{1}^{k} \mathbf{T}_i \tag{1.69}$$

A quadratic function, $\boldsymbol{l}^{\mathrm{T}}\mathbf{B}\boldsymbol{l}$, is said to be a Minimum Norm Quadratic Unbiased Estimator of the linear function,

$$\sum_{1}^{k} p_i \theta_i,$$

with unbiasedness and invariance in **x**, if the quadratic matrix **B** is determined from the following optimization problem (Rao, 1970):

$$\mathrm{Tr}(\mathbf{BTBT}) = \min \tag{1.70}$$

Subject to:

$$\mathbf{BA} = 0 \tag{1.71}$$

$$\mathrm{Tr}\{\mathbf{BT}_i\} = p_i \quad (i = 1, \ldots, k) \tag{1.72}$$

where Tr{ } is the trace operator of a matrix.

Solving Equations (1.70) to (1.72) gives the estimated variance-covariance components as:

$$\theta = \mathbf{S}^{-1}\mathbf{q} \tag{1.73}$$

where

$$\theta = (\theta_1\, \theta_2 \ldots \theta_k)^T \tag{1.74}$$

$$\mathbf{S} = \{s_{ij}\} \text{ with } s_{ij} = \mathrm{Tr}\{\mathbf{RT}_i\mathbf{RT}_j\}\ (i; j = 1, \ldots, k) \tag{1.75}$$

$$\mathbf{q} = \{q_i\} \text{ with } q_i = \boldsymbol{l}^T\mathbf{RT}_i\mathbf{R}\boldsymbol{l}\ (i = 1, \ldots, k) \tag{1.76}$$

$$\mathbf{R} = \mathbf{T}^{-1}\left[\mathbf{I} - \mathbf{A}\left(\mathbf{A}^T\mathbf{T}^{-1}\mathbf{A}\right)^{-}\mathbf{A}^T\mathbf{T}^{-1}\right] \tag{1.77}$$

The covariance matrix of the estimated components is computed by (Chen et al., 1990)

$$\mathbf{C}_\theta = 2\,\mathbf{S}^{-1} \tag{1.78}$$

If some a priori values of the variance-covariance components, i.e., $\theta^0 = (\theta_1^0\, \theta_2^0 \ldots \theta_k^0)^T$, are adopted to initialize the solution procedure, an iterative process, called the iterated MINQUE, or IMINQUE, should be followed (Rao, 1979). Solution equations of IMINQUE are the same as Equations (1.73) to (1.77) except that the matrix **T** in the equations is replaced by the covariance matrix $\mathbf{C}_l$

$$\mathbf{C}_l = \sum_{1}^{k} \theta_i^0 \mathbf{T}_i \tag{1.79}$$

and

$$\theta^{m+1} = \mathbf{S}^{-1}(\theta^m)\,\mathbf{q}(\theta^m) \quad (m = 0, 1, \ldots) \tag{1.80}$$

The iterative process stops when the solution ceases to change.

As discussed in Rao (1979), the appropriateness of the obtained MINQUE results depends heavily on the selection of the error model equation (1.68). An incorrectly assumed error model may lead to negative estimates of variance components and/or give nonsense values for both variance and covariance components. Furthermore, since MINQUE is a statistical method, a sufficient redundancy is required to obtain reliable estimated quantities. After applying the MINQUE method, the appropriateness of the estimated error model and its associated parameters may be justified by statistical techniques as discussed below. Concepts of statistical tests are given in Chapters 3 and 5.

Global Test on the Goodness of Fit

At first, the validity of an estimated error model can be assessed through the global test on the a posterior variance factors $\hat{\sigma}_0^2$ obtained after a least squares network adjustment using the error model of observations estimated by MINQUE. With a correct functional model (i.e., Equation [1.67]), if an estimated error model is appropriate, the estimated variance factor should be statistically equal to 1.0. The testing of the estimated variance factor is done using the following Fisher distribution (cf. Chapter 5):

$$\frac{\hat{\sigma}_0^2}{\sigma_0^2} \sim F(r, \infty) \tag{1.81}$$

where σ_0^2 is the population variance factor with its value being 1.0; r is the redundancy of the adjustment system. Given the significance level α, the estimated error model from MINQUE can be considered appropriate, if the following inequality holds

$$F_{\alpha/2}(r, \infty) \le \hat{\sigma}_0^2 \le F_{1-\alpha/2}(r, \infty) \tag{1.82}$$

Otherwise, a different error model should be postulated and re-evaluated.

Test on the Significance of Individual Parameters

After the estimated error model passed the global test discussed above, the significance of the estimated individual variance-covariance components can also be tested using the following statistic:

$$\frac{\hat{\theta}_i - \mu_{\hat{\theta}_i}}{\hat{\sigma}_{\hat{\theta}_i}} \sim t(r) \tag{1.83}$$

where $\hat{\theta}_i$, $\mu_{\hat{\theta}_i}$, and $\hat{\sigma}_{\hat{\theta}_i}$ are the estimated parameter, its expectation, and its estimated standard deviation, respectively; t(r) represents the student distribution with degrees of freedom of r. $\hat{\sigma}_{\hat{\theta}_i}$ can be extracted from Equation (1.78). Given a certain significance level α, the estimated $\hat{\theta}_i$ is considered statistically significant if the following inequality holds:

$$\left|\hat{\theta}_i\right| > t_{\alpha/2}(r) \cdot \hat{\sigma}_{\hat{\theta}_i} \tag{1.84}$$

An estimated error model of observations from MINQUE may be acceptable if it passed the global test and all the estimated variance-covariance components are significant. For practical applications of the MINQUE method, the interested reader is referred to the given literature.

2 Observation Data Pre-Processing

The purpose of this chapter is to summarize the procedures and formulations for the correction and reduction of raw observation data for network computations. As is well known, raw geodetic measurements are made on the irregular surface of the earth (a physical space) bearing the effects of the earth's geometry and gravity field, whereas network computations must be performed in a geometrical space or on a regular geometrical surface. Therefore, in order for the raw measurements to be usable, they must be corrected for the effects of the earth's geometry and gravity field. In addition, some corrections may also have to be applied to account for the systematic effects of some environmental and instrumental factors, depending on the specific type of measurements being made (cf. Chapter 1). The purpose of observation data pre-processing is, therefore, to apply appropriate corrections to the raw observation data and reduce that data from the surface of the earth to a reference computation surface; for instance, the surface of a reference ellipsoid, or to a conformal mapping plane. The observation data type discussed here includes spatial distances, gyro-azimuths, horizontal directions, horizontal angles, zenith distances, GPS vectors, and spirit-leveled height differences. The raw observation data are first corrected for meteorological effects and instrument calibration, and then subsequently reduced, by applying gravimetrical and/or geometrical corrections, to a reference surface for coordinate calculation.

This chapter starts with a review of the coordinate systems used in geodesy and surveying and a discussion on the choice of a geodetic model in §2.1 that is fundamental in understanding the various gravimetrical and geometrical reductions involved. The appropriate meteorological and instrumental calibration corrections for raw observations are discussed in §2.2 and §2.3. Then presented are the concepts and formulae to reduce observations from the surface of the earth to the Local Geodetic coordinate system, to a reference ellipsoid, and to a conformal mapping plane in §2.4, §2.5, and §2.6, respectively. Finally, §2.7 discusses the reduction of GPS vectors to a conformal mapping plane. The mathematical relationships required to implement the data pre-processing are well documented in survey publications. Emphasis is put here to give a logical flow of the concepts involved and explain the mathematical relationships required. For a more detailed discussion on their theoretical background, the interested reader is referred to the given literature.

2.1 Coordinate Systems and Choice of a Geodetic Model

To define the position of a point on land, at sea, or in space, we need a coordinate reference system. There are a variety of coordinate systems used in geodesy and surveying, and they may in general be classified into the following three categories: terrestrial coordinate systems; celestial coordinate systems; and orbital coordinate systems. Definitions and applications of these various coordinate systems in geodesy were discussed in Krakiwsky and Wells (1971), Vanicek and Krakiwsky (1986), among others. Since this book concentrates on the positioning of a terrestrial geodetic network, terrestrial coordinate systems are needed, and those of major concern to us are the Local Astronomical (LA) system, Conventional Terrestrial (CT) system, Instantaneous Terrestrial (IT) system, Geodetic (G) system, Local Geodetic (LG) system, and Plane Cartesian (PC) system.

2.1.1 Terrestrial Coordinate Systems

First, the Local Astronomical (LA) system is the one in which the conventional geodetic observations are made. This system is centered at the site of the observer on the surface of the earth, and defined through the observer's gravity vector and the direction of the earth's conventional spin axis. The former defines its negative z-axis and together with a parallel to the conventional spin axis they define the xz-plane of the system, and finally the direction of the y-axis is chosen so that the system is left-handed (cf. Vanicek and Krakiwsky, 1986). The gravity vector and the earth's spin axis can be sensed by zenith distance and astronomical azimuth, respectively (see Figure 2.1), and the LA coordinates of a point can be expressed by

$$\begin{pmatrix} x_j \\ y_j \\ z_j \end{pmatrix}^{LA} = s_{ij} \begin{pmatrix} \sin Z_{ij} \cos A_{ij} \\ \sin Z_{ij} \sin A_{ij} \\ \cos Z_{ij} \end{pmatrix} \tag{2.1}$$

where s_{ij}, A_{ij}, and Z_{ij} are the spatial distance, astronomical azimuth, and zenith distance measured from station i to the target point j, respectively. Apparently, the LA system is topocentric and it is spinning as well as revolving with the earth.

Second, the conventional terrestrial system (CT) is a geocentric system with its origin being at the center of mass of the earth. The z-axis of the CT system is

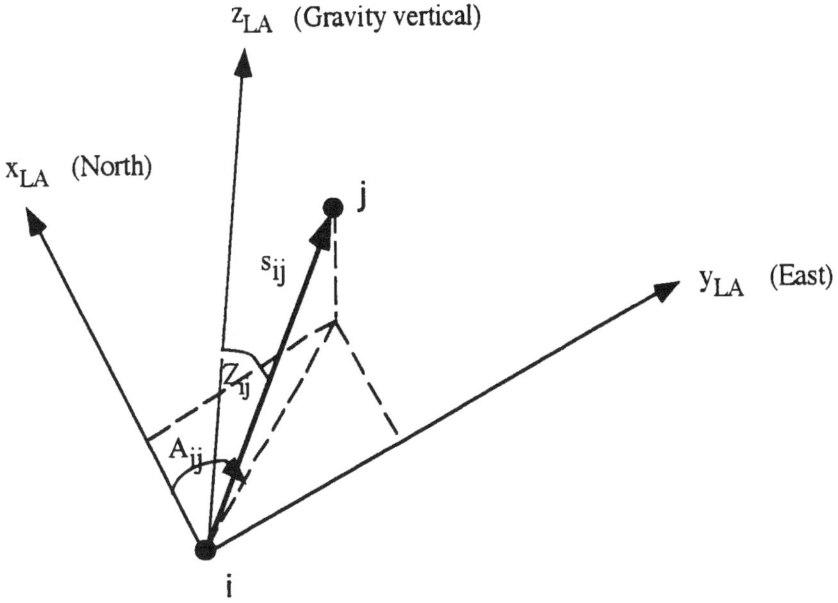

Figure 2.1. Spatial distance s_{ij}, astronomical azimuth A_{ij}, and astronomical zenith distance Z_{ij} measured in the Local Astronomical (LA) system

defined by the vector passing through the CIO (Conventional International Origin) that is defined as the mean position of the instantaneous pole during the period 1900 to 1905. The xz^{CT}-plane contains the mean Greenwich Observatory, and finally, the y^{CT}-axis is selected to make the system right-handed (cf. Vanicek and Krakiwsky, 1986). The CT system is one of the most important coordinate systems in geodesy, and it is in this system that the astronomical latitude Φ and astronomical longitude Λ are measured that define the horizontal position of a point astronomically (see Figure 2.2). The definition of the Instantaneous Terrestrial (IT) system is similar to that of the CT system except that the z^{IT}-axis coincides with the instantaneous spin axis of the earth, instead of the conventional spin axis. In this case, the z^{IT}-axis wobbles around the z^{CT}-axis. It should be noted from Figure 2.2 that the astronomical meridian plane of the observer contains both the gravity vector of the observer and the CIO. It is parallel to the conventional spin axis, but in general may not contain the center of mass of the earth.

The relation between the LA and the CT systems is shown in Figure 2.2, and a rigorous coordinate transformation between these two systems is made by

$$\begin{pmatrix} \Delta x_{ij} \\ \Delta y_{ij} \\ \Delta z_{ij} \end{pmatrix}^{CT} = \mathbf{R}_3(\pi - \Lambda)\,\mathbf{R}_2\left(\frac{\pi}{2} - \Phi\right)\mathbf{P}_2 \begin{pmatrix} \Delta x_{ij} \\ \Delta y_{ij} \\ \Delta z_{ij} \end{pmatrix}^{LA} \tag{2.2}$$

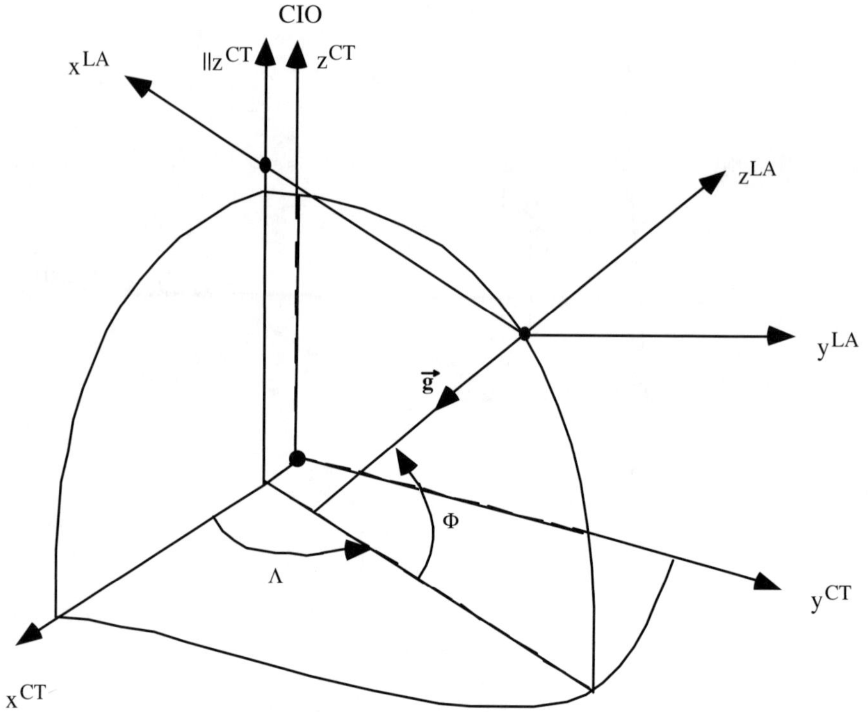

Figure 2.2. Relation between the LA and the CT systems

where $\mathbf{R}_3$ and $\mathbf{R}_2$ are rotation matrices, and $\mathbf{P}_2$ a reflection matrix, and they are defined as follows

$$\mathbf{R}_2(\omega) = \begin{pmatrix} \cos\omega & 0 & -\sin\omega \\ 0 & 1 & 0 \\ \sin\omega & 0 & \cos\omega \end{pmatrix} \tag{2.3}$$

$$\mathbf{R}_3(\omega) = \begin{pmatrix} \cos\omega & \sin\omega & 0 \\ -\sin\omega & \cos\omega & 0 \\ 0 & 0 & 1 \end{pmatrix} \tag{2.4}$$

$$\mathbf{P}_2 = \begin{pmatrix} 1 & 0 & 0 \\ 0 & -1 & 0 \\ 0 & 0 & 1 \end{pmatrix} \tag{2.5}$$

$\mathbf{R}_2(\pi/2 - \Phi)$ and $\mathbf{R}_3(\pi - \Lambda)$ in Equation (2.2) are calculated simply by replacing ω in (2.3) and (2.4) with $\pi/2 - \Phi$ and $\pi - \Lambda$, respectively.

Parallel to the definition of the LA and CT coordinate systems are the LG (Local Geodetic) and G (Geodetic) coordinate systems. The G system is superimposed on a reference ellipsoid chosen to represent the body of the earth. During the late seventeenth and early eighteenth centuries, the earth was determined to be ellipsoidal in shape, and further work in the nineteenth and early twentieth centuries by mathematicians and geodesists showed that the earth's shape is best represented by one of its equipotential surfaces, i.e., the geoid (Thomson, 1976). These historical developments in the determination of the size and shape of the earth, and numerous other physical problems, have led to the traditional splitting of the triplet of coordinates used to describe the positions of terrain points into two horizontal and one vertical component, and the surface of a properly selected biaxial (rotational) ellipsoid and the geoid are used as the reference surfaces for the computation of horizontal and vertical coordinates, respectively. The relation between the reference ellipsoid, the geoid, and the terrain is shown in Figure 2.3 where h and H represent the ellipsoidal and orthometrical height, respectively, and N the geoidal height that serves as a link between the horizontal and vertical geodetic network coordinates. The ellipsoidal height h of a point is defined as its linear distance above the ellipsoid measured along the ellipsoidal normal at the point, while the orthometric height H is defined as the distance above the geoid measured along the plumb line at the point. In practice, one usually relates the ellipsoidal and the orthometrical height by

$$h = H + N \tag{2.6}$$

This formula is precise up to the effect of the curvature of the plumb line that may be neglected in practical surveying applications.

A reference surface used for coordinate computations is called a "Geodetic Datum" in geodesy. In order for an ellipsoid to be used as the datum for horizontal coordinate computation, its size and shape as well as its position and orientation with respect to the earth must be specified. The size and shape of a (biaxial) rotational ellipsoid can be given by the lengths of its semi-major and semi-minor axes or its semi-major axis and flattening. The dimensions of a reference ellipsoid and its position in the earth body are chosen such that it best approximates the geoid either locally (i.e., a local datum) or globally (i.e., a global datum). A detailed discussion of this matter is outside the scope of this book and the interested reader is referred to Heiskanen and Moritz, 1979, Vanicek and Krakiwsky, 1986, among others. A recent example of global geodetic datum is the WGS 84 reference ellipsoid that has a semi-major axis and a flattening of 6,378,137.0 m and 1/298.257223563, respectively. The GPS observations are referred to the WGS 84 coordinate system.

The Geodetic (G) coordinate system is defined with respect to the reference ellipsoid. It is a right-handed system with origin being coincident with the center of the reference ellipsoid. The z^G axis is directed along the semi-minor axis of the ellipsoid, the x^G axis is the intersection of the equatorial plane of the ellipsoid and

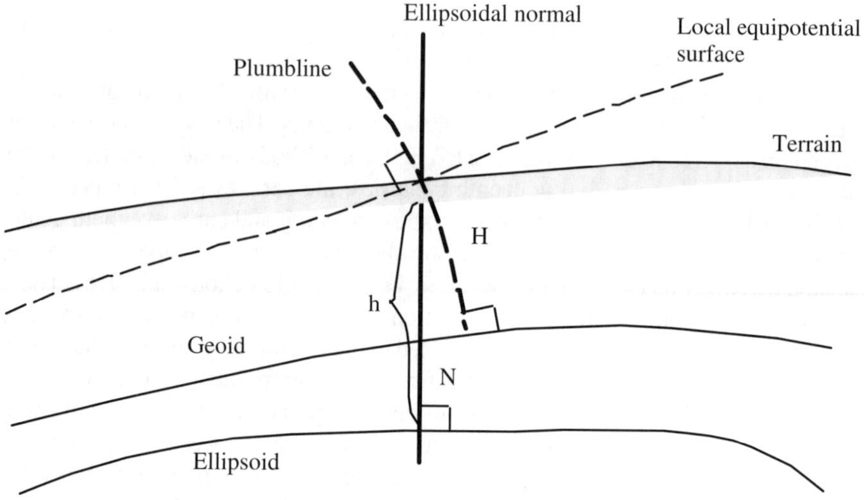

Figure 2.3. Reference surfaces and heights

the Greenwich meridian plane that is defined as the plane containing the semi-minor axis and cutting the surface of the ellipsoid, and finally the y^G axis is chosen to form a right-handed system (see Figure 2.4). In the G system, the position of a terrain point above the reference ellipsoid can be expressed either by the triplet of Cartesian coordinates $(x, y, z)^G$ or by the curvilinear coordinates $(\phi, \lambda, h)^G$. Here, ϕ, λ, and h denote the geodetic latitude, geodetic longitude, and ellipsoidal height, respectively. The geodetic latitude of a point is defined as the acute angular distance between the equatorial plane of the reference ellipsoid and the ellipsoidal normal through the point measured in the meridian plane of the point. The geodetic longitude of a point is defined as the counterclockwise angular distance between the Greenwich meridian plane and the meridian plane of the point, measured in the equatorial plane. It can be shown that the geodetic Cartesian coordinates $(x_i, y_i, z_i)^G$ and their corresponding geodetic curvilinear coordinates $(\phi_i, \lambda_i, h_i)^G$ of a point are related by the following equation:

$$\begin{pmatrix} x_i \\ y_i \\ z_i \end{pmatrix}^G = \begin{pmatrix} \left(N_i + h_i^G\right)\cos\phi_i^G \cos\lambda_i^G \\ \left(N_i + h_i^G\right)\cos\phi_i^G \sin\lambda_i^G \\ \left(N_i \dfrac{b^2}{a^2} + h_i^G\right)\sin\phi_i^G \end{pmatrix} \tag{2.7}$$

where N_i represents the prime vertical radius of curvature at the point, and a and b the semi-major and the semi-minor axes of the reference ellipsoid, respectively.

Another important coordinate system is the Local Geodetic (LG) coordinate system that is centered at the point of interest (i.e., topocentric) and oriented with respect to the chosen reference ellipsoid. The z^{LG} axis is the outward ellipsoid normal passing through the point of interest, the x^{LG} axis points to the geodetic north, and the y^{LG} axis completes a left-handed system. The relation between the LG and G systems are somewhat similar to that between the LA and CT systems shown in Figure 2.2. It can be seen from Figure 2.2 that the LA system is related to the CT system through the astronomical latitude (Φ) and the astronomical longitude (Λ), while in Figure 2.4 the LG system is related to the G system through the geodetic latitude (ϕ) and the geodetic longitude (λ).

The coordinate transformation between the LG and G systems are thus given by

$$\begin{pmatrix} \Delta x_{ij} \\ \Delta y_{ij} \\ \Delta z_{ij} \end{pmatrix}^{G} = \mathbf{R}_3(\pi - \lambda)\mathbf{R}_2\left(\frac{\pi}{2} - \phi\right)\mathbf{P}_2 \begin{pmatrix} \Delta x_{ij} \\ \Delta y_{ij} \\ \Delta z_{ij} \end{pmatrix}^{LG} \tag{2.8}$$

where $\mathbf{R}_3$, $\mathbf{R}_2$, and $\mathbf{P}_2$ mean the same as before.

In the LG system are defined the geodetic azimuth α and the geodetic zenith distance Z' (see Figure 2.5). The LA system is a natural coordinate system that bears the effects of the physical properties of the earth, while the LG system is not. Therefore, the LG system can also serve as a reference system for coordinate computations.

Finally, the position and orientation of the reference ellipsoid within the earth can be described by the translation and rotation of the G coordinate system with respect to the CT coordinate system. If one denotes (x_{e0}, y_{e0}, z_{e0}) and (ω_x, ω_y, ω_z) as the translation components and the misalignment angles of the G system with respect to the CT system, the coordinate transformation between the G and the CT systems follows the rule of the similarity transformation as follows

$$\begin{pmatrix} x_i \\ y_i \\ z_i \end{pmatrix}^{CT} = \mathbf{R}_1(\omega_x)\mathbf{R}_2(\omega_y)\mathbf{R}_3(\omega_z) \begin{pmatrix} x_i \\ y_i \\ z_i \end{pmatrix}^{G} + \begin{pmatrix} x_{e0} \\ y_{e0} \\ z_{e0} \end{pmatrix}^{CT} \tag{2.9}$$

where $\mathbf{R}_1$, $\mathbf{R}_2$, and $\mathbf{R}_3$ are the rotation matrices. $\mathbf{R}_2$ and $\mathbf{R}_3$ are calculated using Equations (2.3) and (2.4), and $\mathbf{R}_1$ is defined as follows

$$\mathbf{R}_1(\omega) = \begin{pmatrix} 1 & 0 & 0 \\ 0 & \cos\omega & \sin\omega \\ 0 & -\sin\omega & \cos\omega \end{pmatrix} \tag{2.10}$$

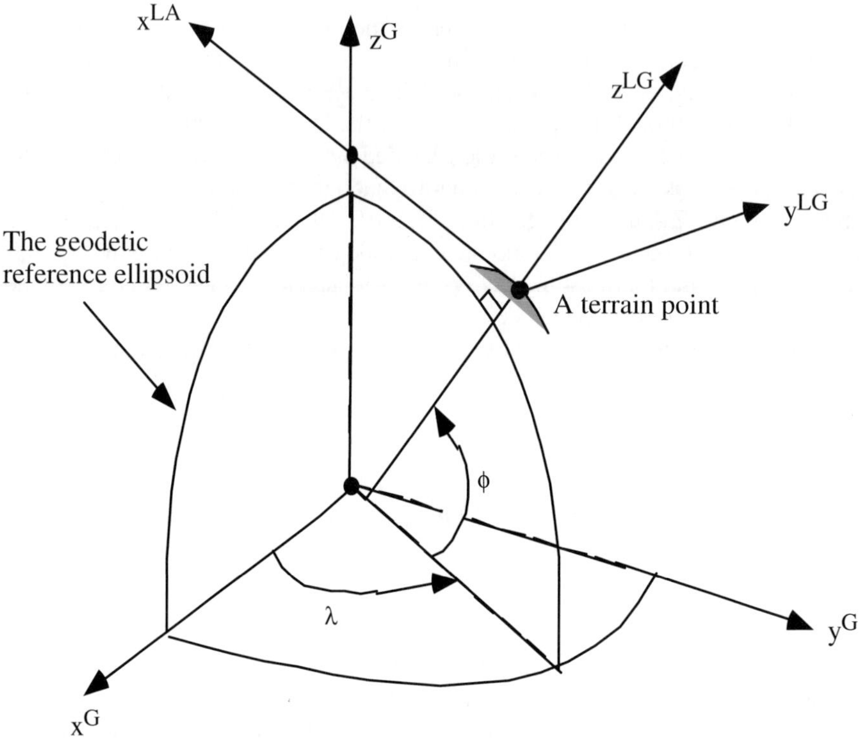

Figure 2.4. The LG and G coordinate systems

In practice, it is advantageous to set the axes of the G and the CT systems parallel to simplify the problem; i.e.,

$$\omega_x = \omega_y = \omega_z = 0 \tag{2.11}$$

Further, for a global geodetic datum (e.g., the WGS 84), the translation components are also set to zeros; i.e.,

$$x_{e0} = y_{e0} = z_{e0} = 0 \tag{2.12}$$

In this case, the reference ellipsoid is referred to as a geocentric ellipsoid. When the G system and the CT system are properly aligned, the relation between the natural astronomical (physically meaningful and observable) quantities (Φ, Λ, A, Z) and their corresponding geodetic quantities (ϕ, λ, α, Z') can be derived as follows (cf. Vanicek and Krakiwsky, 1986):

$$\Delta A = A - \alpha = (\Lambda - \lambda)\sin\phi \tag{2.13}$$

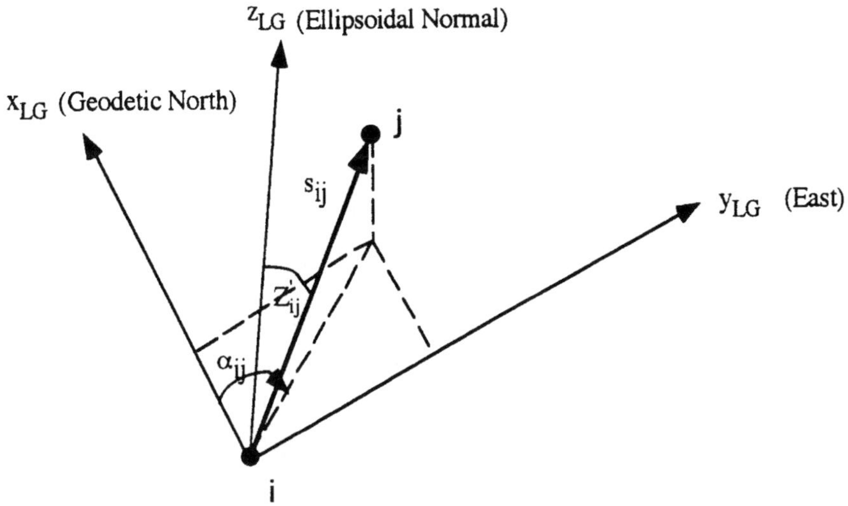

Figure 2.5. Spatial distance s_{ij}, geodetic azimuth α_{ij}, and geodetic zenith distance Z'_{ij} in the Load Geodetic (LG) system

$$\Phi - \phi = \xi \tag{2.14}$$

$$(\Lambda - \lambda)\cos\phi = \eta \tag{2.15}$$

$$Z - Z' = -\xi \cos A - \eta \sin A \tag{2.16}$$

where Equation (2.13) is also called the Laplace equation for azimuth. ξ and η represent the meridian and prime vertical components of the deflection of the vertical, respectively. The relation between the orthometric height H and the ellipsoidal height h has been given by (2.6). These relations serve as the basis for the correction and reduction of raw observations discussed in the following sections. Finally, the relation between the LG and the LA systems is given by

$$\begin{pmatrix} x_i \\ y_i \\ z_i \end{pmatrix}^{LG} = \mathbf{R}_3(\Delta A)\mathbf{R}_2(-\xi)\mathbf{R}_1(\eta)\begin{pmatrix} x_i \\ y_i \\ z_i \end{pmatrix}^{LA} \tag{2.17}$$

where ΔA describes the angle between the x^{LG} and the x^{LA}-axes. For a detailed discussion of some other coordinate systems, and the celestial and orbital coordinate systems as well as their transformations in geodesy, the interested reader is referred to the given literature.

2.1.2 Choice of a Geodetic Model

The Classical 2D Plus 1D Model

As discussed above, due to historical and some practical reasons, the triplet of coordinates used to define the position of a point on the surface of the earth are conventionally separated into two horizontal coordinates and a vertical coordinate, and the computations are done on different reference surfaces. The horizontal coordinates may be geodetic latitude ϕ and geodetic longitude λ calculated on the surface of a reference ellipsoid or Cartesian coordinates X (Easting) and Y (Northing) on a conformal mapping plane. The vertical coordinate is a rigorously defined geodetic height such as dynamic height or orthometric height. When this classical approach is used, the horizontal and vertical control points are often different, especially in the case of national geodetic control networks. That is, it may usually happen that horizontal control points have accurate horizontal coordinates but only approximate heights, which may be scaled on the map, and vertical control points have accurate heights but only approximate horizontal coordinates, which may also be scaled on the map. This happens mainly due to practical reasons. Since the traditional surveying measurements (i.e., azimuth, directions and baselines [horizontal distances]) are only sensitive to the horizontal coordinates and they require intervisibility between the stations, horizontal geodetic control points were in general located on hilltops. The vertical coordinate is determined by precise geodetic leveling that is very difficult to run through mountains. Therefore, one usually designs a totally different set of points as vertical control points and locates them along the roads or public right of way so that the leveling route is easily accessible (Teskey, 1979). It can be said that the classical approach has been in use ever since man first began to establish geodetic control networks, and it is still in use today in all the national geodetic networks over the world and almost exclusively in high-precision surveys for engineering and geoscience projects.

The Modern 3D Model

With the development of modern survey equipment and methods, e.g., EDM (Electromagnetic Distance Meter), GPS (Global Positioning System) and some other extraterrestrial positioning techniques, such as VLBI (Very Long Baseline Interferometry), SLR (Satellite Laser Ranging), etc., over the last two decades, geodesists have started to investigate establishing geodetic control networks in three-dimensional (3D) space. In this case, the position of a point on the terrain is described by the triplet of coordinates (x, y, z) simultaneously. The type of coordinates computed depends on the specific coordinate system chosen, e.g., the CT, G, or LG coordinate system discussed above. Nevertheless, since a rigorous transformation between different coordinate systems can be performed as shown above, equivalent results can be obtained, up to the precision of linearization, whichever coordinate system is used.

There have been many discussions on the advantages and disadvantages of the 3D model. For instance, in the 3D model all three coordinates are determined for every point and since the spatial EDM distances, horizontal directions, zenith distances, and azimuths are actually observed in three-dimensional space, they need only be corrected for refraction effects and instrumental corrections and require no further reduction to some reference surface, except for correcting for gravimetrical effects depending on the accuracy and the extent of the network under study. Also, the 3D model can fully utilize satellite data, fully utilize the output of other systems that operate in 3D space; e.g., photogrammetric systems and inertial systems. The main disadvantage of the 3D model is that because of the variations in the gravity field and the uncertainty associated with vertical refraction, the height component may be less precisely determined than the horizontal components if only trigonometric measurements are made. Spatial distances can be used to determine accurately heights only when the lines of sight are deep.

Finally, the choice of a geodetic model and a coordinate system should be dictated by the purpose and extent of the network, the measuring technique adopted, and the ease of implementing the application. In general, for engineering applications of high accuracy, the classical 2D plus 1D model is still preferred to the modern 3D model, especially for geodetic networks of local extent. This is mainly due to another practical reason: it is a most convenient procedure to set out heights from vertical control and to set out horizontal positions from horizontal control (Tesky, 1979). In the classical model, heights, whether dynamic or orthometric, have a definite physical meaning; that is, points having the same dynamic height lie on the same equipotential surface, and points having the same orthometric height are the same height above the geoid. Even though the modern GPS technique no longer requires intervisibility between network stations, and both the horizontal and vertical coordinates of a point can be determined simultaneously, the height determined by the GPS technique is the height above the ellipsoid, while the height determined by geodetic leveling is the height above the geoid. Therefore, in order to set out an ellipsoidal height from geodetic leveling, an accurate value of the geoid-ellipsoid separation (cf. [2.3]) for the given point must be determined, which is unfortunately unavailable at present for engineering projects of high precision (i.e., 1:100,000 or less) especially in areas where variations in the gravity field are large. Therefore, the GPS technique is currently widely used for horizontal control, and vertical control is still provided by geodetic leveling for high-precision engineering projects.

2.2 Meteorological Corrections for Terrestrial Geodetic Observations

When radiation propagates through the infinite "layers" of the atmosphere, it undergoes some change in both speed and direction due to refractivity (i.e., the refractive index). A brief review of the theory of the propagation of the electromagnetic waves in the atmosphere and the effects of atmospheric refraction on geodetic measurements have been given in §1.2.1. Meteorological corrections are applied to geodetic observations obtained under actual atmospheric conditions to correct for systematic errors. These errors are present due to the fact that the velocity of electromagnetic waves in the atmosphere is not equal to the ideal value that would be obtained in a vacuum. Additionally, the paths followed by electromagnetic waves in the atmosphere are not straight lines. A summary of the various meteorological corrections that have to be applied to conventional terrestrial geodetic measurements will be given below to account for the effects of atmospheric refraction.

2.2.1 EODM (Electro-Optical Distance Meter) Spatial Distances

The meteorological corrections for EODM distances are the First Velocity correction, the Second Velocity correction, and the Wave Path to Chord correction.

First Velocity Correction

The First Velocity correction, if not applied, creates a systematic scale error. This error is present due to the fact that the actual refractive index (n), at the time the measurement is taken, differs from the reference refractive index (n_{ref}) fixed by the manufacturer for the EODM. The First Velocity correction may be written as (cf. Eq. [1.44]):

$$\delta s_1 = \frac{s'(n_{ref} - n)}{n} \tag{2.18}$$

where s′ is the measured distance, and n the actual refractive index that can be calculated, in the case of visible and the infrared spectrum, using Equation (1.11). The refractive indices are computed for both terminals of a line, and the mean of both values is subsequently used to correct the distance; i.e.,

$$n = \frac{1}{n}(n_1 + n_2) \tag{2.19}$$

where n_1 and n_2 are the actual refractive indices calculated at the two terminals of a line.

Second Velocity Correction

Theoretically, if one assumes a spherically layered atmosphere, the mean refractive index at both terminals of the line would be valid for the circular curve with its radius of curvature being the mean radius of curvature of the spheroid (R) along the line (see Figure 2.6).

This, however, is not valid for the actual wave path with its radius of curvature being ρ. Therefore, for the actual wave path, assuming linear vertical gradients of temperature and pressure, the refractive index would be (cf. Rüeger, 1980):

$$n = \frac{1}{2}(n_1 + n_2) + \Delta n \tag{2.20}$$

The Second Velocity correction accounts for the small systematic scale error introduced by Δn, such as:

$$\delta s_2 = -s'\Delta n \tag{2.21}$$

where Δn is calculated by (Rüeger, 1980):

$$\Delta n = (k - k^2)\frac{s'^2}{12R^2} \tag{2.22}$$

with k being the coefficient of lateral refraction defined by Equation (1.16), s' the measured distance with First Velocity correction applied, and R the mean radius of the earth along the line. If the average value of 0.13 is assumed for the coefficient of refraction (k), the Second Velocity correction amounts to 0.1 mm for a distance of 7.6 km and can be neglected for most engineering applications.

Wave Path to Chord Correction

The actual wave path is a curve with approximate radius of curvature being ρ. The Wave Path to Chord correction reduces the measured distance along a curved wave path to the chord distance between the two terminals of the line. The correction is (Rüeger, 1980):

$$\delta s_3 = -k^2 \frac{s_1^3}{24R^2} \tag{2.23}$$

where k is the coefficient of refraction, s_1 the measured distance with the First and Second Velocity corrections applied, and R the mean radius of the earth along the line. Similarly, if the value of 0.13 is assumed for the coefficient of refraction (k), the Wave Path to Chord correction amounts to 0.1 mm for a distance of 17.9 km and can thus be neglected for most engineering applications.

2.2.2 Meteorological Correction for Angular Measurements

As discussed in §1.2, in the atmosphere, an optical path is not a straight line due to the heterogeneous densities of the air. In practice, meteorological corrections are rarely applied to angular measurements due to the difficulties in obtaining the appropriate meteorological data. Nevertheless, a summary of the corrections is

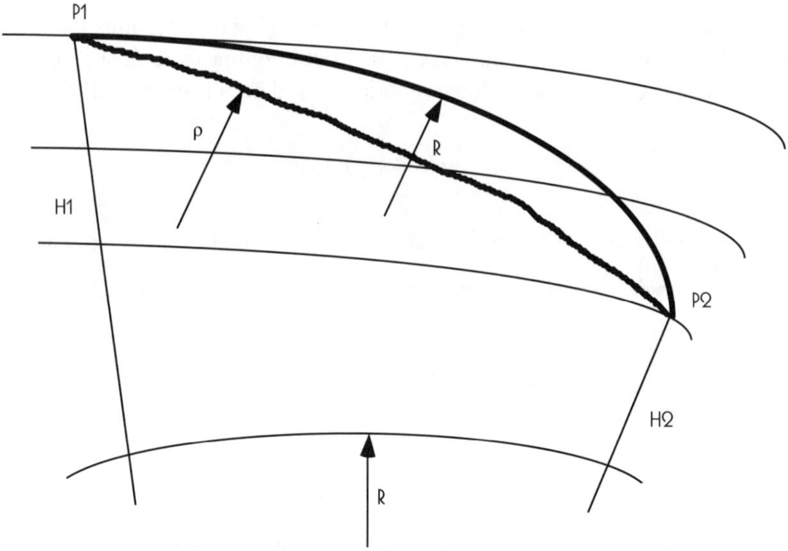

Figure 2.6. Second Velocity correction

given below and it can be used to estimate the expected influences of the atmospheric refraction on these measurements.

Gyro-Azimuth and Horizontal Directions

In practice, to derive the meteorological corrections for angular measurements, the refracted optical path is usually accepted as being a circular curve with the curvature calculated for a mean gradient of n. The meteorological correction for a gyro-azimuth or a horizontal direction is (cf. Eq. [1.17])

$$\delta\alpha_{\text{met.}} = \frac{k_h s}{2R} \tag{2.24}$$

where s is the length of the line of sight, and k_h is the coefficient of lateral refraction of the line of sight.

Horizontal Angles

An observed horizontal angle at a station can be expressed as the difference of two azimuths or directions originating from the same station (see Figure 1.3). Thus, the meteorological correction for an observed horizontal angle reads (cf. Eq. [1.19]):

$$\delta\beta_{\text{met.}} = \frac{1}{2R}\left(k_h^{ik} s_{ik} - k_h^{ij} s_{ij}\right) \tag{2.25}$$

where s_{ij} and s_{ik} are the length of the line of sight along the left and right arms of the angle, respectively, and k_h^{ij}, k_h^{ik} their corresponding coefficient of lateral refraction.

Zenith Distances

Similar to Equation (2.24), the meteorological correction for an observed zenith distance is (cf. Eq. [1.18])

$$\delta z_{met.} = \frac{k_v s}{2R} \tag{2.26}$$

where s is the length of the line of sight, and k_v the coefficient of vertical refraction.

2.3 Instrument Calibration Corrections for Terrestrial Geodetic Observations

The magnitude of instrument errors and constants is kept small by the manufacturers, and their effects are included in the rated accuracy of a type of instrument. The procedure of instrument calibration serves the purpose of verifying the systematic errors and constants of any instrument and deriving corrections to be applied to raw measurements. In the following subsections, the calibration corrections for EODM distances, gyro-azimuths, and spirit-leveled height differences are presented. In general, no calibration corrections are needed for horizontal directions, horizontal angles, and zenith distances because proper observational procedures will be followed to remove the influences of the systematic instrumental errors on these measurements. It should be pointed out that the observation data dealt with in this section are assumed to have been subject to meteorological corrections, when applicable, as discussed in the above section.

2.3.1 Calibration Corrections for EODM Distances

As discussed in §1.2.2, the instrumental calibration corrections for EODM distances are: Cyclic error correction; Zero error correction; and Scale error correction.

Cyclic Error Correction

The cyclic error of an EODM is caused primarily by electric cross-talk within the instrument. The error can be modeled over the defined unit length of the instrument (the period of the cyclic error) by the use of a sine curve, as given by Equation (1.38) as follows

$$s_{mea.} = s_1 + x_{ref.} - A \sin\left[\frac{2\pi}{U}\left(s_{mea.} + B\right)\right] \tag{2.27}$$

where U is the unit wavelength, A the amplitude of the Cyclic error correction, B the phase of the Cyclic error correction; s_1 the distance from the instrument to the first reflector position, $s_{mea.}$ a measured distance to the present reflector position; and finally $x_{ref.}$ is a known distance between the present reflector position and the first reflector position. The Cyclic error correction to a measured distance is then given by

$$\delta c = A \sin\left[\frac{2\pi}{U}(s + B)\right] \tag{2.28}$$

Zero Error Correction

The virtual electro-optical origin, i.e., the zero of an EODM, is usually not coincident with the vertical axis of the instrument. In addition, the zero of the reflector may not coincide with the vertical axis of the centering device. Together, these differences constitute a systematic constant error in a measured distance. The Zero error correction can be determined through a proper calibration procedure. A measured distance is related to the Zero error correction as follows:

$$s^0 = s + u \tag{2.29}$$

where u is the Zero error correction. s^0 and s represent a known distance and its measured value by the EODM, respectively. The Zero error correction can be estimated by averaging a set of redundant observations. The Zero error correction for a measured distance is therefore

$$\delta s_z = u \tag{2.30}$$

Scale Error Correction

The instrumental scale error of an EODM is attributable to the difference between the reference and the actual modulation frequencies. The actual scale of an EODM can be obtained either by measuring the actual modulation frequency with a frequency counter or by standardizing the instrument on a baseline of known length. In the former case, the scale (k_m) of the EODM can be calculated by

$$k_m = \frac{f^0}{f} \tag{2.31}$$

where f^0 and f are the reference modulation frequency and the actual measured modulation frequency of the instrument, respectively. The scale parameter is estimated by averaging a set of redundant observations. The Scale error correction for a measured distance is, therefore, calculated by

$$\delta_s = (k_m - 1)s \tag{2.32}$$

where s is the measured distance after Cyclic and Zero error corrections.

The scale of an EODM can also be determined by standardizing the instrument on a baseline of known length. In this case, the scale of the EODM (k_m) is calculated by

$$k_m = \frac{s^0}{s} \tag{2.33}$$

where s^0 and s represent a known distance and its corresponding measured distance with Cyclic and Zero error corrections applied, respectively. The scale parameter is estimated by averaging a set of redundant observations, and the Scale error correction to a measured distance is calculated using Equation (2.32).

Finally, the Zero error correction and the scale of an EODM may be determined simultaneously. Theoretically, the influences of the Zero error and the scale of an EODM on the measured distances are not separable. Therefore, if the Zero error correction and the scale of an EODM are to be determined by comparing the measured distances to the known distances, a combined solution for the Zero error correction and scale should be made. This can be achieved through the following equation

$$s^0 = k_m(s + u) \tag{2.34}$$

where s^0 is a known distance, and s the measured distance with Cyclic error correction applied, u the Zero error correction, and k_m is the actual scale of the instrument. The unknown parameters, u and k_m, in Equation (2.34) are determined by the method of least squares from a set of redundant observations.

2.3.2 Calibration Corrections for Gyro-Azimuths

The calibration correction for a gyro-azimuth is an unstable constant, usually denoted by E. It is the horizontal angle between the plane of the meridian and the direction of the zero mark of the gyro. The manufacturer cannot always set the E value to zero but attempts to construct the instrument so that the E value is reasonably constant. The value of E must be determined and verified before and after any measurements are made. The calibration correction $\delta\alpha_{cali.}$ for a measured gyro-azimuth can be expressed by

$$\delta\alpha_{cali.} = E \tag{2.35}$$

where E is the calibration constant. The E value can be determined by comparing the measured gyro-azimuth of a line to the known astronomical azimuth as follows:

$$E = A - \alpha_{Gyro} \tag{2.36}$$

where A and α_{Gyro} are the astronomical and gyro-azimuth of the same line, respectively. With a redundant set of observations, an optimum value for E may be obtained by averaging.

2.3.3 Calibration Corrections for Geodetic Leveling

The instrumental calibration corrections to be presented here for geodetic leveling are rod scale error correction and rod index error correction.

Rod Scale Error Correction

If the scale of the rod is not equal to 1.0, a scale error correction must be applied to the leveling data. The scale of the rod can be determined by comparing the length of the Invar strip on the rod to the national length standard. The rods should be calibrated for scale at the beginning and at the end of a survey campaign. The scale error correction for a leveled height difference can be calculated as follows:

$$\delta h_s = [s_b B - s_f F] - [B - F] \tag{2.37}$$

where s_b and s_f are the calibrated scales of the backsight and foresight rods, respectively, and B and F are respectively the backsight and foresight observations.

Rod Index Error Correction

The rod index error is an offset caused by the zero graduation on a leveling stave not representing the bottom of the stave (the BM contact surface). This error can be eliminated through the design of a proper observation procedure, for instance, by using an even number of setups, or one setup with the same rod used for both the backsight and foresight. Otherwise, a correction must be applied as follows

$$\delta h_{inx.} = -[u_b - u_f] \tag{2.38}$$

where u_b and u_f are the rod index errors of the backsight and foresight rods, respectively.

2.4 Gravimetric Corrections: Reduction of Observations to the Local Geodetic Coordinate System

After the raw observations have been subject to meteorological and instrument calibration corrections as discussed in §2.2 and §2.3, gravimetric corrections are applied to geodetic observations to account for the inconsistency between the geoid and the selected reference ellipsoid. Geodetic observations made on the surface of the earth refer to the physical reference elements of the earth. A measured gyro-azimuth, a horizontal direction (angle), and a zenith distance refer to the instantaneous rotation axis of the earth, the local plumb line, and the local astronomical horizon, respectively. For instance, when horizontal directions are measured on the surface of the earth, the instrument is leveled to insure that the vertical axis is coincident with the local gravity vector. Since the local gravity vector and the normal to the ellipsoid at the instrument are in general not coincident, a correction for the deflection of the vertical is needed in order to refer the measured directions on the earth surface to the ellipsoidal normal. With gravimetric corrections applied, geodetic measurements are corrected to refer to the ellipsoidal normal and the local geodetic horizon. A gravimetric correction must be applied to the spirit-leveled height differences to obtain orthometric heights. No gravimetric corrections are required for the measured slope distances.

2.4.1 Gravimetric Corrections for Gyro-Azimuths

A measured gyro-azimuth is the astronomical azimuth referring to the instantaneous rotation axis of the earth. The discrepancy between the astronomical azimuth and the geodetic azimuth are due to the deflection of the vertical at the instrument station. With parallelity of the coordinate axes of the geodetic coordinate system to those of the Conventional Terrestrial (CT) coordinate system, the following gravimetric correction is applied to a measured gyro-azimuth to obtain the geodetic azimuth (Leick, 1990; Mueller, 1977; Vanicek and Krakiwsky, 1986):

$$\delta\alpha_g = c_1 + c_2 \tag{2.39}$$

where

$$c_1 = -\eta_i \tan\phi_i \tag{2.40}$$

$$c_2 = -(\xi_i \sin A_{ij} - \eta_i \cos A_{ij}) \cot Z_{ij} \tag{2.41}$$

ξ_i and η_i are the components of the deflection of the vertical in the meridian and the prime vertical planes, respectively, at the instrument point i, Z_{ij} is the zenith distance from the instrument point to the target point, A_{ij} the measured azimuth of the line ij, and finally, ϕ_i is the latitude of the instrument point.

It is clear that to apply the correction $\delta\alpha_g$, the deflections of the vertical at each point are required. They may be obtained in various ways. The most straightforward approach is to observe the astronomic coordinates (Φ, Λ) at each station and then calculate the meridian and the prime vertical components (ξ, η) of the deflection of the vertical according to Equations (2.14) and (2.15). This is, however, very laborious and costly. Alternatively, one may utilize the results of a contemporary geoid computation technique and compute (ξ, η) at each point (cf. Vanicek and Krakiwsky, 1986).

2.4.2 Gravimetric Corrections for Horizontal Directions and Horizontal Angles

The gravimetric corrections to be applied to the horizontal directions are the same as those given for an azimuth (Vanicek and Krakiwsky, 1986). However, the correction c_1 is optional since it is the same for all the directions at the instrument station and will become a part of the orientation unknown in the adjustment. The gravimetric correction to a measured horizontal angle is given as the difference between the corrections to two horizontal directions as follows (cf. Figure 1.3):

$$\delta\beta_g = (c_2)_{ik} - (c_2)_{ij} \tag{2.42}$$

where c_2 is given by Equation (2.41).

2.4.3 Gravimetric Corrections for Zenith Distances

From Equation (2.16), the gravimetric correction for a measured zenith distance is as follows

$$\delta z_g = (\xi_i \cos A_{ij} + \eta_i \sin A_{ij}) \tag{2.43}$$

where ξ_i, η_i are the components of the deflection of the vertical in the meridian and the prime vertical planes, respectively, at the instrument point i, and A_{ij} is the measured azimuth of the line ij.

2.4.4 Gravimetric Corrections for Geodetic Leveling

The gravimetric correction to be applied to the spirit-leveled height differences is called the orthometric correction that compensates for the nonparallelity of the gravity equipotential surfaces. The correction is derived as follows (cf. Torge, 1980; Vanicek and Krakiwsky, 1986):

$$OC_{ij} = \sum_{k=i}^{j} \frac{\bar{g}_k - g_R}{g_R} \Delta h_k + H_j^o \frac{g_R - \bar{g}_j'}{g_R} - H_i^o \frac{g_R - \bar{g}_i'}{g_R} \tag{2.44}$$

where Δh_k is the leveled height difference between points k and k-1, g_R is the reference gravity, $\bar{g}_i'$ and $\bar{g}_j'$ are the mean gravity along the plumb line at points i and j, respectively, and H_i^o and H_j^o, are the orthometric heights at points i and j, respectively. The computation of the orthometric correction OC_{ij} is done iteratively, since it involves the unknown orthometric heights.

Finally, for a geodetic leveling network of large scale and high precision, the influence of the tidal phenomena, which includes lunar, solar, and ocean tidal forces, may have to be considered. The interested reader is referred to Blazs and Young (1982), Vanicek (1980), Vanicek and Krakiwsky (1986), etc., for details.

2.5 Geometrical Reduction of Observations to the Surface of a Reference Ellipsoid

When two-dimensional geodetic coordinates, i.e., geodetic latitude and longitude (ϕ, λ) are needed, the network computations may be done on the surface of the selected reference ellipsoid. In this case, raw observations made on the surface of the earth must be reduced geometrically to the surface of the reference ellipsoid. This section discusses the mathematical tools needed to perform this geometrical reduction. It should be noted that the observations dealt with in this section must have been corrected for the meteorological, calibration, and gravimetric effects as discussed in the previous Sections 2.2, 2.3, and 2.4, respectively. After effects of the earth's gravity field have been removed, reduction of geodetic measurements from the LG system to the surface of a reference ellipsoid accounts for the effects of the earth's geometry.

2.5.1 Reduction of EODM Distances

A two-phased process may be involved to reduce a spatial EODM distance to the surface of the reference ellipsoid; that is: terrain to the mark-to-mark distance, and then the mark-to-mark distance to the ellipsoid. The former accounts for the effect of the height of the instrument and/or the target, and the latter the effect of the geometry of the earth. At each phase, depending on the observed quantity, different mathematical procedures may be involved.

2.5.1.1 Reduction to the Mark-to-Mark Distances

Reduction Using Station Heights

Assuming no eccentricity of the EODM and the reflector, a measured distance between the center of the EODM and the center of the reflector can be reduced to the distance between the reference survey marks as shown in Figure 2.7. If the ellipsoidal heights of both survey marks are known, the mark-to-mark distance l_{ij}^{g} is related to the measured distance (Δr_{ij}^{EM}) by applying the cosine rule to the two triangles ijc and $i_g j_g c$ as follows:

$$l_{ij}^{g} = \sqrt{a_2^2 + b_2^2 - 2 a_2 b_2 \cos\theta} \tag{2.45}$$

where

$$\cos\theta = \frac{a_1^2 + b_1^2 - \Delta r_{ij}^2}{2\, a_1\, b_1} \tag{2.46}$$

$$a_1 = R_m + h_i + h_{EM} \tag{2.47}$$

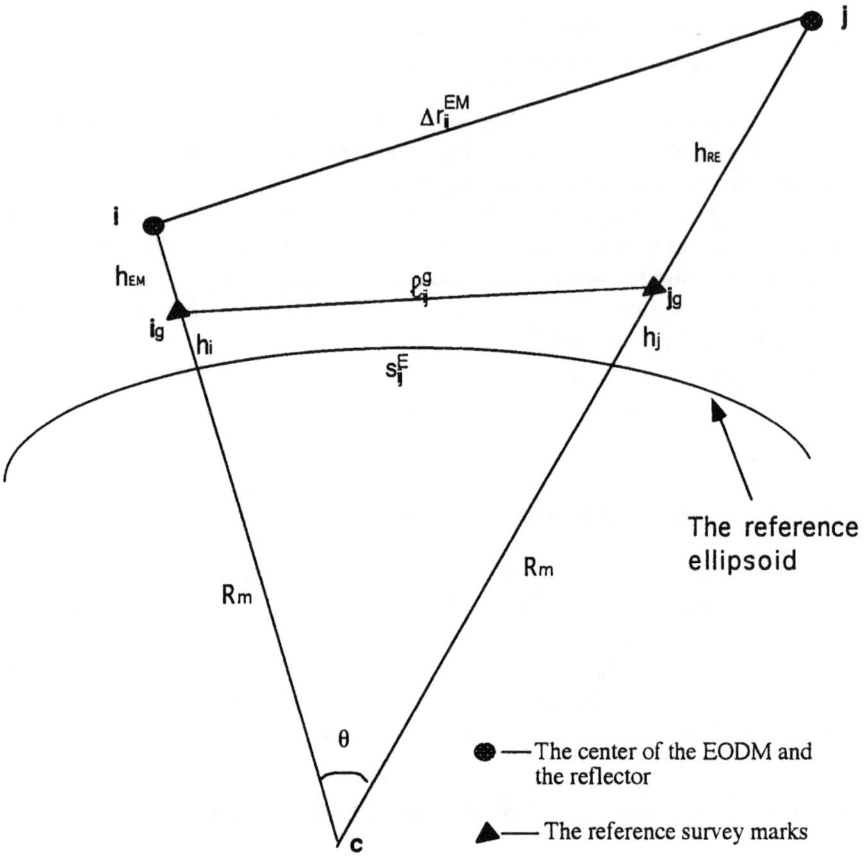

Figure 2.7. Geometrical reduction of distances using station heights

$$b_1 = R_m + h_j + h_{RE} \tag{2.48}$$

$$a_2 = R_m + h_i \tag{2.49}$$

$$b_2 = R_m + h_j \tag{2.50}$$

$$R_m = \frac{1}{2}[R_i(\alpha) + R_j(\alpha)] \tag{2.51}$$

$$R_i(\alpha) = \frac{M_i N_i}{M_i \sin^2\alpha + N_i \cos^2\alpha} \tag{2.52}$$

$$M_i = \frac{a(1-e^2)}{[1-e^2 \sin^2 \phi_i]^{3/2}} \tag{2.53}$$

$$N_i = \frac{a^2}{[a^2 \cos^2 \phi_i + b^2 \sin^2 \phi_i]^{1/2}} \tag{2.54}$$

$$e = \sqrt{\frac{a^2 - b^2}{a^2}} \tag{2.55}$$

h_i = the ellipsoid height of the survey mark at point i_g
h_j = the ellipsoid height of the survey mark at point j_g
h_{EM} = the height of the EODM above the survey mark i_g
h_{RE} = the height of the reflector above the survey mark j_g
ϕ_i = the latitude of a survey mark
R_α = the radius of curvature in the direction of the line
Δr_{ij}^{EM} = the measured slope distance
l_{ij}^{g} = the reduced mark-to-mark slope distance
M_i = the radius of curvature in the meridian plane
N_i = the radius of curvature in the prime vertical plane
a, b = semi-major and semi-minor axes of the reference ellipsoid
e = the first eccentricity of the reference ellipsoid, and finally
α = the azimuth of line ij.

Reduction Using Measured Zenith Distances

If the reduction of a measured EODM distance to the mark-to-mark distance is to be performed using the ellipsoidal height of the survey mark at the instrument station and a measured zenith distance, two steps may be involved. The first step is to correct for the influences of unequal instrument, target, and reflector heights, and the second step is to reduce the corrected distance to the mark-to-mark distance.

First, if the center of the EODM does not coincide with that of the theodolite and/or if the center of the reflector does not coincide with that of the target, a correction has to be made to the measured EODM distance (Δr_{ij}^{EM}), to obtain the chord distance (Δr_{ij}^{TM}) along the optical path of the zenith distance (see Figure 2.8). The correction (δs_{ET}) is derived as follows (Rüeger, 1980):

$$\delta s_{ET} = \Delta h_{ET} \cos Z_{TH} - \frac{\Delta h_{ET}^2}{2\Delta r_{ij}^{EM}} \tag{2.56}$$

where

$$Z_{TH} = Z + \delta z_g + \frac{\Delta r_{ij}^{EM} k}{2R} \tag{2.57}$$

$$\Delta h_{ET} = (h_{EM} - h_{TH} + h_{TR} - h_{RE}) \tag{2.58}$$

Z and δz_g are respectively the measured zenith distance and its corresponding gravimetric correction calculated by Equation (2.23), k the vertical coefficient of refraction, R the mean radius of the earth, h_{EM} and h_{TH} are respectively height of the EODM and the theodolite above the survey mark i_g, and finally, h_{RE} and h_{TR} are respectively height of the reflector and the target above the survey mark j_g.

Second, referring to Figure 2.8, the mark-to-mark distance (l_{ij}^{g}) is obtained from the distance (Δr_{ij}^{TH}) using Equation (2.45) as follows:

$$l_{ij}^{g} = \sqrt{a_2^2 + b_2^2 - 2a_2 b_2 \cos\theta}$$

where a_2 is the same as before, and θ and b_2 are defined by

$$\theta = \arctan\left[\frac{\Delta r_{ij}^{TH} \sin(Z_{TH})}{R_m + h_i + h_{TH} + \Delta r_{ij}^{TH} \cos(Z_{TH})}\right] \tag{2.59}$$

$$b_2 = \left[\frac{\Delta r_{ij}^{TH} \sin(Z_{TH})}{\sin\theta} - h_{TR}\right] \tag{2.60}$$

2.5.1.2 Reduction of the Mark-to-Mark Distances to the Reference Ellipsoid

Reducation Using Station Heights

Referring to Figure 2.7, if the ellipsoidal heights of both survey marks are known, the mark-to-mark distance (l_{ij}^{g}) can be reduced to the distance (S_{ij}^{E}) on the surface of the reference ellipsoid as follows:

$$S_{ij}^{E} = 2R_m \arcsin\left(\frac{l_{ij}^{0}}{2R_m}\right) \tag{2.61}$$

where l_{ij}^{0} is the ellipsoidal chord distance calculated by

$$l_{ij}^{0} = \sqrt{\frac{\left(l_{ij}^{g}\right)^2 - \left(h_j - h_i\right)^2}{\left(1 + h_i / R_m\right)\left(1 + h_j / R_m\right)}} \tag{2.62}$$

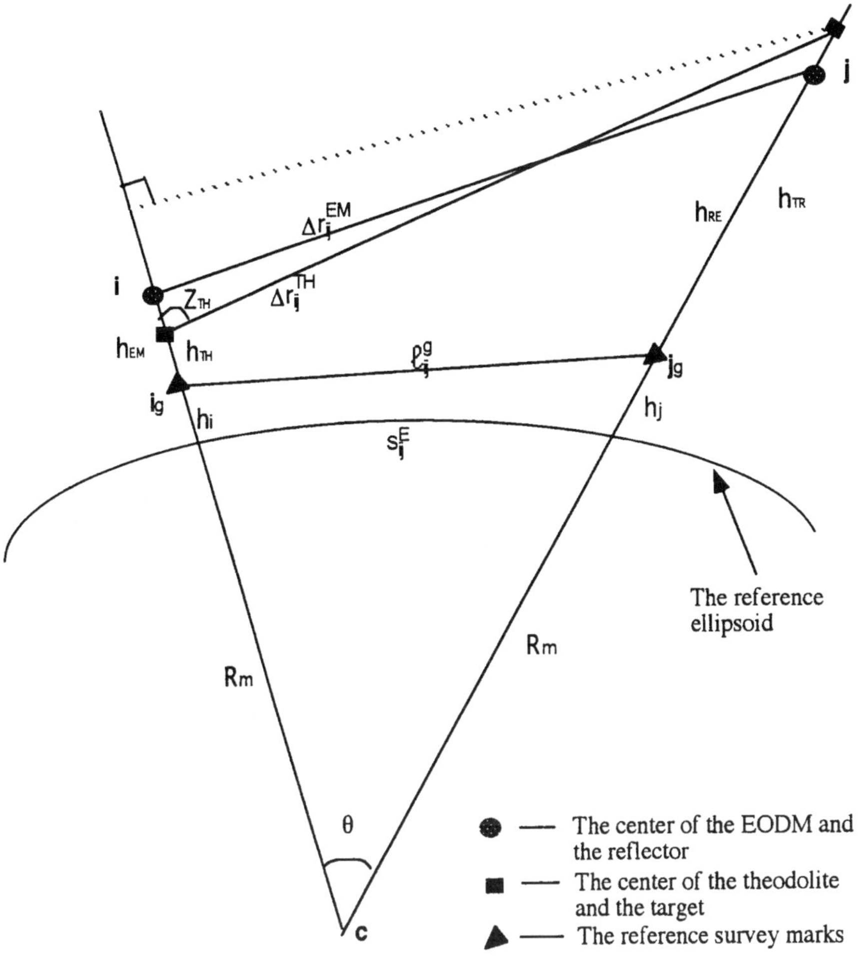

Figure 2.8. Geometrical reduction of distances using zenith distances

with h_i, h_j, and R_m meaning the same as before.

Reduction Using Measured Zenith Distances

Referring to Figure 2.8, if the ellipsoidal height of the survey mark at the instrument station together with a measured zenith distance are known, the reduction of the mark-to-mark distance (l_{ij}^{g}) to the distance (S_{ij}^{E}) on the surface of the reference ellipsoid can also be accomplished by using Equations (2.61) and (2.62). In this case, the ellipsoidal height of the survey mark at the target station can be calculated by:

$$h_j = \left[\frac{\Delta r_{ij}^{TH} \sin(Z_{TH})}{\sin\theta} - h_{TR} - R_m \right] \tag{2.63}$$

where θ is calculated by Equation (2.59), and R_m, h_i, h_{TH}, Δr_{ij}^{TH}, and Z_{TH} mean the same as before.

2.5.2 Reduction of Angular Measurements

After the effects of gravity have been removed, two other geometric effects are to be accounted for that are due to the earth's geometry in order to reduce the angular measurements from the LG system to the surface of a reference ellipsoid. First, a skew-normal correction has to be applied due to the fact that the normals at the instrument and the target points on an ellipsoid are "skewed" to each other; that is, when a target is above the ellipsoid, it is not in the same plane as its normal projection onto the ellipsoid. This correction is called *height of target* reduction, and is calculated by the following Equation (2.65). The second geometric correction is due to the fact that on the surface of an ellipsoid the unique line with the shortest length between any two points is the line called "geodesic" instead of the normal section. The geodesic is therefore used as the reference line for coordinate computations, and all angular measurements must be reduced to refer to this line. This correction is called the normal section to geodesic correction given by Equation (2.66) below.

Geodetic Azimuths

Let α^E be the ellipsoidal azimuth of the geodesic line on the surface of the reference ellipsoid. The discrepancies between the geodetic azimuth (α^G) obtained in §2.4 by applying gravimetric corrections and the ellipsoidal azimuth (α^E) are due to the height of the target, and the inconsistency between the normal section and the geodesic line. α^E is obtained from α^G by applying the following correction:

$$\delta\alpha_E = c_3 + c_4 \tag{2.64}$$

where

$$c_3 = \frac{h_j}{2M_m} e^2 \sin 2\alpha^G \cos^2 \phi_m \tag{2.65}$$

$$c_4 = -\frac{e^2 (S^E)^2 \cos^2 \phi_m \sin 2\alpha^G}{12N_m^2} \tag{2.66}$$

α^G is the geodetic azimuth from the instrument point to the target point, ϕ_m the latitude of the midpoint between the instrument point and the target point, S^E the geodesic length between the two points, h_j the ellipsoidal height of the target point, and finally, M_m and N_m are the radii of curvatures in the meridian and the prime vertical planes at the midpoint, respectively.

Horizontal Directions and Horizontal Angles

As with the reduction of geodetic azimuths, the reduction of direction observations from the Local Geodetic reference system to the ellipsoid can be accomplished through the use of Equation (2.64), and the following correction is applied to reduce a horizontal angle from the LG system to the ellipsoid

$$\delta\beta_E = (c_3 + c_4)_{ik} - (c_3 + c_4)_{ij} \tag{2.67}$$

where c_3 and c_4 are calculated from Equations (2.65) and (2.66), respectively.

2.6 Reduction of Observations to a Conformal Mapping Plane

Finally, when two-dimensional mapping plane coordinates are needed, the observations have to be further reduced from the surface of the reference ellipsoid to a conformal mapping plane to perform network computations, and that is the purpose of this section. Mapping is a general term in mathematics. It means the transformation of information from one surface to another. In surveying and geodesy, a conformal mapping is used that preserves angles, while the linear scale is a function of position only. Depending on the conditions imposed, one may deduce a variety of conformal mapping projections. The most frequently used projections in geodesy are Oblique Mercator projection, Transverse Mercator projection, Universal Transverse Mercator (UTM), Stereographic projection, and Lambert conformal conic projection, etc. For detailed theory and applications of map projections the interested reader is referred to Maling (1973), among others. The choice of a map projection is usually dictated by the location, the shape, and the extent of the area under study to minimize the resulting distortions. In the case of a relatively small area such as that covering an engineering project, we can say that no specific conformal mapping projection may show any distinct advantage. Nevertheless, by choosing an appropriate origin together with a reasonable scale factor at the origin for a conformal mapping projection, the corrections that are needed to reduce observations from the ellipsoid to the mapping plane can be minimized, and that may be very important, especially for geodetic networks established for high-precision engineering projects.

The quantities needed in the reduction of observations from the reference ellipsoid to the mapping plane are the arc-to-chord correction; i.e., the (T-t) correction, the meridian convergence, and the line scale factor. Specific formulae for these quantities depend on the type of projection used (cf. Stem, 1991). Formulae of the Lambert conformal conic projection are used here as examples. For equations of the other map projections, the interested reader is referred to the given literature.

2.6.1 Reduction of Ellipsoidal Distances to a Mapping Plane

To reduce the distance from the ellipsoid to the mapping plane, the grid point scale factor must be known. The grid scale factor is the measure of the linear distortion that has been mathematically imposed on ellipsoid distances so they may be projected onto a plane, and it is defined as the ratio of the length of a linear increment on the grid to the length of the corresponding increment on the ellipsoid (Stem, 1991). The projected distance on a mapping plane (s^M) may thus be calculated rigorously from the following integration

$$s^M = \int k\, ds^E \tag{2.68}$$

where k is the grid point scale factor and the integration is performed along the geodesic. In practice, a mean line scale ($\bar{k}$) is used for distance reduction purposes. The projected distance from the ellipsoid to the mapping plane is calculated as follows (cf. Bomford, 1980)

$$S^M = \bar{k} S^E \tag{2.69}$$

where

$$\bar{k} = (k_1 + 4k_m + k_2)/6 \tag{2.70}$$

k_1 and k_2 are the point scale factors at the endpoints of the line, and k_m is the scale factor at the midpoint of the line.

For the Lambert conformal conic projection, the point scale factor is calculated as follows (Stem, 1991)

$$k_i = W_i R_i \sin\phi_0 / (a \cos\phi_i) \tag{2.71}$$

where

$$W_i = \sqrt{(1 - e^2 \sin^2 \phi_i)} \tag{2.72}$$

$$R_i = K / \exp(Q_i \sin\phi_0) \tag{2.73}$$

$$Q_i = \frac{1}{2}\left[\ln\left(\frac{1+\sin\phi_i}{1-\sin\phi_i}\right) - e\ln\left(\frac{1+e\sin\phi_i}{1-e\sin\phi_i}\right)\right] \tag{2.74}$$

$$K = a\cos\phi_s \exp(Q_s \sin\phi_0) / (W_s \sin\phi_0) \tag{2.75}$$

$$\sin\phi_0 = \frac{\ln(W_n \cos\phi_s) - \ln(W_s \cos\phi_n)}{Q_n - Q_s} \tag{2.76}$$

ϕ_s = the southern standard parallel
ϕ_n = the northern standard parallel

W_s, W_n and Q_s, Q_n are calculated using Equations (2.72) and (2.74) by replacing ϕ_i with ϕ_s and ϕ_n, respectively.

2.6.2 Reduction of Ellipsoidal Azimuths to a Mapping Plane

A geodetic azimuth (α_{ij}^E) on the ellipsoid is reduced to a grid azimuth (t_{ij}) on a mapping plane through application of the following equation:

$$t_{ij} = \alpha_{ij}^E - \gamma_i + \delta_{ij} \tag{2.77}$$

where γ_i and δ_{ij} represent the correction for the convergence of the meridian and the arc-to-chord correction, respectively. Figure 2.9 shows the general relationship among grid azimuth (t), geodetic (ellipsoidal) azimuth (α), meridian convergence (γ), and arc-to-chord correction (δ) at any given point.

For the Lambert conformal conic projection, it is shown that (Stem, 1991)

$$\gamma_i = (\lambda_0 - \lambda_i)\sin\phi_0 \tag{2.78}$$

$$\delta_{ij} = \left(\sin\frac{2\phi_i + \phi_j}{3} - \sin\phi_0\right)\left(\frac{\lambda_i - \lambda_j}{2}\right) \tag{2.79}$$

where λ_0 is the longitude of the origin, and sin ϕ_0 is calculated by Equation (2.76).

2.6.3 Reduction of Horizontal Directions and Horizontal Angles to a Mapping Plane

As with ellipsoidal azimuth, Equation (2.77) is used for the reduction of horizontal directions from the surface of an ellipsoid to a mapping plane, while the reduction of a horizontal angle from the reference ellipsoid to a mapping plane is accomplished as follows (cf. Figure 1.3)

$$\beta^M - \beta^E = \delta_{ik} - \delta_{ij} \tag{2.80}$$

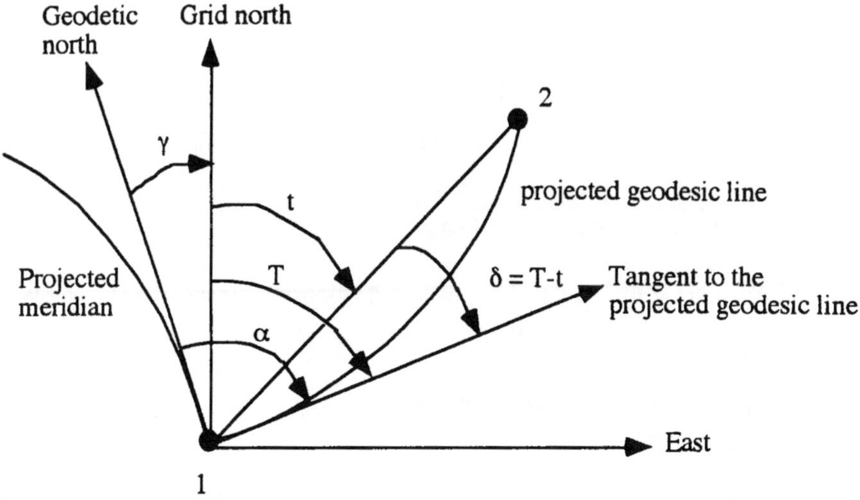

Figure 2.9. Projection of the ellipsoidal azimuth

where δ_{ik} and δ_{ij} are the arc-to-chord corrections calculated from Equation (2.79) for the Lambert conformal conic projection, and β^E and β^M are the horizontal angles on the surface of the ellipsoid and on the mapping plane, respectively.

2.7 Reduction of GPS Coordinates and Baseline Components

The raw GPS observations (e.g., carrier phases and pseudoranges) can be processed through the use of various commercially available software to produce the baseline vectors and their variance-covariance matrices in the WGS 84 system. The original GPS baseline vectors, if they form a network, can be first adjusted in the three-dimensional space producing the 3D geodetic Cartesian coordinates $(x, y, z)^G$ together with their associated variance-covariance matrix C_{xyz}, which may subsequently be transformed into their corresponding geodetic curvilinear coordinates $(\phi, \lambda, \eta)^G$ with their corresponding variance-covariance matrix $C_{\phi\lambda h}$. When it is required to integrate the obtained three-dimensional GPS coordinates or baseline components into the conventional terrestrial geodetic observations to perform a combined adjustment in the two-dimensional space on a particular mapping plane, one may either transform the GPS-obtained horizontal curvilinear coordinates (ϕ, λ) into the mapping plane grid coordinates (X, Y) and using them as weighted stations or transform the GPS-obtained horizontal curvilinear baseline components $(\Delta\phi, \Delta\lambda)$ into the mapping plane grid coordinate differences $(\Delta X, \Delta Y)$ and using them as weighted coordinate differences. In order to apply proper weighting, the most critical task is then to derive the variance-covariance matrix of (X, Y) and $(\Delta X, \Delta Y)$ transformed from (ϕ, λ) and $(\Delta\phi, \Delta\lambda)$, respectively.

2.7.1 Conversion of Adjusted GPS Coordinates to Mapping Plane Coordinates

This conversion depends on the map projection type used. For the Lambert conformal conic projection, the transformation from (ϕ, λ) to grid coordinates (X, Y) is given as follows (Stem, 1991):

$$X = X_0 + R\sin[(\lambda_0 - \lambda)\sin(\phi_0)] \tag{2.81}$$

$$Y = R_b + Y_b - R\cos[(\lambda_0 - \lambda)\sin(\phi_0)] \tag{2.82}$$

where X_0 is the false Easting value at the central meridian, and Y_b the false Northing value for the false origin. R is calculated by (2.73), and

$$R_b = K / \exp(Q_b \sin\phi_0) \tag{2.83}$$

with K being the same as in (2.75).

According to the law of variance-covariance propagation, the variance-covariance matrix C_{XY} of (X,Y) can be expressed as follows

$$\mathbf{C}_{XY} = \mathbf{E}\ \mathbf{C}_{\phi\lambda}\ \mathbf{E}^T \tag{2.84}$$

where **E** is a coefficient matrix calculated by:

$$\mathbf{E} = \begin{pmatrix} \frac{\partial X}{\partial \phi} & \frac{\partial X}{\partial \lambda} \\ \frac{\partial Y}{\partial \phi} & \frac{\partial Y}{\partial \lambda} \end{pmatrix} \tag{2.85}$$

According to Equations (2.81) and (2.82), the elements in Equation (2.85) can be derived as follows:

$$\frac{\partial X}{\partial \phi} = \frac{\partial R}{\partial \phi}\sin[(\lambda_0 - \lambda)\sin(\phi_0)] \tag{2.86}$$

$$\frac{\partial X}{\partial \lambda} = -R\sin(\phi_0)\cos[(\lambda_0 - \lambda)\sin(\phi_0)] \tag{2.87}$$

$$\frac{\partial Y}{\partial \phi} = -\frac{\partial R}{\partial \phi}\cos[(\lambda_0 - \lambda)\sin(\phi_0)] \tag{2.88}$$

$$\frac{\partial Y}{\partial \lambda} = -R\sin(\phi_0)\sin[(\lambda_0 - \lambda)\sin(\phi_0)] \tag{2.89}$$

$$\frac{\partial R}{\partial \phi} = -R \sin\phi_0 \left[\frac{1}{\cos\phi} - \frac{e^2 \cos\phi}{1 - e^2 \sin^2\phi} \right] \tag{2.90}$$

2.7.2 Reduction of GPS Baseline Components to a Mapping Plane

When the original GPS vectors do not form a network, no 3D network adjustment for the GPS vectors can be performed. In this case, in order to integrate the GPS baseline components with the conventional terrestrial geodetic observations to perform a combined network adjustment on the mapping plane, the GPS horizontal baseline components $(\Delta\phi, \Delta\lambda)^G$, with $\mathbf{C}_{(\Delta\phi, \Delta\lambda)}$, must be transformed into the grid coordinate differences $(\Delta X, \Delta Y)$, with a variance-covariance matrix $\mathbf{C}_{(\Delta X, \Delta Y)}$, on the mapping plane. This is accomplished as follows:

$$\phi_j = \phi_i + \Delta\phi_{ij} \tag{2.91}$$

$$\lambda_j = \lambda_i + \Delta\lambda_{ij} \tag{2.92}$$

Given a certain variance-covariance matrix $\mathbf{C}_{(\phi, \lambda)i}$ for (ϕ_i, λ_i), the variance-covariance matrix $\mathbf{C}(\phi, \lambda)_j$ for (ϕ_j, λ_j) is obtained as follows:

$$\mathbf{C}_{(\phi,\lambda)j} = \mathbf{C}_{(\phi,\lambda)i} + \mathbf{C}_{(\Delta\phi,\Delta\lambda)ij} \tag{2.93}$$

From the formulae given previously, (ϕ_i, λ_i) and (ϕ_j, λ_j) can be transformed into the corresponding grid coordinates (X_i, Y_i) and (X_j, Y_j) to obtain the grid coordinate differences as given below:

$$\Delta X_{ij} = X_j - X_i \tag{2.94}$$

$$\Delta Y_{ij} = Y_j - Y_i \tag{2.95}$$

Equations (2.94) and (2.95) can be rewritten in matrix forms as follows

$$\begin{pmatrix} \Delta X_{ij} \\ \Delta Y_{ij} \end{pmatrix} = \begin{pmatrix} -1 & 0 & 1 & 0 \\ 0 & -1 & 0 & 1 \end{pmatrix} \begin{pmatrix} X_i \\ Y_i \\ X_j \\ Y_j \end{pmatrix} \tag{2.96}$$

The variance-covariance matrix $\mathbf{C}(\Delta X, \Delta Y)$ of $(\Delta X, \Delta Y)$ can be obtained from the variance-covariance law as follows

$$\mathbf{C}_{\Delta X \Delta Y} = \mathbf{E}\, \mathbf{C}_{XY}\, \mathbf{E}^T \tag{2.97}$$

where

$$\mathbf{E} = \begin{pmatrix} -1 & 0 & 1 & 0 \\ 0 & -1 & 0 & 1 \end{pmatrix} \tag{2.98}$$

$$\mathbf{C}_{XY} = \begin{pmatrix} \mathbf{C}_{(XY)_i} & \mathbf{C}_{(XY)_i(XY)_j} \\ \mathbf{C}^T_{(XY)_i(XY)_j} & \mathbf{C}_{(XY)_j} \end{pmatrix} \tag{2.99}$$

$\mathbf{C}_{(XY)_i}$, $\mathbf{C}_{(XY)_j}$, and $\mathbf{C}_{(XY)_i(XY)_j}$ are the variance-covariance matrices of (X_i, Y_i), (X_j, Y_j) and the covariance matrix between (X_i, Y_i) and (X_j, Y_j), respectively. These quantities can be calculated using the formulae given before. It should be noted that if several baselines are processed simultaneously, the correlations between the baselines, if available, must be transformed as well.

2.8 Summary and Discussions

Concepts and procedures needed for the correction and reduction of raw observation data for network computations have been addressed in this chapter. Terrestrial raw geodetic observations (e.g., spatial distances, azimuth, horizontal directions and angles, zenith distances) are made on the surface of the earth bearing the effects of environmental and instrumental factors as well as the earth's geometry and gravity field, and therefore they must be pre-processed before they can be used for coordinate computation. This mainly includes meteorological and instrumental calibration corrections, and corrections for the effects of the variations of the earth's gravity field and/or the earth's geometry, depending on the specific coordinate system chosen for coordinate computation.

In general, the position of a terrain point can be expressed either in a 3D model or in a 2D plus 1D model. In the former case, the three-dimensional geodetic position of a terrain point is described in terms of the triplet of coordinates in a three-dimensional coordinate system; e.g., the G and LG systems, while in the latter case it is separated into two horizontal coordinates in a two-dimensional coordinate system plus 1 vertical coordinate; that is, the elevation described by a height system. In the case that 2D horizontal coordinates are needed, the position computations can be done either on the surface of a reference ellipsoid, or on a conformal mapping plane. Practicing surveyors are most exposed to the classical 2D plus 1D model with the conformal mapping plane coordinates and orthometric height. The choice of a geodetic model and a coordinate system is dictated by the type of observation data collected, the kind of coordinates desired, and the extent of the network. Theoretically, however, since rigorous transformations between different coordinate systems can be performed, it should be noted that the computation of coordinates is equivalent, up to the precision of linearization, whichever coordinate system is chosen.

Depending on the coordinate systems chosen, raw observations, after having been subject to meteorological and instrumental corrections, must be reduced to the Local Geodetic (LG) system, to the surface of the reference ellipsoid, or further to a conformal mapping plane. Geodetic leveling data must be reduced to the geoid if orthometric heights are desired. At each reduction phase, one or more reduction steps may be needed, depending on the observed quantity. In certain applications, some of the reduction corrections can be omitted without adversely affecting the relative positioning accuracy, but this should first be shown by analysis of the specific problem. This is especially true of the gravimetric reduction correction that is often omitted only because it is difficult to determine. After raw observations have been subject to corrections and reductions, the variance for the reduced quantities is often taken to be the variance of the observation itself in practice. Theoretically, this is not rigorous but it is practical for most surveying applications. For a precise surveying project, however, the effects of the inaccuracy of both the corrections and reductions applied may have to be considered. This can be done using the law of error propagation as shown in Chapter 1.

Finally, it should be noted that the measurement reduction processes are reversible; that is, a distance, an azimuth, or an angle, which are calculated on the surface of a reference ellipsoid or on a conformal mapping plane, can be reduced up to the terrain simply by adding the negative values of corrections that are required to reduce the corresponding actual measurements on the earth's surface to these reference surfaces. This is very important for a construction layout surveying project where surveyors are faced with the need for terrain values of computed distances and azimuths from pre-designed conformal mapping plane coordinates. In all the reductions to the terrain, however, one should not expect perfect agreement between the computed quantity and the newly observed quantity, since both of these quantities have some statistical fluctuations.

3

Pre-Adjustment Data Screening: Gross Error Detection and Elimination

3.1 Purpose

After raw observation data have been pre-processed according to Chapter 2, they are now ready for coordinate computations. As will be discussed in Chapter 4, geodetic network positioning is often an overdetermined problem; that is, more observations than necessary are available. Due to the effects of the unpredictable observation errors; i.e., random errors, gross errors, and/or systematic errors, the redundant set of observation equations will in general not be consistent, and one may thus derive more than one unique solution for the unknown coordinates using a different subset of observation equations. In order to obtain a best solution that uses all the available observation information, some additional criteria have to be imposed. The most frequently used criterion in geodesy and surveying is the method of least squares that tries to arrive at a best solution by minimizing the sum of squares of the weighted discrepancies among observations.

In principle, a least squares adjustment of the survey data will give the best linear unbiased estimates of the unknown parameters (e.g., station coordinates) only if the true errors of the survey data are randomly distributed. However, as discussed in Chapter 1, due to the influence of the various instrumental, human, and/or environmental factors, one cannot guarantee to obtain gross and/or systematic error-free survey data. Since the least squares minimization includes all observations, random variables, or not, the existence of gross errors would directly degrade the quality of the results. Therefore, in order to produce reliable results, both the pre-adjustment and the post-adjustment data screening techniques have been developed to detect and remove any possible gross and systematic errors in the survey data. The former concentrates on detecting and removing any possible gross errors of *large magnitude* in the survey data prior to the network adjustment, whereas the latter, which will be discussed in Chapter 5, deals with detecting and eliminating gross errors of small (or marginal) magnitude based on statistical testing on the residuals (i.e., corrections to be applied to the survey data) estimated by the least squares process.

The need for pre-adjustment data screening for large gross errors is justified due to the fact that the least squares minimization smooths out gross errors across the entire data set, and any specific large gross error will be smoothed out throughout its neighboring observations. In this case, localization of gross errors based on statistical rejection of residuals available from least squares adjustment is very difficult, since it may point out many non-existing gross errors (Steeves, 1984). The user, therefore, has to search through a large portion of the survey data set that may not contain gross errors, and a "trial and error" approach is required to perform several least squares adjustments by taking out or re-inputting one or more of those observations at a time in order to locate the right one(s) that contains gross error. Furthermore, an estimation problem usually involves a nonlinear mathematical which has to be linearized, and the least squares adjustment has to be accomplished iteratively. A large gross error may damage the linearization process and cause the iteration procedure to diverge, leading to no solution. All these concerns prompt for the need for some method of data snooping before network adjustment is performed.

Pre-adjustment data screening techniques are mainly based on the examination of condition equation misclosures together with the experiences and intuition of the investigator (Kavouras, 1982). For instance, simple checks on triangle closure, geodetic leveling loop closure, etc., have proved to be very useful in detecting and locating large errors. These checks could be done either during or just after the data acquisition. If a large misclosure exists, there is definitely a gross error in one of the circuit observations. The decision can be made with some certain degree of objectivity by examining the misclosures statistically against an assumed statistical distribution. On the other hand, if some observables (e.g., azimuth, direction, angle, distance) are repeatedly measured, each measurement can be tested individually to insure that it is self-consistent. This is done by examining the residuals statistically against an assumed statistical distribution. This chapter continues with a review of the basic concepts and procedures of statistical testing in §3.2. Then, §3.3 and §3.4 discuss the gross error detection techniques based on condition misclosures and repeated measurements, respectively.

3.2 Concepts of Statistical Testing

3.2.1 Null Hypothesis and Alternative Hypothesis

Conceptually, statistical tests serve to determine whether or not anything has gone wrong with some basic postulate; e.g., the postulated probability function for the experiment (sample), or the estimated value of a population parameter such as the mean, the variance, or both. A quantitative statement about the postulated probability density function and its parameters is defined as a statistical hypothesis (Vanicek and Krakiwsky, 1986). A ***null hypothesis*** (H_0) is the one in which the population parameters are postulated to have some particular values. In general, for every null hypothesis there is an infinite number of ***alternative hypotheses*** (H_a) that assume different values for the population parameters. Thus the null hypothesis is the reference level from which any deviation of the different alternative hypothesis has to be detected by statistical tests. Any statistical hypothesis can be tested. A test of a statistical hypothesis H_0 is an algorithm that leads to a statistical decision concerning the validity of H_0 (Vanicek and Krakiwsky, 1986).

Statistical testing is made using a ***statistic,*** which is a special random variable that is a function of one or more random variables. The probability density function, specifically the expectation and the variance of the test statistic, has to be known or derived from the experimental data for the case where the null hypothesis H_0 is true. However, it should be noted that a definite decision about H_0 cannot be made because one does not have complete information about the entire population. Therefore, a decision made based on a finite sample has only a certain confidence (i.e., probability) attached to it. The smaller the probability, the lower the credibility of the null hypothesis. If this probability becomes sufficiently small, the null hypothesis may be rejected (Kavouras, 1982). On the other hand, if the probability is not small enough to sufficiently reduce the credibility of the null hypothesis, it is said that "H_0 is accepted." In this case, it does not imply that H_0 is true but only that the available data cannot convince us that H_0 is false. Therefore, for any statistical test, there are two possible outcomes; i.e., either to accept H_0, or to reject H_0. Similarly, there are two possible outcomes for the test of the alternative hypothesis H_a.

3.2.2 Type I and Type II Errors

Due to the finiteness of the available sample, no definite statistical decision could be made. Therefore, there are always two types of potential errors, identified as Type I and Type II errors, involved in a statistical test. Type I error is defined as the error of rejecting the null hypothesis H_0 when H_0 is actually true. The probability of committing this type of error is called the "significance level" denoted by α, and the probability of making the correct decision is called the "confidence level" ($1-\alpha$). Type II error is defined as the error of accepting the null hypothesis H_0 when it is actually false (i.e., H_a is true), and the probability of committing this type of error is denoted by β. The probability of making the correct decision is called the "power of the test" ($1-\beta$). This is the probability of rejecting the null hypothesis H_0 when it is actually false. Table 3.1 summarizes these possible situations.

Assume that under the null hypothesis H_0 and the alternative hypothesis H_a, the probability density functions of a chosen statistic take the same form but have different values for its population parameters; i.e., the mean and variance as shown in Figure 3.1, the power of the test indicates the smallest difference δ that can be detected if the test has been executed at a significance level α. It is obvious that fixing the smallest difference between the two hypotheses, Type I and Type II errors cannot both be decreased. A decrease in α leads to an increase in β and vice versa.

Table 3.1. Statistical testing of a null hypothesis against an alternative hypothesis

Situation \ Probability \ Decision	Accept H_0	Reject H_0
H_0 is true	Correct decision Confidence level = $1-\alpha$	Type I error Significance level = α
H_0 is false (i.e., H_a is true)	Type II error Probability = β	Correct decision Power of the test = $1-\beta$

It should be noted that selection of the significance level α is more of a subjective matter, and there is in general no objective criteria available for that. In geodesy and surveying, usually the values between 0.001 to 0.05 are selected. A very small significance level expresses a reluctance to reject the null hypothesis unjustly. On the other hand, if the null hypothesis is false the Type II error that is committed is quite large. From Figure 3.1 one can see that if one wishes to decrease the probability of both types of error, δ will increase. That means that the detectable difference between H_0 and H_a would be larger.

3.2.3 Statistical Testing Procedures

A statistical test decision is made by comparing the value of the statistic, which is calculated using the experimental data, to its corresponding boundary value, which is calculated by inverting the probability function of the statistic given the significance level of the test. This situation is depicted by Figure 3.2.

Here we assume that the statistic y is chosen to test the null hypothesis H_0, and that the probability density function of y under the null hypothesis is ϕy. Given the significance level α of the test, the null hypothesis H_0 will be rejected if

$$y > \xi_{1-\alpha} \tag{3.1}$$

where $\xi_{1-\alpha}$ is the upper boundary value of y defined by the following equation

$$\text{Probability } (y > \xi_{1-\alpha}) = \int_{\xi_{1-\alpha}}^{\infty} \phi_y(\xi)d\xi = \alpha \tag{3.2}$$

It should be noted that the above Equations (3.1) and (3.2) account for the case where a false null hypothesis may cause the value of the statistic y, calculated using the experimental data, to be too big. Similarly, in the case that a false null hypoth-

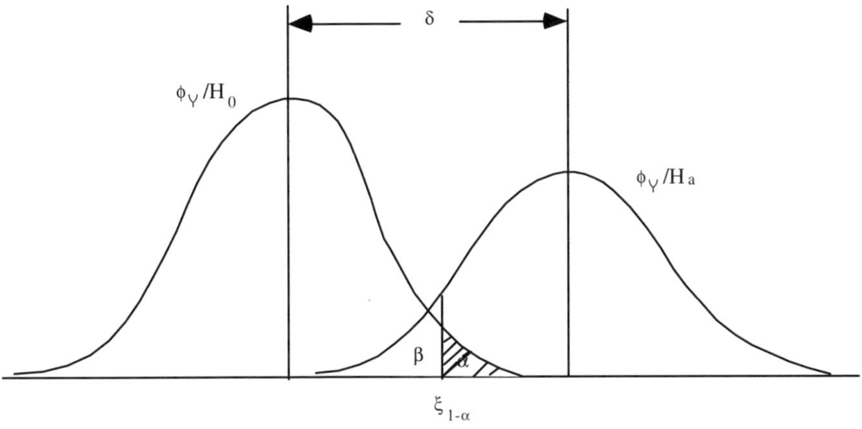

Figure 3.1. Type I error α and Type II Error β

esis H_0 may cause a too small value of the statistic y, the null hypothesis H_0 will be rejected if

$$y < \xi_\alpha \tag{3.3}$$

where ξ_α represents the lower boundary value of y satisfying the following equation (see Figure 3.3)

$$\text{Probability } (y < \xi_\alpha) = \int_{-\infty}^{\xi_\alpha} \phi_y(\xi) d\xi = \alpha \tag{3.4}$$

The above two cases are referred to as *one-tailed test* since only either the upper or the lower value of a chosen statistic is used to make the decision. A *two-tailed test* considers the case that both the upper and lower boundary values are used simultaneously to make the decision. The null hypothesis will be rejected if

$$y > \xi_{1-\alpha/2} \tag{3.5}$$

or

$$y < \xi_{\alpha/2} \tag{3.6}$$

That is, the null hypothesis will be accepted if

$$\xi_{\alpha/2} \le y \le \xi_{1-\alpha/2} \tag{3.7}$$

Here the required probability α is split equally to be contained in the right-hand side tail and the left-hand side tail, respectively (see Figure 3.4). The closed interval $[\xi_{\alpha/2}, \xi_{1-\alpha/2}]$ is also called the confidence interval.

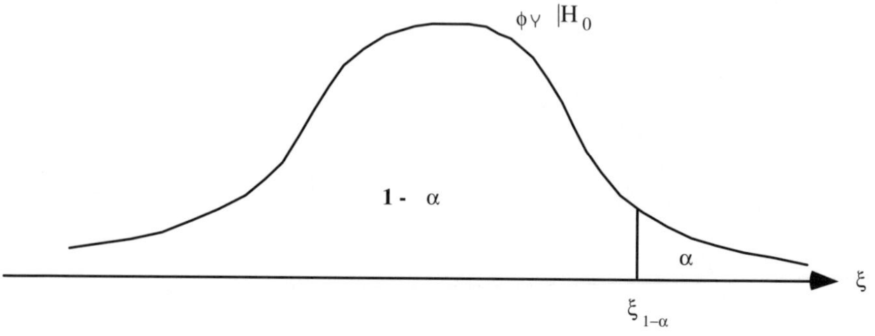

Figure 3.2. One-tailed test: upper bound $\xi_{1-\alpha}$

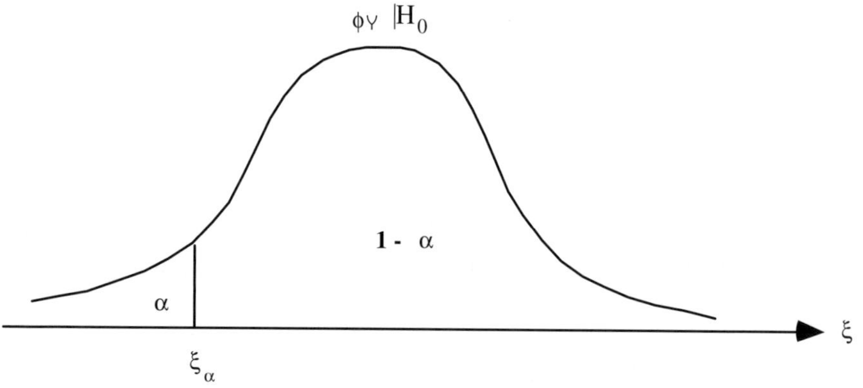

Figure 3.3. One-tailed test: lower bound ξ_α

A two-tailed test is needed when a population parameter is hypothesized to be equal to a specific value. In this case, either a smaller or bigger number is not acceptable. As discussed in Vanicek and Krakiwsky (1986), the advantage of the confidence interval expressed by Equation (3.7) is that when a population parameter is tested, the confidence interval can usually be reformulated to hold for the parameter itself rather than the chosen statistic y. The confidence interval approach is widely used in geodesy and surveying (cf. Chapter 6).

Summarizing the above discussions, the procedures of statistic testing are given below:

Step 1: specify the null hypothesis H_0 and the alternative hypothesis H_a
Step 2: select a statistic y
Step 3: derive the probability density function of y under the null hypothesis H_0

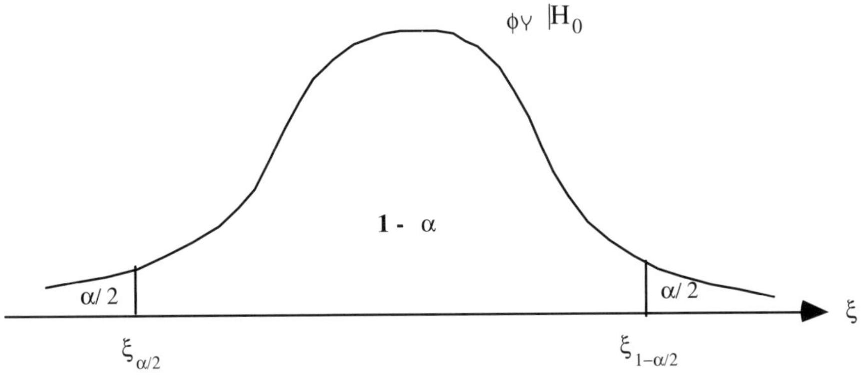

Figure 3.4. Two-tailed test: lower bound $\xi_{\alpha/2}$ and upper bound $\xi_{1-\alpha/2}$

Step 4: choose significance level α for the test
Step 5: choose power of the test $1-\beta$ if necessary
Step 6: calculate the boundary values
Step 7: make decision to accept or reject H_0.

As will be discussed in Chapter 5, selection of the right alternative hypothesis H_a at the time is not a trivial thing since it may appear that more than one H_a may satisfy the tests when H_0 is rejected. The design of H_a requires good knowledge of the collection of the data, experience, and sometimes intuition that are always a challenge to geodesists and surveyors.

3.3 Gross Error Detection from Condition Misclosures

If redundant observations have been made to close a geometrical figure, e.g., a triangle, a leveling loop, etc., the resulting misclosures can be used to test the internal consistency among observations, because statistically if the observation errors are randomly distributed a misclosure is a true error with zero expectation, and therefore it can be tested against the propagated uncertainty at a certain confidence level.

Without loss of generality, let us start with a linear misclosure that is expressed as a linear function of the observations as follows:

$$\Delta_i = c_0 + \sum_{j=1}^{m} c_j l_j = c_0 + \mathbf{c}^T \boldsymbol{l} \tag{3.8}$$

where

$$\mathbf{c} = [c_1\ c_2\ \ldots\ c_m]^T \tag{3.9}$$

$$\boldsymbol{l} = [l_1 \; l_2 \ldots l_m]^T \tag{3.10}$$

c_0 is a constant, and l_j and c_j (j=1, ..., m) are the observations and their corresponding coefficients used to calculate a loop closure, respectively.

The null hypothesis H_0 to be tested statistically is that Δ_i follows the normal distribution with zero mean, i.e.,

$$H_0\text{: } E\{\Delta_i\} = 0 \tag{3.11}$$

where $E\{\ \}$ represents the statistical expectation operator. The test statistic y_i for misclosure Δ_i can be formulated as follows

$$y_i = \frac{\Delta_i}{\sigma_{\Delta_i}} \tag{3.12}$$

where the standard deviation of σ_{Δ_i} can be calculated according to error propagation by

$$\sigma_{\Delta_i} = \sqrt{\mathbf{c}^T \mathbf{C}_l \mathbf{c}} \tag{3.13}$$

if the a priori variance-covariance matrix of the observations is known. In this case, y_i belongs to the standard normal distribution n(0, 1) with zero mean and unit variance. Given the significance level α, the misclosure Δ_i is considered as statistically excessive (i.e., H_0 is rejected) with a confidence level of (1-α), if

$$|y_i| > n_{i-\alpha/2} \tag{3.14}$$

where $n_{1-\alpha/2}$ is the critical value.

If the a priori variance-covariance matrix of observations is not adequately known, it has to be estimated from the observations, and the estimated standard deviation of the misclosure is calculated by

$$\hat{\sigma}_{\Delta_i} = \hat{\sigma}_0 \sqrt{\mathbf{c}^T \mathbf{Q}_l \mathbf{c}} \tag{3.15}$$

where $\mathbf{Q}_l$ is the cofactor matrix of observations, and $\hat{\sigma}_0$ the estimated standard error factor. The standard error factor could be determined either from the same sample or from an independent source. In the former case, the statistic y_i belongs to the tau τ-distribution, and in the latter case, it belongs to the student's t-distribution. The null hypothesis is thus rejected if

$$|y_i| > \tau_{1-\alpha/2}(r) \tag{3.16}$$

or

$$|y_i| > t_{1-\alpha/2}(r) \tag{3.17}$$

depending on how $\hat{\sigma}_0$ is determined, with r being the degrees of freedom of observations used to estimate $\hat{\sigma}_0$.

Critical values for τ-test can be calculated from the student's t-test as follows (Pope, 1976)

$$\tau_{(1-\alpha/2)} = \frac{\sqrt{(n-1)}\, t_{(1-\alpha/2)}}{\sqrt{n-2+t^2_{(1-\alpha/2)}}} \tag{3.18}$$

Finally, if a misclosure is calculated as a nonlinear function of observations as follows

$$\Delta_i = f(l_1, l_2, \ldots, l_m) \tag{3.19}$$

In order to determine the standard deviation of Δ_i, it has to be linearized according to Taylor series as follows

$$\Delta_i = \Delta_i^0 + \sum_{j=1}^{m} \frac{\partial f}{\partial l_j} \delta l_j \tag{3.20}$$

Denote

$$\mathbf{c} = \left[\frac{\partial f}{\partial l_1} \frac{\partial f}{\partial l_2} \cdots \frac{\partial f}{\partial l_m} \right]^T \tag{3.21}$$

σ_{Δ_i} is then calculated in the same way as above, using the law of random error propagation.

Critical values for the n(0, 1), t(r), and τ(r) distributions corresponding to a certain significance level α and degrees of freedom r can be calculated analytically by inverting their corresponding probability functions, and they are usually tabulated in statistics textbooks. Assuming a confidence level of 95%, typical boundary values for the above distributions corresponding to some selected degrees of freedom are given in Table 3.2, from which one can see that the three types of critical values approach the same constant as the degrees of freedom become sufficiently large (e.g., r≥100).

A rule of thumb often used in practice is that all circuits that produce misclosures greater than two times their standard deviations should be flagged as possible areas to be examined for blunders, and consider re-measuring data for circuits having misclosures greater than three times their standard deviations. For a certain type of survey, appropriate limits may be defined through practical experience.

Loop closures in geodetic networks have been used as a basis for detecting possible gross errors in the survey data by surveyors/engineers for many years, and in fact various survey organizations have established criteria of loop closures for various types of surveys as part of the survey specifications. For instance, in the *Standards and Specifications for Geodetic Control Networks* stipulated by Federal

Table 3.2. Critical values for the n(0, 1), t(r), and τ(r) distributions

r	$\tau_{(1-\alpha/2)}$	$t_{(1-\alpha/2)}$	$n_{(1-\alpha/2)}$
2	4.30	1.41	1.96
3	3.18	1.64	1.96
4	2.78	1.76	1.96
5	2.57	1.81	1.96
6	2.45	1.85	1.96
7	2.37	1.87	1.96
8	2.31	1.88	1.96
9	2.26	1.90	1.96
10	2.23	1.90	1.96
15	2.13	1.93	1.96
20	2.09	1.94	1.96
25	2.06	1.94	1.96
30	2.04	1.94	1.96
35	2.03	1.95	1.96
40	2.02	1.95	1.96
50	2.01	1.95	1.96
60	2.00	1.95	1.96
80	1.99	1.96	1.96
100	1.98	1.96	1.96

Geodetic Control Subcommittee (FGCS) of the United States (cf. FGCS, 1984), loop closure criteria for various types of surveys, e.g., triangulation, traverse, inertial surveying, geodetic leveling, and Global Positioning System (GPS), of different orders of accuracy, have been stipulated. For instance, for geodetic leveling of First Order Class I it is required that the loop closure be less than $4\sqrt{S}$ mm (with s being the total length of the leveling loop under study in kilometers).

In order to efficiently detect the presence of large errors in the survey data prior to network adjustment, all the possible conditions in the network should be checked, and this prompts for the need for a systematic procedure. However, before the advent of powerful computers this would be a tremendously large amount of human work even for networks of moderate size. Furthermore, it is very difficult, if not impossible, for a human to isolate all the possible non-trivial set of loops (or conditions), especially in the case of large networks with mixed types of observations. For a long time there has been no efficient algorithm, suitable for automation on an electronic computer, for the detection of blunders in the raw survey data in terms of misclosures before network adjustment.

With the development of modern computers, some studies have been undertaken by geodesists to develop methodologies to automate the checks of condition misclosures by computers. For instance, Steeves (1984) developed an approach to isolate a fundamental set of circuits for two-dimensional geodetic networks using the graph theory; that is, computers are used to generate the loops, compute misclosures, and perform statistical analysis. Any misclosure exceeding a certain limit can be flagged automatically for human examination. Nevertheless, his approach is only applicable to some special cases. Recent developments in this field are characterized by my work (1994b) in which I developed a new algorithm, using matrix algebra that allows the computers to identify all the possible non-trivial loops and compute loop closures for geodetic leveling networks fully automatically. Circuits need not be defined manually, and any loop closure that exceeds a desired precision criterion will be flagged and this serves as the basis for surveyors to check blunders in the survey data. The developed methodology is universally applicable and therefore it can be extended to two- or three-dimensional networks with mixed types of observations. The algorithm developed in this program requires input of the raw survey data only. The computer calculates misclosures and statistical confidence intervals for each of the fundamental set of circuits and prints out the results for human analysis. The algorithm does not necessarily locate the circuits with the minimum lengths, but it does locate all the possible non-trivial circuits, the total number of which is equal to the redundancy of the network.

With the automated algorithm for loop closure computation, a human's effort has been reduced to examining observations associated with loops that have excessive misclosures and locating the gross error. It should be noted that since each misclosure relates to more than one observation, a single condition equation with a large misclosure does not indicate which observation(s) contains a gross error. It is, therefore, not always possible to pinpoint the exact location of a blunder by examining circuit misclosures, even if one assumes that only one blunder exists. It is certainly more involved in the case of multi-gross errors. Obviously, the fewer the observations involved in each of the condition equations, the easier the localization of possible gross errors. Nevertheless, the amount of data that has to be checked for possible blunders is limited to the data associated with the common edges of contiguous circuits having large misclosure. The amount of data that has to be examined manually is therefore minimized.

3.4 Testing of Consistency of Repeated Observations

If an observable (e.g., a direction, azimuth, angle, or distance, etc.) has been repeatedly measured, resulting in a data series, l_i (i=1, ..., n), any individual observation l_i can be tested statistically against the rest of the observations to see whether they are statistically compatible. The general null hypothesis H_0 can thus be stated as follows:

H_0: l_i is compatible with the rest of the observations

Possible choices of test statistic for the test of H_0 are given in Table 3.3, from which one can see that a specific test statistic is selected depending on whether the population mean (μ) and variance (σ^2) are known or not. Intuitively, if the population mean is known, the true value of the observable is known; otherwise, it is unknown. Similarly, if the population variance is known, the accuracy of the instrument being used for the measurements is known, and vice versa.

Table 3.3. Choice of statistic for statistical testing of repeated observations

σ \ μ (y)	μ known	μ unknown
σ known	$y_1 = \dfrac{l_i - \mu}{\sigma}$	$y_2 = \dfrac{l_i - \bar{l}}{\sqrt{\dfrac{n-1}{n}}\sigma}$
σ unknown	$y_3 = \dfrac{l_i - \mu}{s}$	$y_4 = \dfrac{l_i - \bar{l}}{\sqrt{\dfrac{n-1}{n}}s}$

In Table 3.3, $\bar{l}$ and s represent the sample mean and sample variance, respectively, and they are calculated as follows:

$$\bar{l} = \frac{1}{n}\sum_{i=1}^{n} l_i \tag{3.22}$$

$$s = \frac{1}{n-1}\sum_{i=1}^{n}(l_i - \bar{l})^2 \tag{3.23}$$

When the population variance σ^2, i.e., the accuracy of the instrument is known, both the statistic y_1 and y_2 belong to the standard normal distribution n(0, 1) with zero mean and unit variance. Therefore, given the significance level of the test α, an observation l_i (i = 1, ..., n) is considered as being consistent with the rest observations l_j (j≠i; j=1, ..., n) if

$$|y_1| \text{ or } |y_2| \le n_{1-\alpha/2} \tag{3.24}$$

The statistic y_1 is used when the population mean μ, i.e., the true value of the observable is known. Otherwise, the statistic y_2 is used.

On the other hand, if the population variance is unknown, i.e., the accuracy of the instrument is not adequately known, either statistic y_3 or y_4 is used, depending on whether the population mean μ is known or not. Theoretically, if the sample variance s^2 is estimated from an independent source, both statistic y_3 and y_4 belong to the student's t-distribution with degrees of freedom being (n-1). In this case, H_0 is accepted at (1-α) confidence level if

$$|y_3| \text{ or } |y_4| \le t_{1-\alpha/2}(n-1) \tag{3.25}$$

The statistic y_3 is used when the population mean μ, i.e., the true value of the observable is known. Otherwise, the statistic y_4 is used.

On the other hand, if the population variance σ^2 is unknown, and the sample variance s^2 is estimated from the same sample, both the statistic y_3 and y_4 belong to the Tau (τ-) distribution (Pope, 1976), and H_0 is accepted at (1-α) confidence level if

$$|y_3| \text{ or } |y_4| \leq \tau_{1-\alpha/2}(n-1) \tag{3.26}$$

Some typical critical values for the n(0, 1), t(r), and τ(r) distributions under a given confidence level of 95% have been listed in Table 3.2.

4 Least Squares Network Adjustment

After raw measurements have been pre-processed and screened for large gross errors (Chapters 2 and 3), they are now ready to be used for coordinate computations. As is well known, physical parameters can be determined either directly by taking measurements of those parameters themselves (if possible), or indirectly by taking measurements of another set of parameters that are related to the unknown parameters under study by a known mathematical model. As discussed before, physical quantities can never be measured perfectly due to the limiting human senses, the limiting resolution of the measuring instrument, and the uncontrollable effects of environmental factors, etc., which will always introduce observational errors in a measuring process. In order to reduce the effect of measurement inconsistencies it has become a routine procedure to make redundant measurements that, due to the effects of observational errors will, however, not agree with one another (i.e., they will not be consistent), and therefore each minimum subset of measurements would yield a different result, leading to no unique result unless an additional criterion is introduced.

The method of least squares, which arrives at a best solution by minimizing the sum of squares of weighted discrepancies between measurements, is one of the most frequently used methods to obtain unique estimates for physical parameters from a set of redundant measurements. The concepts of least squares estimation and curve fitting were introduced in the early 1800s by Legendre and Gauss mainly for the purpose of reducing physical and astronomical data. Statistically, application of the least squares principle does not require a priori knowledge of the distribution associated with the observations. It can be shown, however, that if the weight matrix **P** is chosen to be the inverse of the variance-covariance matrix of the observations, the least squares estimate is the *minimum variance* estimate, and if the observation errors have a normal distribution, the least squares estimate is the *maximum likelihood* estimate.

Deriving a least squares algorithm starts with a basic mathematical model that is defined as a functional relationship between unknown parameters under study (e.g., coordinates of unknown points) and the available observations (e.g., azimuth, distances, angles, GPS measurements). In geodetic science, the mathematical models that relate observations and unknown parameters have been classified into the following three types:

1. Combined model
2. Parametric model
3. Condition model

In the combined model, the observations and parameters are related through an implicit function, while in the parametric model observations are explicitly expressed as functions of the parameters, and finally, in the condition model formulated are conditions between measurements. In principle, parametric and condition models are special cases of the combined model. Choice of a mathematical model is dictated by the specific problem under study, and the ease in formulating mathematical equations. In geodesy, the parametric model has been proven to be much more advantageous over the condition model due to the ease of formulating parametric equations. It is in general difficult to formulate condition equations that were mainly used before the advent of modern computers to save manual computation work. The combined model is used when observations cannot be explicitly expressed as functions of unknown parameters.

Depending on the specific type of mathematical model in use, a least squares network adjustment is classified as a combined adjustment, a parametric adjustment, or a condition adjustment. Mathematically, the combined model can be converted into the parametric model by using a set of pseudo-observations. Many problems can be solved by either method, and solutions from different methods are equivalent. This chapter starts with the parametric modeling of geodetic measurements in §4.1. §4.2 discusses the network datum problem, and finally, the principle of parametric least squares network adjustment is dealt with in §4.3. For discussions on the combined and condition network adjustment and some other related issues, the interested reader is referred to Vanicek and Krakiwsky (1986), Koch (1987), Caspary (1987), and Leick (1990), among others.

4.1 Observational Modeling

As discussed in Chapter 2, geodetic network positioning can be made in terms of either the modern 3D model or the classical 2D plus 1D model. In the former case, the position of a terrain point is described in a three-dimensional coordinate system, while in the latter case it is separated into two horizontal coordinates in a two-dimensional coordinate system plus 1 vertical coordinate; that is, the elevation described by a height system. Obviously, the specific coordinates to be sought in network computation depend on the coordinate system chosen to work with. In three-dimensional space, they can be either the Geodetic (G) or Local Geodetic (LG) coordinate systems. In two-dimensional space, the horizontal coordinates are either (ϕ, λ) calculated on a reference ellipsoid or (N, E) (i.e., Northing, Easting) calculated on a mapping plane. Finally, for the vertical coordinate, the most frequently used height system is the orthometric height. From the computational point of view, however, since rigorous transformation between different coordinate systems can be made, equivalent results will be achieved regardless of which coordinate system is chosen to work with. The main factors governing the choice of a specific coordinate system are the purpose and the extent of the network. In practice, an overwhelming majority of practicing surveyors have been well exposed to the classical approach and prefer to use conformal mapping plane coordinates as horizontal control, and orthometric heights as vertical control.

In this section, in order to facilitate the discussion, observation equations in both the three-dimensional Local Geodetic (LG) coordinate system and the two-dimensional mapping plane coordinate system are given. The former is suitable for geodetic networks of local extent established for engineering purposes; for instance, the micronetworks used in industrial metrology, and networks for monitoring local crustal deformations, etc., and the latter is suitable for networks of local or regional scales. Observation modeling for geodetic leveling is not singled out since it is the same as modeling of coordinate difference observations, as discussed.

4.1.1 Observation Modeling in a Right-Handed Local Geodetic (RLG) System

Since LG is the system that is closest to the natural Local Astronomical (LA) coordinate system in which the conventional terrestrial geodetic observations are actually made (see Figure 2.1), the formulation of observation equations is more intuitive and can be greatly simplified in this system. As shown in §2.1, the difference between LG and LA is due to the effects of the earth's gravity field. The LA system is defined as being left-handed with origin at the station and orientation defined by the station's astronomical latitude, longitude, and a tangent to the local gravity vector, while the LG system is defined as being left-handed with origin at the station and orientation defined by the station's geodetic latitude, longitude, and the ellipsoidal normal. Thus, the raw observations have to be corrected for the effects of the earth's gravity field before they can be used, in addition to meteorological and instrumental corrections (cf. §2.4).

Since the North American convention for a coordinate system is right-handed, a Right-handed Local Geodetic (RLG) coordinate system can be obtained by a reflection of the y-axis and a rotation around the z-axis to switch the locations of the x and y axis (see Figure 4.1).

The physical meaning of the RLG system is not changed except for the description of the horizontal axes. The LG and RLG coordinates are related by the following similarity transformation

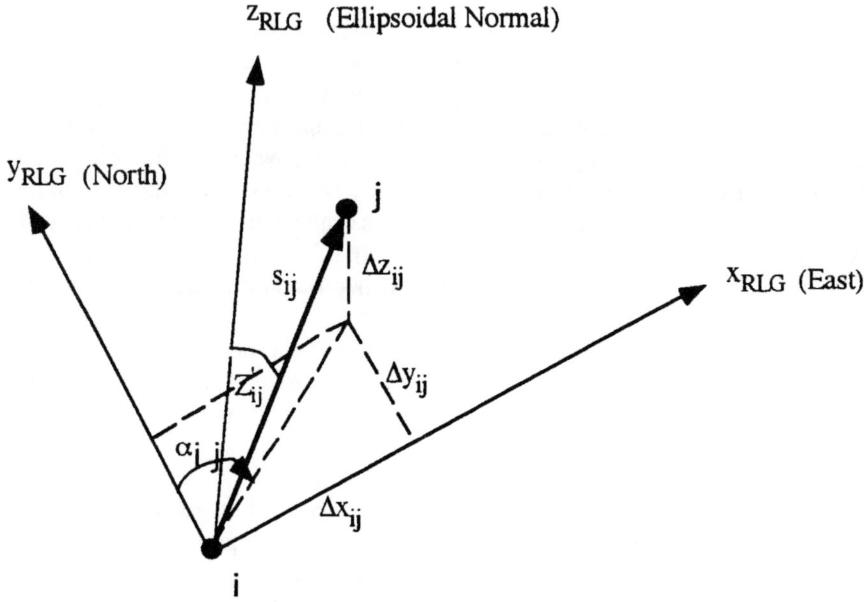

Figure 4.1. Spatial distances s_{ij}, geodetic azimuth α_{ij}, and geodetic zenith distance Z'_{ij} in the Right-handed Local Geodetic (RLG) system

$$\begin{pmatrix} x_i \\ y_i \\ z_i \end{pmatrix}^{RLG} = \mathbf{R}_3\left(-\frac{\pi}{2}\right)\mathbf{P}_2 \begin{pmatrix} x_i \\ y_i \\ z_i \end{pmatrix}^{LG} \tag{4.1}$$

where $\mathbf{R}_3$ is a rotation matrix, and $\mathbf{P}_2$ a reflection matrix calculated by Equation (2.4) and (2.5), respectively. The position of a terrain point in the RLG system is expressed by

$$\begin{pmatrix} x_i \\ y_i \\ z_i \end{pmatrix}^{RLG} = s_{ij} \begin{pmatrix} \sin Z'_{ij} \ \sin\alpha_{ij} \\ \sin Z'_{ij} \ \cos\alpha_{ij} \\ \cos Z'_{ij} \end{pmatrix} \tag{4.2}$$

where s_{ij}, α_{ij}, and Z'_{ij} represent a spatial distance, a geodetic azimuth, and a geodetic zenith distance, respectively (cf. Figure 4.1). It should be noted that theoretically every observing station in the network has its own RLG system. Since the directions of the ellipsoidal normal at each point are practically the same for networks of local extent, one may constrain one of the RLG systems from moving (or alternatively a fictitious mean system) and to solve for the translation components (station coordinates) of each of the other systems with respect to this fixed system.

As discussed in Chapter 2, in order for raw observations to be usable in the RLG system, they must first be corrected, whenever applicable, for the meteorological and instrumental effects, and then reduced to the local geodetic system by applying gravimetrical corrections to angular measurements (i.e., azimuth, horizontal directions and angles, and zenith distances). Spatial distances need not be corrected for gravimetrical effects. Therefore, in the following equations, the observation quantities are assumed to have been obtained from the pre-processed raw measurements according to Chapter 2.

4.1.1.1 Nonlinear Observation Equations

Distance

After pre-processing, the originally measured raw distances have been reduced to the straight line spatial distances, i.e., the chords, between two points. According to the above Equation (4.1), the distance is related to the coordinates of the two terminal points as follows:

$$s_{ij} = \sqrt{\left(x_j - x_i\right)^2 + \left(y_j - y_i\right)^2 + \left(z_j - z_i\right)^2} \tag{4.3}$$

where s_{ij} is the measured raw distance after meteorological and instrumental corrections, and (x_i, y_i, z_i) and (x_j, y_j, z_j) are the RLG coordinates of the instrument station i and target station j, respectively.

Azimuth

A measured azimuth can be seen as an "absolute" direction that is measured relative to the true north and thus provides orientation to a geodetic network. An azimuth can be measured either by an astronomical method or using a gyro instrument. After reduced to the local geodetic system by pre-processing, a measured raw azimuth becomes a spatial azimuth referring to the normal section of the reference ellipsoid, passing through the terrain points in question. It is related to the horizontal coordinates of the terminal points as follows (cf. [4.1])

$$\alpha_{ij} = \arctan\left[\left(x_j - x_i\right) / \left(y_j - y_i\right)\right] \tag{4.4}$$

where a_{ij} is the measured raw azimuth reduced to the local geodetic system, and (x_i, y_i) and (x_j, y_j) are the RLG horizontal coordinates of instrument station i and target station j, respectively.

Horizontal Direction

Horizontal direction observations are measured relative to the reference direction of the theodolite; e.g., the "zero" of the horizontal circle. In order to relate the

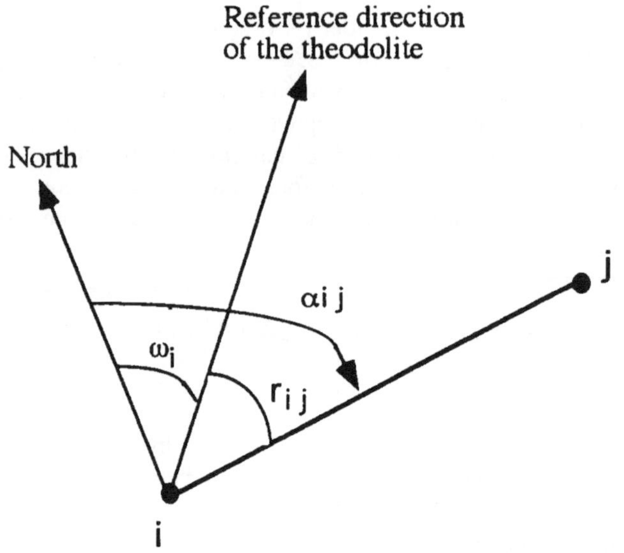

Figure 4.2. Orientation of a horizontal direction

location of the zero to the north direction, an unknown "nuisance" parameter, called the orientation unknown ω_i, must be added to the directions and be solved for by the adjustment along with the unknown coordinates. The direction observations are usually arrived at from numerous sightings from a given point to other points. The relationship between an azimuth α_{ij} and a direction r_{ij} is thus given by (see Figure 4.2)

$$a_{ij} = r_{ij} + \omega_i \tag{4.5}$$

The observation equation for a direction in the RLG coordinate system is therefore

$$r_{ij} = \arctan\left[\left(x_j - x_i\right) / \left(y_j - y_i\right)\right] - \omega_i \tag{4.6}$$

where r_{ij} is the measured raw direction reduced to the local geodetic system, ω_i is the orientation unknown, and (x_i, y_i) and (x_j, y_j) are the RLG horizontal coordinates of the instrument station i and target station j, respectively.

Horizontal Angle

Referring to Figure 4.3, a horizontal angle can be seen as simply the difference between two azimuths or two directions. Thus, the observation equation for a horizontal angle can be obtained by differencing two direction or two azimuth observation equations referring to lines ij and ik; i.e.,

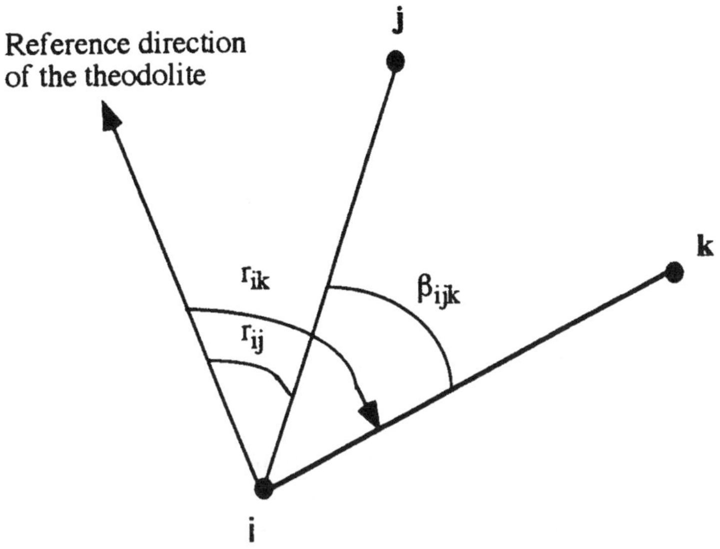

Figure 4.3. Observation of a horizontal angle

$$\beta_{ijk} = \arctan\left[(x_k - x_i)/(y_k - y_i)\right] - \arctan\left[(x_j - x_i)/(y_j - y_i)\right] \tag{4.7}$$

where β_{ijk} is the measured raw angle reduced to the local geodetic system, and (x_i, y_i), (x_j, y_j) and (x_k, y_k) are the RLG horizontal coordinates of instrument station i and target stations j and k, respectively.

Zenith Distance

Referring to Figure 4.1, a geodetic zenith distance obtained from a measured astronomical zenith distance by gravimetrical correction is measured relative to the ellipsoidal normal. The observation equation reads

$$Z_{ij} = \arctan\left[\frac{\sqrt{(x_j - x_i)^2 + (y_j - y_i)^2}}{(z_j - z_i)}\right] \tag{4.8}$$

where Z_{ij} is the measured raw zenith distance reduced to the local geodetic system, and (x_i, y_i, z_i) and (x_j, y_j, z_j) are the RLG coordinates of the instrument station i and target station j, respectively.

Coordinates and Coordinate Differences

When the RLG coordinates of a point or the coordinate differences between two points are measured, they are simply related to their corresponding unknowns as follows

$$\begin{pmatrix} x_i \\ y_i \\ z_i \end{pmatrix}^m = \begin{pmatrix} x_i \\ y_i \\ z_i \end{pmatrix} \tag{4.9}$$

$$\begin{pmatrix} \Delta x_{ij} \\ \Delta y_{ij} \\ \Delta z_{ij} \end{pmatrix}^m = \begin{pmatrix} x_j \\ y_j \\ z_j \end{pmatrix} - \begin{pmatrix} x_i \\ y_i \\ z_i \end{pmatrix} \tag{4.10}$$

where $(x_i, y_i, z_i)^m$ and $(\Delta x_{ij}, \Delta y_{ij}, \Delta z_{ij})^m$ are the measured RLG coordinates and coordinate differences, respectively, and (x_i, y_i, z_i) and $(\Delta x_{ij}, \Delta y_{ij}, \Delta z_{ij})$ their corresponding unknowns. Coordinates or coordinate differences that are directly measured by some extraterrestrial positioning techniques (e.g., GPS) are usually expressed in a global geodetic coordinate system. They may be transformed to the RLG system through the procedures discussed in §2.1 and formulae (2.8) and (4.1).

4.1.1.2 Linearized Observation Equations

For network adjustment computations, the above nonlinear observation equations have to be linearized, and an iterative solution approach is used. Assuming that the approximate coordinates of the network stations are available, the above nonlinear observation equations can be approximated by the linear term of their respective Taylor series as follows:

Distance

$$\begin{aligned} s_{ij} + v_{s_{ij}} = s_{ij}^0 &- \frac{\Delta x_{ij}^0}{s_{ij}^0}\delta x_i - \frac{\Delta y_{ij}^0}{s_{ij}^0}\delta y_i - \frac{\Delta z_{ij}^0}{s_{ij}^0}\delta z_i \\ &+ \frac{\Delta x_{ij}^0}{s_{ij}^0}\delta x_j + \frac{\Delta y_{ij}^0}{s_{ij}^0}\delta y_j + \frac{\Delta z_{ij}^0}{s_{ij}^0}\delta z_j \end{aligned} \tag{4.11}$$

Azimuth

$$\begin{aligned} \alpha_{ij} + v_{\alpha_{ij}} = \alpha_{ij}^0 &- \frac{\Delta y_{ij}^0}{\left(L_{ij}^0\right)^2}\delta x_i + \frac{\Delta x_{ij}^0}{\left(L_{ij}^0\right)^2}\delta y_i + \\ &\frac{\Delta y_{ij}^0}{\left(L_{ij}^0\right)^2}\delta x_j - \frac{\Delta x_{ij}^0}{\left(L_{ij}^0\right)^2}\delta y_j \end{aligned} \tag{4.12}$$

Horizontal Direction

$$r_{ij} + v_{r_{ij}} = r_{ij}^0 - \frac{\Delta y_{ij}^0}{\left(L_{ij}^0\right)^2}\delta x_i + \frac{\Delta x_{ij}^0}{\left(L_{ij}^0\right)^2}\delta y_i + \frac{\Delta y_{ij}^0}{\left(L_{ij}^0\right)^2}\delta x_j - \frac{\Delta x_{ij}^0}{\left(L_{ij}^0\right)^2}\delta y_i - \delta\omega_i \tag{4.13}$$

Horizontal Angle

$$\beta_{ijk} + v_{\beta ijk} = \beta_{ijk}^0 + \left(\frac{\Delta y_{ij}^0}{\left(L_{ij}^0\right)^2} - \frac{\Delta y_{ik}^0}{\left(L_{ik}^0\right)^2}\right)\delta x_i + \left(\frac{\Delta x_{ik}^0}{\left(L_{ik}^0\right)^2} - \frac{\Delta x_{ij}^0}{\left(L_{ij}^0\right)^2}\right)\delta y_i - \frac{\Delta y_{ij}^0}{\left(L_{ij}^0\right)^2}\delta x_j + \frac{\Delta x_{ij}^0}{\left(L_{ij}^0\right)^2}\delta y_j + \frac{\Delta y_{ik}^0}{\left(L_{ik}^0\right)^2}\delta x_k - \frac{\Delta x_{ik}^0}{\left(L_{ik}^0\right)^2}\delta y_k \tag{4.14}$$

Zenith Distance

$$Z_{ij} + v_{Z_{ij}} = Z_{ij}^0 - \frac{\Delta x_{ij}^0 \Delta z_{ij}^0}{L_{ij}^0\left(s_{ij}^0\right)^2}\delta x_i - \frac{\Delta y_{ij}^0 \Delta z_{ij}^0}{L_{ij}^0\left(s_{ij}^0\right)^2}\delta y_i + \frac{L_{ij}^0}{\left(s_{ij}^0\right)^2}\delta z_i + \frac{\Delta x_{ij}^0 \Delta z_{ij}^0}{L_{ij}^0\left(s_{ij}^0\right)^2}\delta x_j + \frac{\Delta y_{ij}^0 \Delta z_{ij}^0}{L_{ij}^0\left(s_{ij}^0\right)^2}\delta y_j - \frac{L_{ij}^0}{\left(s_{ij}^0\right)^2}\delta z_j \tag{4.15}$$

Coordinates and Coordinate Differences

$$\begin{pmatrix} x_i \\ y_i \\ z_i \end{pmatrix}^m + \begin{pmatrix} v_{x_i} \\ v_{y_i} \\ v_{z_i} \end{pmatrix} = \begin{pmatrix} x_i^0 \\ y_i^0 \\ z_i^0 \end{pmatrix} + \begin{pmatrix} \delta x_i \\ \delta y_i \\ \delta z_i \end{pmatrix} \tag{4.16}$$

$$\begin{pmatrix} \Delta x_{ij} \\ \Delta y_{ij} \\ \Delta z_{ij} \end{pmatrix}^m + \begin{pmatrix} v_{\Delta x_{ij}} \\ v_{\Delta y_{ij}} \\ v_{\Delta z_{ij}} \end{pmatrix} = \begin{pmatrix} \Delta x_{ij}^0 \\ \Delta y_{ij}^0 \\ \Delta z_{ij}^0 \end{pmatrix} + \begin{pmatrix} \delta x_j \\ \delta y_j \\ \delta z_j \end{pmatrix} - \begin{pmatrix} \delta x_i \\ \delta y_i \\ \delta z_i \end{pmatrix} \tag{4.17}$$

where

$$\Delta x_{ij}^0 = x_j^0 - x_i^0; \tag{4.18}$$

$$\Delta y_{ij}^0 = y_j^0 - y_i^0; \tag{4.19}$$

$$\Delta z_{ij}^0 = z_j^0 - z_i^0; \tag{4.20}$$

$$\Delta x_{ik}^0 = x_k^0 - x_i^0 \tag{4.21}$$

$$\Delta y_{ik}^0 = y_k^0 - y_i^0 \tag{4.22}$$

$$\Delta z_{ik}^0 = z_k^0 - z_i^0 \tag{4.23}$$

x^0, y^0, z^0 are the approximate coordinates of the network points; $v_{s_{ij}}$, $v_{\alpha_{ij}}$, $v_{r_{ij}}$, $v_{\beta_{ij}}$, $v_{Z_{ij}}$ are the differential corrections to an observed spatial distance, an azimuth, a horizontal direction, a horizontal angle, and a zenith distance, respectively; δx_i, δy_i, δz_i, δx_j, δy_j, δz_j, δx_k, δy_k, δz_k are the differential changes in the initially given coordinates of the instrument station i, and targets stations j and k, respectively; and s_{ij}^0, α_{ij}^0, r_{ij}^0, β_{ijk}^0, and Z_{ij}^0 are the approximate values of a spatial distance, an azimuth, a horizontal direction, a horizontal angle, and a zenith distance, calculated from the given approximate coordinates, respectively, as follows:

$$s_{ij}^0 = \sqrt{\left(x_j^0 - x_i^0\right)^2 + \left(y_j^0 - y_i^0\right)^2 + \left(z_j^0 - z_i^0\right)^2} \tag{4.24}$$

$$\alpha_{ij}^0 = \arctan\left[\left(x_j^0 - x_i^0\right) / \left(y_j^0 - y_i^0\right)\right] \tag{4.25}$$

$$r_{ij}^0 = \arctan\left[\left(x_j^0 - x_i^0\right) / \left(y_j^0 - y_i^0\right)\right] - \omega_i^0 \tag{4.26}$$

$$\beta_{ijk}^0 = \arctan\left[\left(x_k^0 - x_i^0\right) / \left(y_k^0 - y_i^0\right)\right] - \arctan\left[\left(x_j^0 - x_i^0\right) / \left(y_j^0 - y_i^0\right)\right] \tag{4.27}$$

$$Z_{ij}^0 = \arctan\left[\frac{\sqrt{\left(x_j^0 - x_i^0\right)^2 + \left(y_j^0 - y_i^0\right)^2}}{\left(z_j^0 - z_i^0\right)}\right] \tag{4.28}$$

$$L_{ij}^0 = \sqrt{\left(\Delta x_{ij}^0\right)^2 + \left(\Delta y_{ij}^0\right)^2} \tag{4.29}$$

$$L_{ik}^0 = \sqrt{\left(\Delta x_{ik}^0\right)^2 + \left(\Delta y_{ik}^0\right)^2} \tag{4.30}$$

Denote

$$\delta \mathbf{p}_{ij} = (\delta x_i \ \delta y_i \ \delta z_i \ \delta x_j \ \delta y_j \ \delta z_j)^T \tag{4.31}$$

$$\delta \mathbf{p}_{ik} = (\delta x_i \ \delta y_i \ \delta z_i \ \delta x_k \ \delta y_k \ \delta z_k)^T \tag{4.32}$$

$$\mathbf{a}_{sij}^T = \left(\frac{\Delta x_{ij}^0}{s_{ij}^0} \ \frac{\Delta y_{ij}^0}{s_{ij}^0} \ \frac{\Delta z_{ij}^0}{s_{ij}^0}\right) \tag{4.33}$$

$$\mathbf{a}_{\alpha ij}^T = \left(\frac{\Delta y_{ij}^0}{\left(L_{ij}^0\right)^2} \ \ -\frac{\Delta x_{ij}^0}{\left(L_{ij}^0\right)^2} \ \ 0\right) \tag{4.34}$$

$$\mathbf{a}_{\alpha ik}^T = \left(\frac{\Delta y_{ik}^0}{\left(L_{ik}^0\right)^2} \ \ -\frac{\Delta x_{ik}^0}{\left(L_{ik}^0\right)^2} \ \ 0\right) \tag{4.35}$$

$$\mathbf{a}_{Zij}^T = \left[\frac{\Delta x_{ij}^0 \Delta z_{ij}^0}{L_{ij}^0\left(s_{ij}^0\right)^2} \ \ \frac{\Delta y_{ij}^0 \Delta z_{ij}^0}{L_{ij}^0\left(s_{ij}^0\right)^2} \ \ -\frac{L_{ij}^0}{\left(s_{ij}^0\right)^2}\right] \tag{4.36}$$

Then the above Equations (4.11) to (4.15) can be expressed more compactly in matrix and vector format as follows:

Distance:

$$v_{s_{ij}} = \mathbf{a}_{s_{ij}}^{T}\mathbf{U}\delta\mathbf{p}_{ij} + \left(s_{ij}^{0} - s_{ij}\right) \tag{4.37}$$

Azimuth:

$$v_{\alpha_{ij}} = \mathbf{a}_{\alpha_{ij}}^{T}\mathbf{U}\delta\mathbf{p}_{ij} + \left(\alpha_{ij}^{0} - \alpha_{ij}\right) \tag{4.38}$$

Horizontal Direction:

$$v_{r_{ij}} = \mathbf{a}_{\alpha_{ij}}^{T}\mathbf{U}\delta\mathbf{p}_{ij} - \delta\omega_{i} + \left(r_{ij}^{0} - r_{ij}\right) \tag{4.39}$$

Horizontal Angle:

$$v_{\beta_{ijk}} = \mathbf{a}_{\alpha_{ik}}^{T}\mathbf{U}\delta\mathbf{p}_{ik} - \mathbf{a}_{\alpha_{ij}}^{T}\mathbf{U}\delta\mathbf{p}_{ij} + \left(\beta_{ijk}^{0} - \beta_{ijk}\right) \tag{4.40}$$

Zenith Angle:

$$v_{Z_{ij}} = \mathbf{a}_{Z_{ij}}^{T}\mathbf{U}\delta\mathbf{p}_{ij} + \left(Z_{ij}^{0} - Z_{ij}\right) \tag{4.41}$$

where

$$\mathbf{U} = [-\mathbf{I} \quad \mathbf{I}] \tag{4.42}$$

with **I** being a 3 by 3 unit matrix.

4.1.2 Observation Modeling on a Conformal Mapping Plane

In order to perform network computations on a mapping plane, raw measurements must first be corrected for meteorological, instrumental, and gravimetrical effects, and then reduced to the reference ellipsoid, and finally reduced to the chosen mapping plane following the data pre-processing procedures discussed in Chapter 2. Figure 4.4 shows the relation between the plane coordinates and the grid distance and grid azimuth that are denoted by l_{ij} and t_{ij}, respectively, to distinguish them from their corresponding spatial distance s_{ij} and azimuth α_{ij}.

The basic observables on a conformal mapping plane include the grid distance, grid azimuth, horizontal directions, and horizontal angles reduced to the mapping plane. From Figure 4.4, the nonlinear observation equations can be established as follows

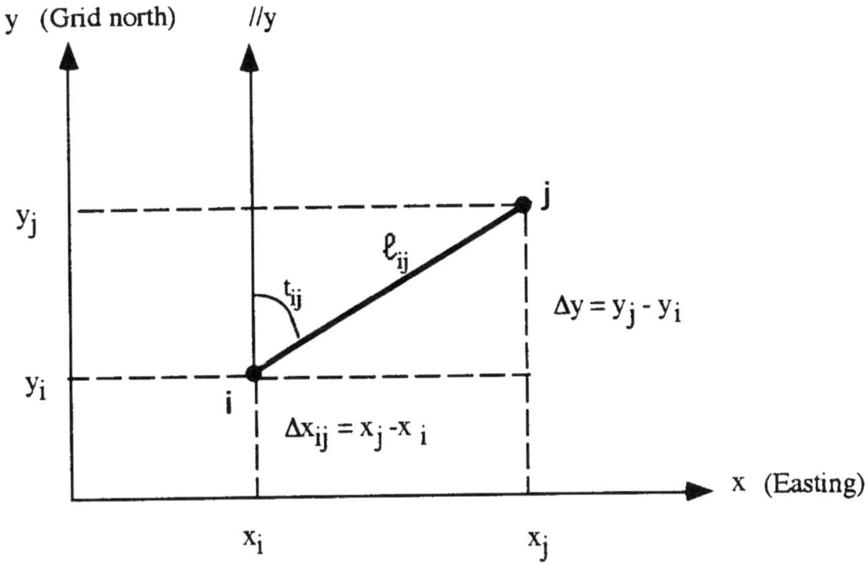

Figure 4.4. Grid azimuth t_{ij} and grid distance l_{ij} in a conformal mapping plane coordinate system

Grid Distance:

$$l_{ij} = \sqrt{(x_j - x_i)^2 + (y_j - y_i)^2} \tag{4.43}$$

Grid Azimuth:

$$t_{ij} = \arctan[(x_j - x_i)/(y_j - y_i)] \tag{4.44}$$

Horizontal Direction:

$$r_{ij} = \arctan[(x_j - x_i)/(y_j - y_i)] - \omega_i \tag{4.45}$$

Horizontal Angle:

$$\beta_{ijk} = [(x_k - x_i)/(y_k - y_i)] - \arctan[(x_j - x_i)/(y_j - y_i)] \tag{4.46}$$

where l_{ij}, t_{ij}, r_{ij}, and β_{ijk} represent respectively the grid distance, grid azimuth, grid horizontal direction, and grid horizontal angle that are obtained from their corresponding measured quantities through pre-processing; (x_i, y_i), (x_j, y_j), and (x_k, y_k) are the Easting and Northing coordinates for the instrument station i, and target points j and k, respectively; and ω_i is the orientation unknown at station i.

Similarly, the linearized form of the observation equations are as follows:

Grid Distance:

$$v_{l_{ij}} = \mathbf{a}_{l_{ij}}^{T}\delta\overline{\mathbf{p}}_{ij} + \left(l_{ij}^{0} - l_{ij}\right) \tag{4.47}$$

Grid Azimuth:

$$v_{t_{ij}} = \mathbf{a}_{t_{ij}}^{T}\delta\overline{\mathbf{p}}_{ij} + \left(t_{ij}^{0} - t_{ij}\right) \tag{4.48}$$

Horizontal Direction:

$$v_{r_{ij}} = \mathbf{a}_{t_{ij}}^{T}\delta\overline{\mathbf{p}}_{ij} - \delta\omega_{i} + \left(r_{ij}^{0} - r_{ij}\right) \tag{4.49}$$

Horizontal Angle:

$$v_{\beta_{ijk}} = \mathbf{a}_{t_{ik}}^{T}\delta\overline{\mathbf{p}}_{ik} - \mathbf{a}_{t_{ij}}^{T}\delta\overline{\mathbf{p}}_{ij} + \left(\beta_{ijk}^{0} - \beta_{ijk}\right) \tag{4.50}$$

where x^0 and y^0 are the approximate coordinates of the network points, and

$$\delta\overline{\mathbf{p}}_{ij} = \begin{pmatrix}\delta_{x_i} & \delta_{y_i} & \delta_{x_j} & \delta_{x_j}\end{pmatrix}^{T} \tag{4.51}$$

$$\delta\overline{\mathbf{p}}_{ik} = \begin{pmatrix}\delta_{x_i} & \delta_{y_i} & \delta_{x_k} & \delta_{y_k}\end{pmatrix}^{T} \tag{4.52}$$

$$\mathbf{a}_{l_{ij}}^{T} = \begin{pmatrix} -\frac{\Delta x_{ij}^{0}}{l_{ij}^{0}} & -\frac{\Delta y_{ij}^{0}}{l_{ij}^{0}} & \frac{\Delta x_{ij}^{0}}{l_{ij}^{0}} & \frac{\Delta y_{ij}^{0}}{l_{ij}^{0}} \end{pmatrix} \tag{4.53}$$

$$\mathbf{a}_{t_{ij}}^{T} = \begin{pmatrix} -\frac{\Delta y_{ij}^{0}}{\left(l_{ij}^{0}\right)^{2}} & \frac{\Delta x_{ij}^{0}}{\left(l_{ij}^{0}\right)^{2}} & \frac{\Delta y_{ij}^{0}}{\left(l_{ij}^{0}\right)^{2}} & -\frac{\Delta x_{ij}^{0}}{\left(l_{ij}^{0}\right)^{2}} \end{pmatrix} \tag{4.54}$$

$$\mathbf{a}_{t_{ik}}^{T} = \begin{pmatrix} -\frac{\Delta y_{ik}^{0}}{\left(l_{ik}^{0}\right)^{2}} & \frac{\Delta x_{ik}^{0}}{\left(l_{ik}^{0}\right)^{2}} & \frac{\Delta y_{ik}^{0}}{\left(l_{ik}^{0}\right)^{2}} & -\frac{\Delta x_{ik}^{0}}{\left(l_{ik}^{0}\right)^{2}} \end{pmatrix} \tag{4.55}$$

$$l_{ij}^{0} = \sqrt{\left(\Delta x_{ij}^{0}\right)^{2} + \left(\Delta y_{ij}^{0}\right)^{2}} \tag{4.56}$$

$$t_{ij}^0 = \arctan\left[\left(x_j^0 - x_i^0\right) / \left(y_j^0 - y_i^0\right)\right] \tag{4.57}$$

$$r_{ij}^0 = \arctan\left[\left(x_j^0 - x_i^0\right) / \left(y_j^0 - y_i^0\right)\right] - \omega_i^0 \tag{4.58}$$

$$\beta_{ijk}^0 = \arctan\left[\left(x_k^0 - x_i^0\right) / \left(y_k^0 - y_i^0\right)\right] - \arctan\left[\left(x_j^0 - x_i^0\right) / \left(y_j^0 - y_i^0\right)\right] \tag{4.59}$$

$$\Delta x_{ij}^0 = x_j^0 - x_i^0 \tag{4.60}$$

$$\Delta y_{ij}^0 = y_j^0 - y_i^0 \tag{4.61}$$

$$\Delta x_{ik}^0 = x_k^0 - x_i^0 \tag{4.62}$$

$$\Delta y_{ik}^0 = y_k^0 - y_i^0 \tag{4.63}$$

4.1.3 Replacing the Independent Directions Observed at a Station by Correlated Angles

The observation of horizontal directions introduces orientation parameters. In order to avoid the orientation parameters, the set of directions, denoted by $\boldsymbol{l}_{di}$, observed at station i with observational matrix $\mathbf{P}_{l_{di}}$ can be replaced rigorously by a set of correlated angles, denoted by $\boldsymbol{l}_{ai}$, corresponding to the same station with observational matrix $\mathbf{P}_{l_{ai}}$. Furthermore, since it is impractical to use different observational weights for each individual direction observed at the same station, the same observational weight is therefore assumed for all the directions observed at the same station. The relation between $\boldsymbol{l}_{di}$ and $\boldsymbol{l}_{ai}$, and the relation between $\mathbf{P}_{l_{di}}$ and $\mathbf{P}_{l_{ai}}$ are derived below.

Referring to Figure 4.5, one assumes that if a set of n directions, i.e., r_1, r_2, ..., and r_n, are observed at station i, the vector $\boldsymbol{l}_{di}$ is therefore denoted as

$$\boldsymbol{l}_{di} = \left(r_1 \;\; r_2 \;\; \dots \;\; r_n\right)^T \tag{4.64}$$

Assuming that all the directions at the station will be measured with the same observational weight p_i, then the observational weight matrix $\mathbf{P}_{l_{di}}$ for $\boldsymbol{l}_{di}$ can be expressed as

$$\mathbf{P}_{l_{di}} = \mathbf{I} \cdot p_i \tag{4.65}$$

where $\mathbf{I}$ is a n by n identity matrix.

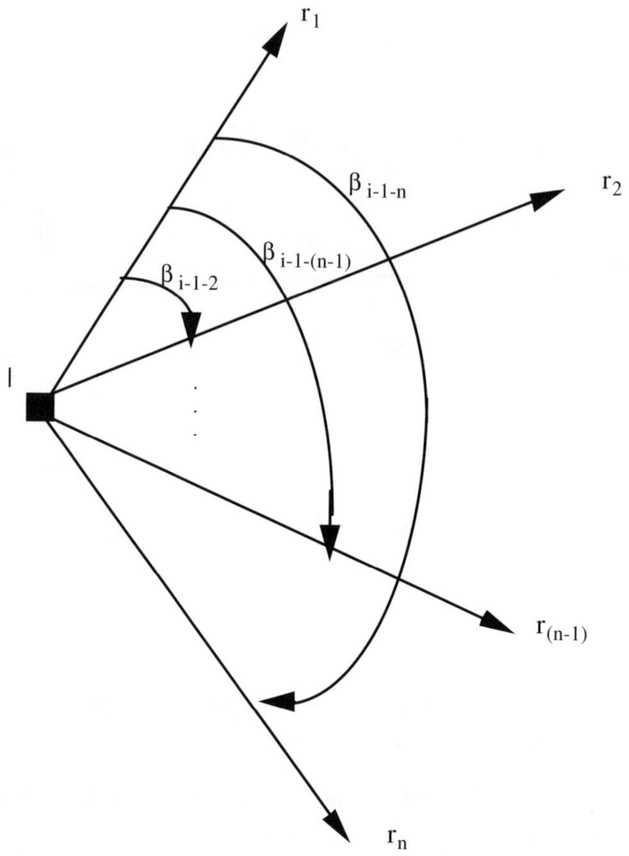

Figure 4.5. Observation of horizontal directions at a network station

The set of independent directions l_{di} can be replaced rigorously by a set of correlated angles, i.e., $\beta_{i\text{-}1\text{-}2}$, $\beta_{i\text{-}1\text{-}3}$, ..., $\beta_{i\text{-}1\text{-}(n\text{-}1)}$, and $\beta_{i\text{-}1\text{-}n}$, defined as follows

$$\beta_{i-1-2} = r_2 - r_1 \tag{4.66}$$

$$\beta_{i-1-3} = r_3 - r_1 \tag{4.67}$$

...

$$\beta_{i-1-(n-1)} = r_{(n-1)} - r_1 \tag{4.68}$$

$$\beta_{i-1-n} = r_n - r_1 \tag{4.69}$$

Denoting $\boldsymbol{l}_{ai}$ by

$$\boldsymbol{l}_{ai} = (\beta_{i-1-2} \ \beta_{i-1-3} \ \cdots \ \beta_{i-1-n})^T \tag{4.70}$$

one then obtains the relation between $\boldsymbol{l}_{ai}$ and $\boldsymbol{l}_{di}$ as follows

$$\boldsymbol{l}_{ai} = \mathbf{B}\, \boldsymbol{l}_{di} \tag{4.71}$$

where **B** is a (n-1) by n coefficient matrix derived from Equations (4.66) to (4.69) as follows

$$\mathbf{B} = \begin{pmatrix} -1 & 1 & 0 & 0 & \cdots & 0 \\ -1 & 0 & 1 & 0 & \cdots & 0 \\ \vdots & \vdots & \vdots & \ddots & \vdots & \vdots \\ -1 & 0 & \cdots & 0 & 1 & 0 \\ -1 & 0 & \cdots & 0 & 0 & 1 \end{pmatrix} \tag{4.72}$$

By applying the law of variance-covariance propagation, matrix $\mathbf{P}_{l_{ai}}$ reads

$$\begin{aligned} \mathbf{P}_{l_{ai}} &= \left(\mathbf{B}\left(\mathbf{P}_{l_{di}}\right)^{-1}\mathbf{B}^T\right)^{-1} \\ &= \left(\mathbf{B}\,\mathbf{B}^T\right)^{-1} p_i \end{aligned} \tag{4.73}$$

By replacing $\boldsymbol{l}_{di}$ with $\boldsymbol{l}_{ai}$ the orientation parameter ω_i at station i has been removed through differencing and the configuration matrix for an angle $\beta_{i\text{-}j\text{-}k}$ can be obtained from Equations (4.40) and (4.50) for the RLG and mapping plane coordinate systems, respectively.

4.2 Geodetic Network Datum

4.2.1 Definitions and Parameters

The datum of a geodetic network is defined as the basic (minimum) parameters needed to define the network in space or to position the network relative to a pre-defined coordinate system. This problem emerges from the fact that the conventional geodetic measurements, e.g., distance, horizontal directions, and/or angles, are internal measurements made between or among network points, and therefore they can only define the relative positions (i.e., relative coordinates) of the network points, while the "absolute" coordinates of a network point are external quantities, i.e., they are measured relative to a pre-defined coordinate system. Thus, with the conventional measurements alone, position computations cannot be initiated unless some other basic information, i.e., the network datum, is supplied.

A majority of practicing surveyors are still not well exposed to the term "geodetic network datum." Theoretically, for decades, this problem was bypassed in the theory of network adjustment computations when the method of condition adjustment was dominantly used, which would greatly reduce the amount of hand computation work before the advent of modern computers. Practically, surveyors took it for granted that some points in a newly established network have to be known in order to compute the coordinates of unknown points. This is especially true even today among the practicing surveyors. For instance, in the United States, whenever a local horizontal or vertical network of a certain order of accuracy is established to provide geodetic control for a mapping and/or civil construction project, it is required according to the FGCS (Federal Geodetic Control Subcommittee) specifications that the new network be tied to a certain number of appropriately distributed existing national geodetic control points of a certain order of accuracy (cf. FGCS, 1984), and the coordinates of the existing control points are then kept fixed in the network adjustment and thus inherently provide datum information (usually more than necessary) for the newly established network. One, therefore, does not have to deal with the datum problem.

The datum definition for national geodetic control networks was done by astronomic measurements together with making some theoretic assumptions when the first observations of the First-order control were processed. All subsequent observations extending or densifying the national networks are related to the national datum by observations connecting new points with already fixed ones. Similar situations exist in every country around the world. Some limitations and problems with this conventional procedure for geodetic network datum definition must be noted here. First of all, the old positions of the national geodetic control points were usually obtained a long time ago when the accuracy of geodetic measurements was lower than that available today with the modern geodetic instruments. Especially with the development of GPS and other extraterrestrial positioning system, e.g., SLR (Satellite Laser Ranging), VLBI (Very Long Baseline Interferometry), the positioning accuracy available today could be one order (or more) better than before. Unless the national geodetic control is continuously updated with the newer and more accurate observations available, a newly established network with better internal accuracy could be distorted by the fixed old positions. Even though the national geodetic control is updated over time, there will always be a gap, between two consecutive updatings, during which the coordinates of the national control networks are not up to date. Second, there are cases in which the required accuracy of a newly established network, e.g., networks established for high-precision engineering and/or deformation monitoring purposes, is much higher than that of the existing national control. In this case, adopting the national datum would therefore degrade the results, and independent networks are required, and the datum is chosen in an optimal way depending on the purpose of a particular network. Thirdly, the conventional procedure usually overconstrains the network; that is, more coordinates are fixed than necessary to provide the new network datum. In this case, the geometry of the network is affected by the strains imposed on the network, and therefore the internal accuracy of the network, i.e., the actual precision of the geodetic measurements, has to be estimated separately. One important condition of the datum definition for a geodetic network is that it must not affect the geometry of the network; i.e., the relative position of the points shall be defined solely by the geodetic observations (Caspary, 1987).

With the development of modern computers that allow the method of parametric network adjustment, due to the ease in formulating observation equations and computer programming, as a standard procedure for data processing in surveying and geodesy, the definition of a geodetic network datum has become an important part in the context of network adjustment theory, and it has been well addressed by geodesists over the last decade. As discussed before, since the geodetic network datum definition is associated with defining a coordinate system, the basic datum parameters can be identified as the following three types: translation, rotation (i.e., orientation), and scale.

Depending on whether a network is established in one-, two-, or three-dimensional space and the geodetic measurements made within the network, the types and total number of datum parameters for different types of geodetic networks are given in Table 4.1. Mathematically, the need for the datum parameters of a network means that, in addition to having at least as many observations as unknowns, a network must be supplied with the above minimum information in order to proceed to position computations. As a minimum, a network must have scale, orientation, and one known position. If two or more points are known, the scale and orientation are inherent as well. Otherwise, the normal equations matrix as discussed in §4.3 below will be singular, leading to no definite network solution. The various approaches to defining network datum parameters will be discussed below.

4.2.2 Datum Definition by External Observations

The geodetic network datum parameters may be provided by external observations. Table 4.2 gives a list of the geodetic/astronomical measurements and the datum parameters they may provide in a three-dimensional space. It is clear that horizontal directions/angles do not provide any datum parameters. A distance measurement in a network can provide the network scale only to the extent that the contributing biases, which are dependent on the measuring techniques, can be sufficiently accounted for. An azimuth, which can be measured either by astronomical method, or by a gyro instrument, or by GPS, provides the orientation of a network with respect to the astronomical or geodetic north. Raw zenith distances provide rotations of a network about the astronomical north and astronomical east. When reduced to the Local Geodetic coordinate system, the geodetic zenith distances provide rotations of a network about the geodetic north and geodetic east. The measured astronomical positions, i.e., the astronomical latitude Φ, and astronomical longitude Λ, provide horizontal translations of a network with respect to the origin of the Conventional Terrestrial coordinate system (cf. §2.1).

Table 4.1. Datum parameters of geodetic networks

Network Type		Datum Parameters		Total Number
Three-dimensional Network (with no distances)		3 translations	t_x = translation in x direction t_y = translation in y direction t_z = translation in z direction	7
		3 rotations	ω_x = rotation about x direction ω_y = rotation about y direction ω_z = rotation about z direction	
		1 scale	s = scale	
Horizontal Network	triangulation network	2 translations	t_x = translation in x direction t_y = translation in y direction	4
		1 rotation	ω_z = rotation about z direction	
		1 scale	s = scale	
	trilateration network	2 translations	t_x = translation in x direction t_y = translation in y direction	3
		1 rotation	ω_z = rotation about z direction	
Vertical Network		1 translation	t_z = translation in z direction	1

Table 4.2. Datum contents of geodetic measurements

	Datum Parameters						
	Translation			Rotation			Scale
Observable	t_x	t_y	t_z	ω_x	ω_y	ω_z	s
Distances	–	–	–	–	–	–	√
Horizontal directions/angles	–	–	–	–	–	–	–
Azimuth	–	–	–	–	–	√	–
Zenith distances	–	–	–	√	√	–	–
Astronomical positions	√	√	–	–	–	√	√
GPS positions	√	√	√	√	√	√	√
3D position differences by GPS or INS, etc.	–	–	–	√	√	√	√
2D position differences	–	–	–	–	–	√	√
Height differences	–	–	–	–	√	√	√

Some clarifications about the datum content of the GPS positions should be made here. The GPS positioning computations are carried out in the geocentric geodetic coordinate system WGS 84. The original GPS observations are either pseudoranges or carrier beat phases. Theoretically, as with the conventional terrestrial geodetic networks, a stand-alone three-dimensional GPS network also has seven datum parameters to be defined; i.e., three translation, three orientation, and one scale parameters. The distance content of the GPS measurements (e.g., pseudoranges or carrier beat phases) may provide the datum scale to the extent that the contributing biases can be sufficiently accounted, and the missing translation and orientation parameters must be provided by some constraints on the coordinates. For most engineering surveying applications, this is done by fixing the satellite position vectors that are given by the adopted broadcast or precise ephemerides of the GPS satellites used in the data reduction. The GPS ephemerides are originally generated by a set of global tracking stations whose coordinates are determined by some method independent of GPS; e.g., VLBI (Very Long Baseline Interferometry) or SLR (Satellite Laser Ranging). The GPS datum is then transferred to the stations of the survey network via these ephemerides without the need of direct observations between the survey and the global tracking stations. Thus, the datum definition for a stand-alone GPS network is usually complex and overconstrained (cf. also Delikaraoglou and Lahaye 1989, Wells et al. 1987). For practicing surveyors, the results of a GPS survey are in the form of three-dimensional positions or position differences. Therefore, we can assume that the three coordinates of one GPS station provide the translation of the network with respect to the origin of the geodetic system WGS 84. If the GPS coordinates of three or more points are known, the scale and rotations of the network are also defined. However, since at present the relative positions obtained by GPS are much more accurate than the absolute positions, GPS provides weak translation parameters, but highly accurate orientation and scale parameters.

4.2.3 Datum Definition by Minimum Constraints

Although some datum parameters of a geodetic network may be provided by external observations, it should be noted that some external observations, e.g., astronomical positions and azimuth, are very costly to obtain, and GPS provides very weak translation parameters. Another approach to define a network datum is by specifying some constraints on the coordinates. Theoretically, the constraints can be in the form of fixed constraints; i.e., specified "known" positions; conditional constraints, i.e., through additional relationships between certain points; or weighted constraints, i.e., assigning weights to some or all station coordinate values involved in the network adjustment. However, as discussed before, when specifying constraints one has to remember that the datum definition must not affect the geometry of the network, i.e., the relative position of the points shall be defined solely by the geodetic observations. The datum shall not impose any strains on the network; that is, no more constraints are specified than necessary to provide the network datum. This is called the approach of minimum constraints.

First, let us assume that a geodetic network of m stations is to be established in the RLG Cartesian coordinate system as shown in Figure 4.6. If distances are measured, the number of datum parameters of the network will be six, i.e., three translations plus three rotations. The conventional approach to provide minimum constraints for its datum definition is to fix the three RLG coordinates of one station to their approximate values, and fix one LG azimuth together with two LG zenith distances from the same station to another station(s) at their approximate values computed from the approximate RLG coordinates. Alternatively, one may also fix six RLG coordinates divided over at least three points not on the same line to their approximate values. If distances are not measured or if they are measured but are not good enough to be used to provide scale for the network, an additional distance in the network must be fixed, or fix seven RLG coordinates divided over at least three points not on the same line to their approximate values.

Mathematically, specifying a constraint means defining a mathematical equation. For instance, referring to the linearized observation modeling as given in §4.1.1.2, fixing a coordinate of a station to its approximate value means that the differential correction to be sought for the given approximate value will be zero. Similarly, fixing an azimuth or a distance calculated from the given approximate coordinates means that the differential correction to be sought for the calculated azimuth or distance will be zero. In three-dimensional space, considering the case that the seven datum parameters are defined by fixing the coordinates of points #1, fixing the azimuth from point #1 to #2, fixing the distance from point #1 to #2, and fixing the two zenith distances from point #1 to #2, and from point #1 to #3 (see Figure 4.6), the following seven constraint equations can be formulated:

$$\delta x_1 = 0 \tag{4.74}$$

$$\delta y_1 = 0 \tag{4.75}$$

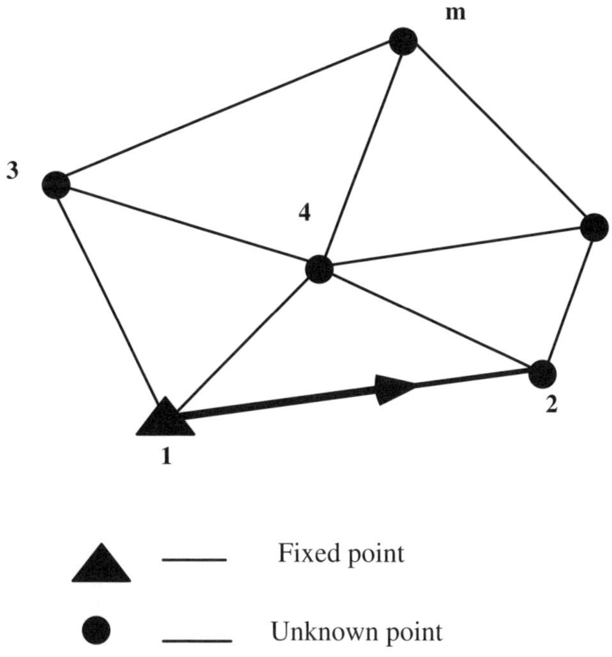

Figure 4.6. Minimum datum constraints of a geodetic network

$$\delta z_1 = 0 \tag{4.76}$$

$$v_{\alpha_{12}} = \mathbf{a}_{\alpha_{12}}^{T} \mathbf{U} \delta \mathbf{p}_{12} = 0 \tag{4.77}$$

$$v_{Z_{12}} = \mathbf{a}_{Z_{12}}^{T} \mathbf{U} \delta \mathbf{p}_{12} = 0 \tag{4.78}$$

$$v_{Z_{13}} = \mathbf{a}_{Z_{13}}^{T} \mathbf{U} \delta \mathbf{p}_{13} = 0 \tag{4.79}$$

$$v_{s_{12}} = \mathbf{a}_{s_{12}}^{T} \mathbf{U} \delta \mathbf{p}_{12} = 0 \tag{4.80}$$

If we denote vector $\mathbf{x}$ as

$$\mathbf{x} = [\delta x_1 \;\; \delta y_1 \;\; \delta z_1 \;\; \delta x_2 \;\; \delta y_2 \;\; \delta z_2 \;\; \ldots \;\; \delta x_m \;\; \delta y_m \;\; \delta z_m]^T \tag{4.81}$$

the above Equations (4.74) to (4.80) can be expressed more compactly in matrix and vector form as follows:

$$\mathbf{D}^{\mathrm{T}}\mathbf{x} = 0 \tag{4.82}$$

where

$$\mathbf{D}^{\mathrm{T}} = \begin{pmatrix} 1 & 0 & 0 & 0 & 0 & 0 & 0 & 0 & 0 & 0 & \cdots & 0 \\ 0 & 1 & 0 & 0 & 0 & 0 & 0 & 0 & 0 & 0 & \cdots & 0 \\ 0 & 0 & 1 & 0 & 0 & 0 & 0 & 0 & 0 & 0 & \cdots & 0 \\ a_1 & a_2 & a_3 & -a_1 & -a_2 & -a_3 & 0 & 0 & 0 & 0 & \cdots & 0 \\ b_1 & b_2 & b_3 & -b_1 & -b_2 & -b_3 & 0 & 0 & 0 & 0 & \cdots & 0 \\ c_1 & c_2 & c_3 & 0 & 0 & 0 & -c_1 & -c_2 & -c_3 & 0 & \cdots & 0 \\ d_1 & d_2 & d_3 & -d_1 & -d_2 & -d_3 & 0 & 0 & 0 & 0 & \cdots & 0 \end{pmatrix} \tag{4.83}$$

with

$$(a_1\ a_2\ a_3) = \mathbf{a}_{\alpha 12}^{\mathrm{T}} \tag{4.84}$$

$$(b_1\ b_2\ b_3) = \mathbf{a}_{z12}^{\mathrm{T}} \tag{4.85}$$

$$(c_1\ c_2\ c_3) = \mathbf{a}_{Z13}^{\mathrm{T}} \tag{4.86}$$

$$(d_1\ d_2\ d_3) = \mathbf{a}_{s12}^{\mathrm{T}} \tag{4.87}$$

Depending on the geodetic observations made in the network, some rows in the above datum matrix $\mathbf{D}^{\mathrm{T}}$ may be deleted. For instance, if an azimuth is measured, the fourth row of $\mathbf{D}^{\mathrm{T}}$ can be deleted. If zenith distances are measured, the fifth and sixth rows can be deleted. If distances are measured, the seventh row can be deleted, and finally, if 3D positions are measured, rows one to three can be deleted.

Similar results can be obtained for a two-dimensional geodetic network. In this case, if distances are measured (e.g., a trilateration or triangulateration network), the number of datum parameters of the network will be three, i.e., two translations plus one rotation. The network datum can be defined by fixing the two grid coordinates of one station at their approximate values plus one grid azimuth from the same station to another at its approximate value computed from the approximate grid coordinates. Alternatively, one may also fix three grid coordinates over two points to their approximate values. If distances are not measured or if they are measured but are not good enough to be used to provide scale for the network, an additional distance in the network must be fixed, or fix the four grid coordinates of two points at their approximate values. Considering the case in which, point #1 is fixed, and the azimuth and distance from point #1 to #2 are fixed, the constraint equations are as follows:

$$\delta x_1 = 0 \tag{4.88}$$

$$\delta y_1 = 0 \tag{4.89}$$

$$v_{t12} = \mathbf{a}_{t12}^T \delta\overline{\mathbf{p}}_{12} = 0 \tag{4.90}$$

$$v_{l12} = \mathbf{a}_{l12}^T \delta\overline{\mathbf{p}}_{12} = 0 \tag{4.91}$$

If we denote vector **x** as

$$\mathbf{x} = \begin{bmatrix}\delta x_1 & \delta y_1 & \delta x_2 & \delta y_2 & \cdots & \delta x_m & \delta y_m\end{bmatrix}^T \tag{4.92}$$

the above constraint Equations (4.88) to (4.91) can be expressed in the same form as Equation (4.82) with the datum matrix $\mathbf{D}^T$ as follows:

$$\mathbf{D}^T = \begin{pmatrix} 1 & 0 & 0 & 0 & 0 & \cdots & 0 \\ 0 & 1 & 0 & 0 & 0 & \cdots & 0 \\ e_1 & e_2 & -e_1 & -e_2 & 0 & \cdots & 0 \\ f_1 & f_2 & -f_1 & -f_2 & 0 & \cdots & 0 \end{pmatrix} \tag{4.93}$$

where

$$\begin{pmatrix} e_1 & e_2 & -e_1 & -e_2 \end{pmatrix} = \mathbf{a}_{t12}^T \tag{4.94}$$

$$\begin{pmatrix} f_1 & f_2 & -f_1 & -f_2 \end{pmatrix} = \mathbf{a}_{l12}^T \tag{4.95}$$

Finally, in the case of a geodetic leveling network (one-dimensional network), only height differences between benchmarks are measured. If the orthometric height system is chosen, it is referred to the geoid, and one thus has to fix the orthometric height of one benchmark at its approximate value in order to provide the translation datum parameter. The scale of a leveling network is provided by height differences. The scale of the level rods used must be regularly calibrated in order to provide proper scale. In the case that the orthometric height of point #1 is fixed, the constraint equation will be

$$\delta H_1 = 0 \tag{4.96}$$

Denote

$$\mathbf{x} = [\delta H_1 \quad \delta H_2 \quad \dots \quad \delta H_m]^T \tag{4.97}$$

The datum equation matrix for a geodetic leveling network is then

$$\mathbf{D}^T = (1 \quad 0 \quad 0 \quad \dots \quad 0) \tag{4.98}$$

4.2.4 Datum Definition by Inner Constraints

Inner constraints are another set of minimum constraints that are formulated based on relationships between some or all station coordinate values involved in the network adjustment. In the context of network adjustment theory, this is also called "free network adjustment."

Physically, instead of fixing the coordinates of an actual station, an actual azimuth, an actual distance, and actual zenith distances in a network, inner constraints are to fix the coordinates of some fictitious station, and fix some fictitious azimuth, distance, and zenith distances. For instance, in three-dimensional space, inner constraints require that before and after network adjustment the following conditions hold: 1) the coordinates of the centroid of the network, i.e., the average coordinates of the network points, stay unchanged; 2) the network does not rotate with respect to the centroid around the x, y, and z-axes; and 3) the average distance from the centroid to each point in the network remains unchanged. Similarly, for two-dimensional networks, inner constraints require that the coordinates of the center of the network, and the average azimuth and the average distance from the center to each point in the network stay unchanged. Finally, for one-dimensional vertical control networks, an inner constraint requires that the average height of the network points remains unchanged.

Mathematically, the constraint equations for inner constraints can be derived from similarity transformation equations by requiring that the sum of the squares of the differential corrections to the given approximate coordinates of all network points be a minimum, i.e.,

$$\mathbf{x}^T\mathbf{x} = \text{minimum} \tag{4.99}$$

where **x** is defined by the above Equations (4.81), (4.92), or (4.97), depending on the network type. To illustrate how the constraint equations are derived, let us consider a geodetic network in two-dimensional space. In the case that the four datum parameters of the network, i.e. two translations (t_x, t_y), one rotation (ω_z) plus one scale (s), are not defined, and given the initial coordinates (x_i^0, y_i^0) of a point i, its new coordinates (x_i, y_i) can be related to their corresponding initial (old) coordinates by similarity transformation as follows

$$\begin{pmatrix} x_i \\ y_i \end{pmatrix} = \begin{pmatrix} t_x \\ t_y \end{pmatrix} + s\begin{pmatrix} \cos\omega_z & -\sin\omega_z \\ \sin\omega_z & \cos\omega_z \end{pmatrix}\begin{pmatrix} x_i^0 \\ y_i^0 \end{pmatrix} (i = 1,\dots,m) \tag{4.100}$$

To solve for the four unknowns t_x, t_y, ω_z, and s, the above equations may be linearized under the approximate values of $s^0 = 1.0$ and $\omega_z^0 = 0.0$ leading to

$$\begin{pmatrix} \delta x_i \\ \delta y_i \end{pmatrix} = \begin{pmatrix} 1 & 0 & -y_i^0 & x_i^0 \\ 0 & 1 & x_i^0 & y_i^0 \end{pmatrix} \begin{pmatrix} t_x \\ t_y \\ \delta\omega_z \\ \delta s \end{pmatrix} \quad (i = 1, \ldots, m) \tag{4.101}$$

where $\delta\omega_z$ and δs are the differential rotation and scale change defined by

$$\delta\omega_z = \omega_z - \omega_z^0 \tag{4.102}$$

$$\delta s = s - s^0 \tag{4.103}$$

Under the requirement of Equation (4.99), one obtains

$$\sum_1^m \delta x_i = 0 \tag{4.104}$$

$$\sum_1^m \delta y_i = 0 \tag{4.105}$$

$$\sum_1^m -y_i^0 \delta x_i + x_i^0 \delta y_i = 0 \tag{4.106}$$

$$\sum_1^m x_i^0 \delta x_i + y_i^0 \delta y_i = 0 \tag{4.107}$$

If expressed in matrix form, Equations (4.104) to (4.107) become

$$\mathbf{H}^T \mathbf{x} = 0 \tag{4.108}$$

where

$$\mathbf{H}^T = \begin{pmatrix} 1 & 0 & 1 & 0 & \cdots & 1 & 0 \\ 0 & 1 & 0 & 1 & \cdots & 0 & 1 \\ -y_1^0 & x_1^0 & -y_2^0 & x_2^0 & \cdots & -y_m^0 & x_m^0 \\ x_1^0 & y_1^0 & x_2^0 & y_2^0 & \cdots & x_m^0 & y_m^0 \end{pmatrix} \tag{4.109}$$

The first two rows of matrix $\mathbf{H}^T$ define the translations of the network, and the third and fourth rows define the orientation and the scale of the network, respectively. If some of the network datum parameters have been resolved by external observations, the corresponding rows in matrix $\mathbf{H}^T$ may be deleted. For instance, if distances are measured, i.e., for a trilateration or triangulateration network, matrix $\mathbf{H}^T$ will be

$$\mathbf{H}^T = \begin{pmatrix} 1 & 0 & 1 & 0 & \cdots & 1 & 0 \\ 0 & 1 & 0 & 1 & \cdots & 0 & 1 \\ -y_1^0 & x_1^0 & -y_2^0 & x_2^0 & \cdots & -y_m^0 & x_m^0 \end{pmatrix} \tag{4.110}$$

If both azimuth(s) and distance(s) are measured; that is, both the network orientation and scale are defined, matrix $\mathbf{H}^T$ will be:

$$\mathbf{H}^T = \begin{pmatrix} 1 & 0 & 1 & 0 & \cdots & 1 & 0 \\ 0 & 1 & 0 & 1 & \cdots & 0 & 1 \end{pmatrix} \tag{4.111}$$

The $\mathbf{H}^T$ matrix corresponding to any other cases in which external observations have been made to define part of the network datum parameters can be derived by the same logic. Obviously, in the case that all the network datum parameters have been defined by external observations, there will be no need to introduce any additional datum constraints; that is, matrix $\mathbf{H}^T$ will not be needed.

Similarly, for a three-dimensional network with seven datum parameters, we can derive matrix $\mathbf{H}^T$ as follows

$$\mathbf{H}^T = \begin{pmatrix} 1 & 0 & 0 & 1 & 0 & 0 & \cdots & 1 & 0 & 0 \\ 0 & 1 & 0 & 0 & 1 & 0 & \cdots & 0 & 1 & 0 \\ 0 & 0 & 1 & 0 & 0 & 1 & \cdots & 0 & 0 & 1 \\ 0 & -z_1^0 & y_1^0 & 0 & -z_2^0 & y_2^0 & \cdots & 0 & -z_m^0 & y_m^0 \\ z_1^0 & 0 & -x_1^0 & z_2^0 & 0 & -x_2^0 & \cdots & z_m^0 & 0 & -x_m^0 \\ -y_1^0 & x_1^0 & 0 & -y_2^0 & x_2^0 & 0 & \cdots & -y_m^0 & x_m^0 & 0 \\ x_1^0 & y_1^0 & z_1^0 & x_2^0 & y_2^0 & z_2^0 & \cdots & x_m^0 & y_m^0 & z_m^0 \end{pmatrix} \tag{4.112}$$

where the first three rows define the translation parameters, the fourth, fifth, and sixth rows define the network rotations, and the seventh row provides the network scale. In the case that some external observations have been made to define part of the network datum parameters, some rows in matrix $\mathbf{H}^T$ that correspond to the defined parameters can be removed. For instance, distance may remove the seventh row of $\mathbf{H}^T$, a 3D position may remove the first three rows of $\mathbf{H}^T$, an azimuth may remove the sixth row of $\mathbf{H}^T$, two or more height differences may remove the fourth, fifth, and seventh rows of $\mathbf{H}^T$, and so on. If all the seven datum parameters are defined by external observations, matrix $\mathbf{H}^T$ becomes a nuisance since no datum constraints will be needed (cf. also Cooper, 1987).

For a one-dimensional vertical network, the inner constraint datum equation matrix is

$$\mathbf{H}^{T} = \begin{pmatrix} 1 & 1 & 1 & \ldots & 1 \end{pmatrix} \tag{4.113}$$

Obviously, if a height is given in the network, this constraint can be deleted.

Finally, an important relation between the network configuration matrix $\mathbf{A}$ and the inner constraint matrix $\mathbf{H}$ is

$$\mathbf{AH} = 0 \tag{4.114}$$

Equation (4.114) is crucial in simplifying least squares network adjustment equations as discussed in §4.3.3 below.

4.3 Modeling and Solution of Least Squares Network Adjustment

As discussed before, a least squares network adjustment is classified as a combined adjustment, a parametric adjustment, or a condition adjustment, depending on the specific type of observation mathematical model in use. Choice of a specific adjustment model depends on the nature of the problem under study and the ease in formulating observation equations. Mathematically, however, one type of adjustment problem can be converted to another, and the solutions from different types of adjustment models are equivalent. In this section are discussed the theory and algorithm of the parametric least squares network adjustment that is the most frequently used adjustment model in geodetic science. For instance, the observation equations derived in §4.1 above are all of the parametric type. For discussions on the combined and condition least squares network adjustment and some other related issues, the interested reader is referred to the given literature.

4.3.1 Uniquely Defined Versus Overdetermined Positioning Problem

After a geodetic network has been measured, one obtains a set of observation equations that can be used to solve for the unknown coordinates of network stations. Mathematically, a set of equations may be classified into an underdetermined problem, a uniquely determined problem, or an overdetermined problem. An underdetermined problem means that the total number of independent equations is less than the total number of independent unknown parameters to be solved for. A uniquely determined problem means that the total number of independent equations is equal to the total number of independent unknown parameters. And finally, in an overdetermined problem, the total number of independent equations is larger than the total number of independent unknown parameters to be solved for. A set of equations are considered as being independent if none of them can be expressed as a linear combination of the rest equations; that is, none of them can be derived from all the rest equations. Similarly, a set of unknown parameters is considered as being independent if none of them can be derived from the rest parameters by linear combinations of them.

From the practical point of view, in geodetic positioning, we consider only the uniquely or overdetermined cases; that is, the total number of independent equations must be equal to or larger than the total number of independent unknown parameters. Typical examples of the uniquely determined positioning problem are polar method (i.e., to measure one distance and one azimuth), azimuth or distance intersection, angle resection, and open traverses, etc. These simple methods of positioning are widely used in land survey and construction layout, among others. The danger with the uniquely defined positioning methods is that since no geometrical check in the form of a closed geometrical figure is available, any gross errors in the observations may directly degrade the quality of the results. Therefore, redundant measurements are always recommended in order to allow for the detection of both gross errors and systematic effects, and that leads to an overdetermined mathematical problem. Here the term "redundant measurements" means that more observations are available than necessary to determine a set of unknown parameters. Some examples of overdetermined positioning problems include, for instance, a closed traverse, and a network with a closed geometrical figure.

In order to solve for the unknown station coordinates from observations, one used to derive a nonlinear formula of closed form for the case of a uniquely defined positioning problem. With the advent of modern computers, the linearized observation equations derived in §4.1 can be used to arrive at the unknown coordinates by solving a set of linear equations iteratively. That may greatly simplify the computation procedures since there is no need to develop different computation rules and formulae for the different positioning problems of intersection, resection, and traverse, etc. Even though a different positioning problem is accomplished by a different combination of measurements of azimuth, distance, and directions or angles, the linearized observation equations of different observables can be treated in the same way mathematically.

In an overdetermined positioning problem, due to the effects of the unpredictable observation errors, i.e., random, gross, and/or systematic errors, the redundant set of observation equations will not be consistent, and one may thus derive more than one unique solution for the unknown coordinates using a different subset of observation equations. In order to obtain a best solution that uses all the available observation information, some additional criteria have to be imposed. The most frequently used criterion in surveying is the method of least squares that tries to arrive at a best solution by minimizing the sum of squares of the weighted observational residuals. It should be kept in mind that prior to being used in network computations, raw observations have to be pre-processed and screened for any possible gross errors of large magnitude using the theory and procedures discussed in Chapters 2 and 3, respectively.

4.3.2 The Gauss-Markov Model

In general, the stochastic model of a geodetic network can be expressed in the following form

$$E\{\bar{l}\} = \mathbf{A}_1\mathbf{x} + \mathbf{A}_2\xi \tag{4.115}$$

$$D\{\bar{l}\} = \bar{\mathbf{P}}^{-1}\sigma_0^2, \tag{4.116}$$

with datum constraints

$$\mathbf{D}^{\mathrm{T}}\mathbf{x} = 0 \tag{4.117}$$

where $E\{\cdot\}$ and $D\{\cdot\}$ are respectively the statistical expectation and dispersion operators. The rank of matrix $\mathbf{D}$ is equal to the number of rank defects of $\mathbf{A}_1$, and the other quantities are defined as follows

- $\bar{l}$ = the n-vector of observations
- $\mathbf{x}$ = the u-vector of unknown coordinates
- ξ = the m_r-vector of unknown orientation parameters (m_r is the total number of stations with horizontal direction observations)
- $\mathbf{A}_1$ = the n by u configuration matrix
- $\mathbf{A}_2$ = the n by m_r constant coefficient matrix
- $\bar{\mathbf{P}}$ = the n by n weight matrix of observations, and finally
- σ_0^2 = the a priori variance factor.

Equation (4.115) is the linearized form of a nonlinear mathematical model. The nonlinear observation equations are linearized around the Taylor point $\mathbf{x}^0$ (i.e., the approximate coordinate vector), and thus the vector $\mathbf{x}$ is actually the vector of corrections to the initially adopted approximate coordinate vector, and vector $\bar{l}$ represents the misclosure vector. Equations (4.115) and (4.116) are usually called the Gauss-Markov model for geodetic network adjustment.

As discussed in §4.1.3, the direction orientation unknowns can be removed by replacing the directions by correlated angles. In this case, elimination of the orientation parameters ξ allows for Equations (4.115) and (4.116) to be expressed in terms of only the coordinate vector as follows

$$E\{l\} = \mathbf{A}\mathbf{x} \tag{4.118}$$

$$D\{l\} = \mathbf{P}^{-1}\sigma_0^2 \tag{4.119}$$

where l and $\mathbf{P}$ are derived from $\bar{l}$ and $\bar{\mathbf{P}}$ through the elimination of the nuisance parameters ξ. Finally, in least squares network adjustment, Equation (4.118) is written as

$$\boldsymbol{l} + \mathbf{v} = \mathbf{A}\mathbf{x} \tag{4.120}$$

where **v** is called the vector of observational residuals.

4.3.3 Least Squares Network Adjustment Equations

Equations (4.119), (4.120), and (4.117) describe the statistical properties of the observations and the relationship existing among the unknown coordinates and residuals to be solved for. The system of Equations (4.120) and (4.117) is underdetermined since there are more unknowns than equations. The method of least squares network adjustment tries to solve for an optimal estimate of both the coordinates and residuals by minimizing the sum of squares of the weighted residuals, leading to the following optimization problem

Minimize

$$\mathbf{v}^{\mathrm{T}}\mathbf{P}\mathbf{v} \tag{4.121}$$

Subject to:

$$\boldsymbol{l} + \mathbf{v} = \mathbf{A}\mathbf{x}$$

$$\mathbf{D}^{\mathrm{T}}\mathbf{x} = 0$$

This optimization problem can be easily solved by using the Lagrange method. The variation function ϕ relating the unknown quantities **x** and **v** to the known quantities **A**, **P**, *l*, and **D** is constructed by

$$\phi = \mathbf{v}^{\mathrm{T}}\mathbf{P}\mathbf{v} + 2\mathbf{k}_1^{\mathrm{T}}(\boldsymbol{l} + \mathbf{v} - \mathbf{A}\mathbf{x}) + 2\mathbf{k}_2^{\mathrm{T}}\left(\mathbf{D}^{\mathrm{T}}\mathbf{x} - 0\right) \tag{4.122}$$

To find the minimum of the variation function, the derivatives of ϕ with respect to **x** and **v** are set to zeros; that is,

$$\frac{1}{2}\frac{\partial\phi}{\partial\mathbf{v}} = \mathbf{P}\,\mathbf{v} + \mathbf{k}_1 = 0 \tag{4.123}$$

$$\frac{1}{2}\frac{\partial\phi}{\partial\mathrm{x}} = \mathbf{A}^{\mathrm{T}}\mathbf{k}_1 + \mathbf{D}\mathbf{k}_2 = 0 \tag{4.124}$$

Equations (4.123), (4.124), and (4.120), (4.117) together form the following least squares ***normal equations*** system

$$\mathbf{P}\,\mathbf{v} + \mathbf{k}_1 = 0$$

$$-\mathbf{A}^{\mathrm{T}}\mathbf{k}_1 + \mathbf{D}\mathbf{k}_2 = 0$$

$$\boldsymbol{l} + \mathbf{v} = \mathbf{Ax}$$

$$\mathbf{D}^{\mathrm{T}}\mathbf{x} = 0$$

These equations may be written in a hypermatrix form as follows

$$\begin{pmatrix} \mathbf{P} & \mathbf{I} & \mathbf{0} & \mathbf{0} \\ \mathbf{0} & -\mathbf{A} & \mathbf{D} & \mathbf{0} \\ \mathbf{I} & \mathbf{0} & \mathbf{0} & -\mathbf{A} \\ \mathbf{0} & \mathbf{0} & \mathbf{0} & \mathbf{D}^{\mathrm{T}} \end{pmatrix} \begin{pmatrix} \mathbf{v} \\ \mathbf{k}_1 \\ \mathbf{k}_2 \\ \mathbf{x} \end{pmatrix} = \begin{pmatrix} \mathbf{0} \\ \mathbf{0} \\ -\boldsymbol{l} \\ \mathbf{0} \end{pmatrix} \tag{4.125}$$

Solving matrix Equation (4.125), using the theory of generalized inverse, leads to the optimal estimates $\hat{\mathbf{x}}$, $\hat{\mathbf{v}}$ and $\hat{\boldsymbol{l}}$ for the coordinates, residuals, and observations, respectively, as follows

$$\hat{\mathbf{x}} = \left(\mathbf{N} + \mathbf{DD}^{\mathrm{T}}\right)^{-1} \mathbf{A}^{\mathrm{T}}\mathbf{P}\boldsymbol{l} \tag{4.126}$$

$$\hat{\mathbf{v}} = \left[\mathbf{A}\left(\mathbf{N} + \mathbf{DD}^{\mathrm{T}}\right)^{-1} \mathbf{A}^{\mathrm{T}}\mathbf{P} - \mathbf{I}\right]\boldsymbol{l} \tag{4.127}$$

$$\hat{\boldsymbol{l}} = \mathbf{A}\left(\mathbf{N} + \mathbf{DD}^{\mathrm{T}}\right)^{-1} \mathbf{A}^{\mathrm{T}}\mathbf{P}\boldsymbol{l} \tag{4.128}$$

where

$$\mathbf{N} = \mathbf{A}^{\mathrm{T}}\mathbf{PA} \tag{4.129}$$

Associated with the above estimates are their corresponding accuracy estimates. At first, the a posterior variance factor can be estimated by

$$\hat{\sigma}_0^2 = \frac{\hat{\mathbf{v}}^{\mathrm{T}}\mathbf{P}\hat{\mathbf{v}}}{\mathrm{r}} \tag{4.130}$$

where r is the total redundancy of the system. The a priori variance-covariance matrix $\mathbf{C}_{\hat{x}}$, $\mathbf{C}_{\hat{v}}$ and $\mathbf{C}_{\hat{l}}$ and the a posterior variance-covariance matrix $\hat{\mathbf{C}}_{\hat{x}}$, $\hat{\mathbf{C}}_{\hat{v}}$ and $\hat{\mathbf{C}}_{\hat{l}}$ for $\hat{\mathbf{x}}$, $\hat{\mathbf{v}}$ and $\hat{\boldsymbol{l}}$ respectively are derived according to the law of error propagation as follows

$$\mathbf{C}_{\hat{x}} = \sigma_0^2 \mathbf{Q}_{\hat{x}} \tag{4.131}$$

$$\mathbf{C}_{\hat{v}} = \sigma_0^2 \mathbf{Q}_{\hat{v}} \tag{4.132}$$

$$\mathbf{C}_{\hat{l}} = \sigma_0^2 \mathbf{Q}_{\hat{l}} \tag{4.133}$$

$$\hat{\mathbf{C}}_{\hat{x}} = \hat{\sigma}_0^2 \mathbf{Q}\hat{x} \tag{4.134}$$

$$\hat{\mathbf{C}}_{\hat{v}} = \hat{\sigma}_0^2 \mathbf{Q}_{\hat{v}} \tag{4.135}$$

$$\hat{\mathbf{C}}_{\hat{l}} = \hat{\sigma}_0^2 \mathbf{Q}_{\hat{l}} \tag{4.136}$$

where $\mathbf{Q}_{\hat{x}}$, $\mathbf{Q}_{\hat{v}}$ and $\mathbf{Q}_{\hat{l}}$ are called the co-factor matrix for $\hat{\mathbf{x}}$, $\hat{\mathbf{v}}$ and $\hat{\boldsymbol{l}}$ respectively, and they are calculated by

$$\mathbf{Q}_{\hat{x}} = \left(\mathbf{N} + \mathbf{D}\mathbf{D}^{\mathrm{T}}\right)^{-1} - \mathbf{H}\left(\mathbf{H}^{\mathrm{T}}\mathbf{D}\mathbf{D}^{\mathrm{T}}\mathbf{H}\right)^{-1}\mathbf{H}^{\mathrm{T}} \tag{4.137}$$

$$\mathbf{Q}_{\hat{v}} = \mathbf{P}^{-1} - \mathbf{A}\left(\mathbf{N} + \mathbf{D}\mathbf{D}^{\mathrm{T}}\right)^{-1}\mathbf{A}^{\mathrm{T}} \tag{4.138}$$

$$\mathbf{Q}_{\hat{l}} = \mathbf{A}\left(\mathbf{N} + \mathbf{D}\mathbf{D}^{\mathrm{T}}\right)^{-1}\mathbf{A}^{\mathrm{T}} \tag{4.139}$$

4.3.4 Solution of Overconstrained Network Adjustment

The above solution Equations (4.126) to (4.128) and (4.137) to (4.139) are derived only for the case of minimum or inner constraints; that is, the total number of constraint equations must be equal to the datum defects of the network or the rank defects of the configuration matrix A. In the case that more constraints than necessary to remove the network datum defects have been defined; for instance, fixing two or more stations for a two-dimensional trilateration network, a new set of solution equations must be derived.

Assume that the number of constraints defined and the number of network datum defects are r_c and r_d, respectively. For the overconstrained case, we have

$$r_c > r_d \tag{4.140}$$

In general, let us assume that the constraint equation is described by

$$\mathbf{R}^{\mathrm{T}}\mathbf{x} = \mathbf{c} \tag{4.141}$$

where $\mathbf{R}^{\mathrm{T}}$ is a r_c by u matrix, and c a r_c by one constant vector.

The best approach to solve for the least squares network adjustment problem in this case is to eliminate the constraint Equation (4.141) by reducing the dimension of the unknown coordinate vector $\mathbf{x}$ by r_c. At first, let us arrange the sequence of the unknown parameters in $\mathbf{x}$ and decompose the left-hand side of Equation

(4.141) so that the first r_c unknowns can be expressed by a linear combination of the rest unknowns; that is,

$$\begin{pmatrix}\mathbf{R}_1^T & \mathbf{R}_2^T\end{pmatrix}\begin{pmatrix}\mathbf{x}_1\\ \mathbf{x}_2\end{pmatrix} = \mathbf{c} \tag{4.142}$$

where $\mathbf{x}_1$ and $\mathbf{x}_2$ are two sub-vectors of dimension r_c and $(u\text{-}r_c)$, respectively. $\mathbf{R}_1$ is a r_c by r_c nonsingular matrix, and R_2 is a $(u\text{-}r_c)$ by r_c matrix. $\mathbf{x}_1$ is therefore expressed by $\mathbf{x}_2$ as follows

$$\mathbf{x}_1 = \left(\mathbf{R}_1^T\right)^{-1}\left(\mathbf{c} - \mathbf{R}_2^T\mathbf{x}_2\right) \tag{4.143}$$

Similarly, the right-hand side of Equation (4.120) can be decomposed according to $\mathbf{x}_1$ and $\mathbf{x}_2$ as follows

$$\boldsymbol{l} + \mathbf{v} = \mathbf{B}_1\mathbf{x}_1 + \mathbf{B}_2\mathbf{x}_2 \tag{4.144}$$

where $\mathbf{B}_1$ and $\mathbf{B}_2$ is a n by r_c and n by $(u\text{-}r_c)$ matrix, respectively. Substituting Equation (4.143) into (4.144), one obtains

$$\boldsymbol{l} + \mathbf{v} = \mathbf{B}_1\left(\mathbf{R}_1^T\right)^{-1}\mathbf{c} + \left[\mathbf{B}_2 - \mathbf{B}_1\left(\mathbf{R}_1^T\right)^{-1}\mathbf{R}_2^T\right]\mathbf{x}_2 \tag{4.145}$$

Denote

$$\tilde{\mathbf{A}} = \mathbf{B}_2 - \mathbf{B}_1\left(\mathbf{R}_1^T\right)^{-1}\mathbf{R}_2^T \tag{4.146}$$

$$\tilde{\boldsymbol{l}} = \boldsymbol{l} - \mathbf{B}_1\left(\mathbf{R}_1^T\right)^{-1}\mathbf{c} \tag{4.147}$$

Equation (4.145) then becomes

$$\tilde{\boldsymbol{l}} + \mathbf{v} = \tilde{\mathbf{A}}\ \mathbf{x}_2 \tag{4.148}$$

From Equation (4.148), one can see that the dimension of the original unknown coordinate vector $\mathbf{x}$ has been reduced by r_c. The least squares network adjustment problem is thus posed as

Minimize

$$\mathbf{v}^T\mathbf{P}\mathbf{v}$$

Subject to:

$$\tilde{\boldsymbol{l}} + \mathbf{v} = \tilde{\mathbf{A}}\,\mathbf{x}_2$$

Similar to the above section, this problem can be solved using the Lagrange method, and the variation function ϕ relating the unknown quantities $\mathbf{x}_2$ and $\mathbf{v}$ to the known quantities $\tilde{\mathbf{A}}$, $\mathbf{P}$, and $\tilde{l}$ reads

$$\phi = \mathbf{v}^T\mathbf{P}\mathbf{v} + 2\mathbf{k}^T\left(\tilde{l} + \mathbf{v} - \tilde{\mathbf{A}}\mathbf{x}_2\right) \tag{4.149}$$

By setting the derivatives of ϕ with respect to $\mathbf{x}_2$ and $\mathbf{v}$ to zeros, we then obtain the system of least squares ***normal equations*** as follows

$$\mathbf{P}\,\mathbf{v} + \mathbf{k} = 0 \tag{4.150}$$

$$\tilde{\mathbf{A}}\,\mathbf{k} = 0 \tag{4.151}$$

$$\tilde{\mathbf{l}} + \mathbf{v} = \tilde{\mathbf{A}}\,\mathbf{x}_2$$

or written in hypermatrix form as

$$\begin{pmatrix} \mathbf{P} & \mathbf{I} & \mathbf{0} \\ \mathbf{0} & \tilde{\mathbf{A}}^T & \mathbf{0} \\ \mathbf{I} & \mathbf{0} & -\tilde{\mathbf{A}} \end{pmatrix} \begin{pmatrix} \mathbf{v} \\ \mathbf{k} \\ \mathbf{x}_2 \end{pmatrix} = \begin{pmatrix} \mathbf{0} \\ \mathbf{0} \\ -\tilde{l} \end{pmatrix} \tag{4.152}$$

Parallel to Equations (4.126) to (4.139), we can derive, from the above matrix Equation (4.152), the optimal estimates for the coordinates, residuals, and observations and their associated variance-covariance matrices as follows

$$\hat{\mathbf{x}}_2 = \left(\tilde{\mathbf{A}}^T\mathbf{P}\tilde{\mathbf{A}}\right)^{-1}\tilde{\mathbf{A}}^T\mathbf{P}\tilde{l} \tag{4.153}$$

$$\hat{\mathbf{v}} = \left[\tilde{\mathbf{A}}\left(\tilde{\mathbf{A}}^T\mathbf{P}\tilde{\mathbf{A}}\right)^{-1}\tilde{\mathbf{A}}^T\mathbf{P} - \mathbf{I}\right]\tilde{l} \tag{4.154}$$

$$\hat{\tilde{l}} = \tilde{\mathbf{A}}\left(\tilde{\mathbf{A}}^T\mathbf{P}\tilde{\mathbf{A}}\right)^{-1}\tilde{\mathbf{A}}^T\mathbf{P}\tilde{l} \tag{4.155}$$

$$\hat{\sigma}_0^2 = \frac{\hat{\mathbf{v}}^T\mathbf{P}\hat{\mathbf{v}}}{r} \tag{4.156}$$

$$\mathbf{C}_{\hat{x}_2} = \sigma_0^2\mathbf{Q}_{\hat{x}_2} \tag{4.157}$$

$$\mathbf{C}_{\hat{v}} = \sigma_0^2\mathbf{Q}_{\hat{v}} \tag{4.158}$$

$$\mathbf{C}_{\hat{\tilde{l}}} = \sigma_0^2 \mathbf{Q}_{\hat{\tilde{l}}} \tag{4.159}$$

$$\hat{\mathbf{C}}_{\hat{x}_2} = \hat{\sigma}_0^2 \mathbf{Q}_{\hat{x}_2} \tag{4.160}$$

$$\hat{\mathbf{C}}_{\hat{v}} = \hat{\sigma}_0^2 \mathbf{Q}_{\hat{v}} \tag{4.161}$$

$$\hat{\mathbf{C}}_{\hat{\tilde{l}}} = \hat{\sigma}_0^2 \mathbf{Q}_{\hat{\tilde{l}}} \tag{4.162}$$

with

$$\mathbf{Q}_{\hat{x}_2} = \left(\tilde{\mathbf{A}}^{\mathrm{T}} \mathbf{P} \tilde{\mathbf{A}}\right)^{-1} \tag{4.163}$$

$$\mathbf{Q}_{\hat{v}} = \mathbf{P}^{-1} - \tilde{\mathbf{A}} \mathbf{Q}_{\hat{x}_2} \tilde{\mathbf{A}}^{\mathrm{T}} \tag{4.164}$$

$$\mathbf{Q}_{\hat{\tilde{l}}} = \tilde{\mathbf{A}} \mathbf{Q}_{\hat{x}_2} \mathbf{A}^{\mathrm{T}} \tag{4.165}$$

Finally, the estimate of $\mathbf{x}_1$ and its a priori and a posterior variance-covariance matrices are

$$\mathbf{x}_1 = \left(\mathbf{R}_1^{\mathrm{T}}\right)^{-1} \left(\mathbf{c} - \mathbf{R}_2^{\mathrm{T}} \hat{\mathbf{x}}_2\right) \tag{4.166}$$

$$\mathbf{C}_{\hat{x}_1} = \left(\mathbf{R}_1^{T}\right)^{-1} \mathbf{R}_2^{T} \mathbf{C}_{\hat{x}_2} \mathbf{R}_2 \mathbf{R}_1^{-1} \tag{4.167}$$

$$\hat{\mathbf{C}}_{\hat{x}_1} = \left(\mathbf{R}_1^{T}\right)^{-1} \mathbf{R}_2^{T} \hat{\mathbf{C}}_{\hat{x}_2} \mathbf{R}_2 \mathbf{R}_1^{-1} \tag{4.168}$$

This approach has the advantage of leading to a reduced order of the matrix of normal equations and thus saving computer time and storage. Although this approach is proposed for overconstrained network adjustment, it can also be used for the minimum constrained network adjustment.

It should be mentioned that if the network datum is defined by fixing station coordinates and observations, an alternative to introducing constraint equations for the fixed quantities is to assign a very small standard error to these fixed quantities. These assigned standard errors should be so small as to be beyond the resolution of the survey instruments being used; for instance, to assign a standard error of 0.01 seconds of arc for a fixed azimuth, or to assign a standard error of 0.01 mm for a fixed distance or station coordinates.

Finally, it should be noted that in the case of a constrained network adjustment it is often to solve for station coordinates together with the network datum parameters simultaneously. When this occurs, usually differential network scale change (with respect to 1.0) and differential rotations are solved. This is done by introducing the following constraints on the coordinates with known values

$$\mathbf{x}_i^k = \mathbf{x}_i^0 + \delta\mathbf{x}_i + \mathbf{H}_i\mathbf{t} \tag{4.169}$$

where $\mathbf{x}_i^k$ and x_i^0 are the vectors of known coordinates and approximate coordinates for station i, respectively. $\delta\mathbf{x}_i$ is the vector of differential corrections to the approximate station coordinates. $\mathbf{H}_i$ is the inner constraint matrix discussed in §4.2.4, and **t** is the vector of similarity transformation parameters with differential scale change and rotations. For instance, in the two-dimensional space, matrix $\mathbf{H}_i$ and vector **t** are respectively as follows [cf. Equation (4.101)]

$$\mathbf{H}_i = \begin{pmatrix} 1 & 0 & -y_i^0 & x_i^0 \\ 0 & 1 & x_i^0 & y_i^0 \end{pmatrix} \tag{4.170}$$

$$\mathbf{t} = \left(t_x \; t_y \; \delta\omega_z \; \delta s\right)^T \tag{4.171}$$

where t_x and t_y are the two translation components, and $\delta\omega_z$ and δs the differential rotation and differential scale change of the network, respectively, while in the three-dimensional space, $\mathbf{H}_i$ and **t** are respectively as follows

$$\mathbf{H}_i = \begin{pmatrix} 1 & 0 & 0 & 0 & z_i & -y_i^0 & x_i^0 \\ 0 & 1 & 0 & -z_i^0 & 0 & x_i^0 & y_i^0 \\ 0 & 0 & 1 & y_i^0 & -x_i^0 & 0 & z_i^0 \end{pmatrix} \tag{4.172}$$

$$\mathbf{t} = \left(t_x \quad t_y \quad t_z \quad \delta\omega_x \quad \delta\omega_y \quad \delta\omega_z \quad \delta s\right)^T \tag{4.173}$$

where t_x, t_y and t_z are the three translation parameters, $\delta\omega_x$, $\delta\omega_y$, and $\delta\omega_z$ the three differential rotations, and δs is the differential scale change. Depending on the total number of known coordinates, part or all of the above transformation parameters may be solved.

4.3.5 S-Transformations

The variance-covariance matrix of the estimated station coordinates strongly depends on the selection of points whose coordinates serve as datum parameters. These points are sometimes called the zero-variance computational base of the model. All elements of the covariance matrix $\mathbf{C}_x$ are to be considered relative variances in respect to the computational base. The S-transformation has been developed to relate variance-covariance matrices that are calculated under different zero-variance computational bases (Baarda, 1973).

Assume that the variance-covariance matrix $\mathbf{C}_x^1$ is calculated under a network datum described by datum matrix $\mathbf{D}_1$, then its corresponding variance-covariance matrix $\mathbf{C}_x^2$ under another network datum matrix $\mathbf{D}_2$ can be expressed by

$$\mathbf{C}_x^2 = \mathbf{S}\,\mathbf{C}_x^1\,\mathbf{S}^T \tag{4.174}$$

where matrix **S** is the so-called S-transformation matrix defined by

$$\mathbf{S} = \left[\mathbf{I} - \mathbf{H}\left(\mathbf{D}_2^T\mathbf{H}\right)^{-1}\mathbf{D}_2^T\right] \tag{4.175}$$

with **I** being a u by u identity matrix, and matrix **H** meaning the same as before. It should be noted that the S-transformation works for the transformation of variance-covariance matrices that are calculated under the minimum or inner datum constraints only.

5 Post-Adjustment Data Screening: Outlier Detection and Gross Error Localization

After a least squares network adjustment, a set of station coordinates and some relevant quantities are obtained. However, as discussed in §3.1, the theory of the least squares estimation has been founded on the basic assumption that all the gross errors and systematic effects have been eliminated before the adjustment is performed, and only random errors affect the data. Therefore, before we can use the results from our estimation or forward these results to our clients for use, both the observations and the mathematical model have to be assessed to determine whether the results that we have obtained are reliable. And that is the purpose of post-adjustment data screening and analysis.

Post-adjustment data screening concentrates on the detection and elimination of gross errors of ***small magnitude,*** assuming that gross errors of large magnitude have to be detected and eliminated through pre-designed techniques and procedures either during the data collection process or at the stage of pre-adjustment data screening (cf. Chapter 3) or both. Due to their local effects, large gross errors are in general easier to recognize when they occur in a network adjustment because the least squares estimation tends to smooth a large error throughout its neighboring observations, producing big residuals in a large portion of the data unless their magnitude is such as to destroy the linearization of the observation equations leading to no solution; i.e., the iterative solution process does not converge.

Unlike the pre-adjustment data screening techniques that are based on testing of condition misclosures, post-adjustment data screening is based on statistical testing of the estimated observational residuals. Although the estimated residuals are, in principle, indicative of the behavior of both the observation and the mathematical model, it is very difficult, if not possible, to separate the two since mathematically either a bad geometrical model or a bad observation(s) will affect the estimated residuals in the same way. Therefore, both the observations and the model are in fact tested simultaneously. Fortunately, since in geodetic network positioning the mathematical modeling of each type of observations is well defined based on simple geometrical and physical laws, we could in general attribute failure of statistical testing on the estimated residuals to bad observations, i.e., observations contaminated by gross errors.

This chapter starts with a discussion on the relationship between residuals, outliers, and gross errors in §5.1. §5.2 defines the null hypothesis and alternative hypothesis to be statistically tested. §5.3 and §5.4 discuss two most frequently used post-adjustment data screening techniques, and finally §5.5 summarizes the philosophy behind and outlines a general strategy for post-adjustment data screening.

5.1 Residuals, Outliers, and Gross Errors

In least squares network adjustment, an observational ***residual*** is defined as the difference between the estimated value of the observation and its corresponding measured value. From §4.1, the least squares network adjustment provides the estimated residual vector as follows [cf. Equation (4.127)]

$$\begin{aligned}\hat{\mathbf{v}} &= \hat{\boldsymbol{l}} - \boldsymbol{l} \\ &= \left[\mathbf{A}\left(\mathbf{A}^{\mathrm{T}}\mathbf{P}\,\mathbf{A} + \mathbf{D}\mathbf{D}^{\mathrm{T}}\right)^{-1}\mathbf{A}^{\mathrm{T}}\,\mathbf{P} - \mathbf{I}\right]\boldsymbol{l} \\ &= -\mathbf{R}\,\boldsymbol{l}\end{aligned} \tag{5.1}$$

where

$$\mathbf{R} = \mathbf{I} - \mathbf{A}\left(\mathbf{A}^{\mathrm{T}}\,\mathbf{P}\,\mathbf{A} + \mathbf{D}\mathbf{D}^{\mathrm{T}}\right)^{-1}\mathbf{A}^{\mathrm{T}}\,\mathbf{P} \tag{5.2a}$$

Matrices **A**, **D**, and **P** represent the parameter configuration matrix, the network datum matrix, and the observational weight matrix, respectively. $\boldsymbol{l}$ and $\hat{\boldsymbol{l}}$ *are* the observation vector and its corresponding estimate, respectively. The vector of residuals $\hat{\mathbf{v}}$ plays an important role in data analysis after the adjustment process. In some applications, for instance, curve fitting, the adequacy of the functional model is analyzed based on the magnitudes of the elements of $\hat{\mathbf{v}}$, and this analysis may even lead to discarding a chosen model completely and to remodeling the problem.

Denote matrix **R** by

$$\mathbf{R} = \begin{pmatrix} r_1 & r_{12} & \cdots & r_{1i} & \cdots & r_{1n} \\ r_{21} & r_2 & \cdots & r_{2i} & \cdots & r_{2n} \\ \vdots & \vdots & \vdots & \vdots & \vdots & \vdots \\ r_{i1} & r_{i2} & \cdots & r_i & \cdots & r_{in} \\ \vdots & \vdots & \vdots & \vdots & \vdots & \vdots \\ r_{n1} & r_{n2} & \cdots & r_{ni} & \cdots & r_n \end{pmatrix} \tag{5.2b}$$

where r_i (i = 1, ..., n) and r_{ij} ($i \neq j$; i, j = 1, 2, ..., n) are the diagonal and off-diagonal elements of matrix **R**, respectively.

An individual residual $\hat{\mathbf{v}}_i$ (i = 1, ..., n) can therefore be expressed from Equation (5.1) as follows

$$\hat{v}_i = \hat{l}_i - l_i = -\sum_{j=1}^{n} r_{ij}\, l_j \quad (i = 1, \ldots, n) \tag{5.3}$$

The a priori variance-covariance matrix of $\hat{\mathbf{v}}$ is [cf. (4.132)]

$$\mathbf{C}_{\hat{v}} = \sigma_0^2\, \mathbf{Q}_{\hat{v}} = \sigma_0^2\, \mathbf{RP} \tag{5.4}$$

with σ_0^2 being the a priori given variance factor.

Observational residuals are in theory assumed to have been caused by random errors in the observations only. A ***gross error*** is a large disturbance in the observation data due to malfunctioning of either the instrument or the surveyor or both. If one assumes that the observation vector $\boldsymbol{l}$ contains a gross error vector $\nabla \boldsymbol{l}$, the influence of $\nabla \boldsymbol{l}$ on the estimated residual vector would be [cf. (5.1)]

$$\nabla \hat{\mathbf{v}} = -\mathbf{R}\, \nabla \boldsymbol{l} \tag{5.5}$$

and

$$\nabla \hat{v}_i = -\sum_{j=1}^{n} r_{ij}\, \nabla l_j \quad (i = 1, \ldots, n) \tag{5.6}$$

And the total residuals due to the effects of both random and gross errors in the observation data would be

$$\tilde{v}_i = \hat{v}_i + \nabla \hat{v}_i \quad (i = 1, \ldots, n) \tag{5.7}$$

Therefore, the existence of gross errors would cause the magnitude of residuals to grow, and an ***outlier*** is defined as a residual that according to some statistical test rule exceeds some boundary value that is established based on some assumptions on the stochastic properties of the observations (Caspary, 1987). In fact, a residual is in general the effect of a mixture of all error types (i.e., random error, gross error, and systematic error) that may exist in the observation. In our context we consider an outlier as being caused by a gross error contained in the measurement.

From the variance-covariance matrix of the estimated residuals [Equation (5.4)] one can see that even when the raw observations are uncorrelated, i.e., matrix **P** is diagonal, the estimated residuals are in general correlated since matrix **R** is generally not a diagonal matrix. That is, matrix **R** is in general a full matrix and it contains the full information on the geometry of the network design. From Equations (5.3) and (5.6), one can see that each individual residual is a linear combination of all the observations, and the higher the correlation among the residuals the more intricate the linear dependence of the residuals on the observations. Therefore, a variation in any observation of the set may affect a specific residual, or on the other hand, a change in one specific observation would influence more than one residual. Only in the case of zero correlation, i.e., matrix **R** is a diagonal matrix that would produce diagonal $\mathbf{C}_{\hat{v}}$ matrix, a variation of one observation would affect only one residual; that is, its corresponding residual. Similarly, if the residuals are correlated, a change in one observation would influence more than one unknown parameter, otherwise only one unknown parameter would be affected. Due to the existing correlations among the estimated residuals, even when an outlier is detected it is difficult for us to pinpoint the erroneous observation since its gross error may have been spread over all the residuals. That is why in this chapter we distinguish outlier detection and gross error localization as two totally different subjects.

In order to examine more closely the relationship between an outlier and gross errors, let us assume the special case that only one observational value l_i (i=1, ..., n) is erroneous by ∇l_i at the time and introduce the vector

$$\nabla_i \boldsymbol{l} = \left(0, 0, \ldots, \nabla l_i, 0, \ldots, 0\right)^T \tag{5.8}$$

that contains zeros except for ∇l_i at the i-th position. The observation vector $\boldsymbol{l}$ is then changed by $\nabla_i \boldsymbol{l}$. If the vector $\nabla_i \boldsymbol{l}$ is full it describes a systematic effect on the observations. According to Equation (5.5), the effect of $\nabla_i \boldsymbol{l}$ in the observations on the estimated residuals will be:

$$\nabla \hat{v} = -R \nabla_i l \tag{5.9}$$

From Equation (5.6), one can see that the effect $\nabla_i v_i$ of an error ∇li in observation l_i onto its corresponding residual is determined by the i-th diagonal element of **R**, i.e.,

$$\nabla_i \hat{v}_i = -r_i \nabla l_i \tag{5.10}$$

with

$$r_i = (\mathbf{R})_{ii} \tag{5.11}$$

Here r_i (i=1, ..., n) are diagonal elements of matrix **R** [cf. (5.2b)] and they are called as the redundancy numbers. It can be proven that each r_i is always between 0 and 1, and they sum up to the total redundancy of the system, i.e.,

$$0 \le r_i \le 1 \tag{5.12}$$

$$\sum_{i=1}^{n} r_i = r \tag{5.13}$$

with r being the total redundancy of the system. r_i can be seen as the contribution of the observation l_i to the total redundancy r of the system (Förstner, 1986).

In the meantime, the effect $\nabla_i\hat{v}_j$ ($j\neq i$) of the gross error ∇l_i in observation l_i onto the other residuals $\hat{v}_j$ ($j\neq i$, $j=1, ..., n$) is determined by the off-diagonal elements of **R** as follows:

$$\nabla_i \hat{v}_j = r_{ji} \, \nabla l_i \quad (j \neq i, \; j = 1, \ldots, n) \tag{5.14}$$

where

$$r_{ji} = (\mathbf{R})_{ji} \quad (j \neq i, \; j = 1, \ldots, n) \tag{5.15}$$

From Equation (5.10), it is clear that since r_i always is between 0 and 1, only possibly a small part of a gross error shows up in the residuals and the rest of it will be absorbed in the determination of the unknown parameters. A gross error ∇l_i in an observation that has large redundancy number r_i will affect more the corresponding residual v_i. Therefore, the larger the r_i, the better the control on the ith observation. On the other hand, due to the correlations among the estimated residuals, Equation (5.14) shows that the effect of the gross error ∇l_i in observation l_i might have been spread over all the residuals. Therefore, in order to pinpoint the erroneous observation l_i through the examination of its corresponding residual $\hat{v}_i$ the absolute value of $\nabla_i\hat{v}_i$ must be larger than that of $\nabla_i\hat{v}_j$ ($j\neq i$, $j=1, ..., n$), or equivalently

$$r_i > |r_{ji}| \, (j \neq i, \; j = 1, \ldots, n) \tag{5.16}$$

In this case, if we have an indication that observation l_i is erroneous, an estimate ∇l_i for the size of the gross error may be obtained from the estimated residual as follows [assuming $\hat{v}_i = \nabla_i\hat{v}_i$ in Equation (5.10) in the presence of gross errors]

$$\hat{\nabla} l_i = -\frac{\hat{v}_i}{r_i} \tag{5.17}$$

In the case that

$$r_i \leq \max\{|r_{ji}| \, (j \neq i, \; j = 1, \ldots, n)\} \tag{5.18}$$

localization of the gross error is difficult.

There are two extreme cases for a redundancy number r_i. The first case is the ideal case in which the redundancy number $r_i = 1$. This happens, however, only when a measurement is made of a known quantity; for instance, a measured distance between two fixed points. In this case 100% of any gross error will be revealed in the residual $\hat{v}_i$, and it will not have any effect on the determination of the unknown parameters. The second case is the worst case in which the redundancy number $r_i = 0$. In this case ∇l_i does not affect the residuals at all; therefore it cannot be detected, and it will directly be transferred into the estimated unknown parameters. For instance, in the case of intersection by measuring two angles from two known stations to determine uniquely the position of an unknown point, the redundancy numbers r_i for both observations are zeros, and there will be no control at all on angular measurements. Any gross errors in the angles will affect the position computation. The average redundancy of a network is calculated by

$$\bar{r} = \frac{r}{n} \tag{5.19}$$

If a network is not properly designed, the individual (local) redundancy numbers r_i may vary significantly and do not remain close to $\bar{r}$ (Pope, 1976). That means that the controllability is not the same for all the observations. Thus by revealing how high or low the controllability is in the different parts of the network, the redundancy numbers reflect its geometrical strength (Förstner, 1986). In practice, it is desirable to have a network with relatively large and uniform redundancy numbers so that the ability to detect gross errors will be the same in every part of the network. Since the redundancy numbers depend on only the geometry of the network and have nothing to do with the actual measurements, they can be considered during the design of the network as discussed later in Chapters 9 to 11.

5.2 Defining Null Hypotheses and Alternative Hypothesis for Outlier Detection

Basic concepts and procedures of statistical testing have been discussed in §3.2. The first step of statistical testing is to define the basic postulates, i.e., the so-called null hypothesis and alternative hypothesis, to be tested. Although least squares estimation does not require that the observations be normally distributed, for the purpose of parametric statistical testing, this is required. As discussed before, the method of least squares has been founded on the assumption that the raw observations are free of blunders and systematic effects. Therefore, according to the Gauss-Markov model of network adjustment given in §4.3.2, this basic assumption can be posed as follows: the observations are normally distributed with the expected value of $\mathbf{Ax}$ and variance-covariance matrix $\mathbf{C}_l$. The null hypothesis for statistical testing is thus stated in general as follows

$$H_0 : \boldsymbol{l} \in n\left(\mathbf{Ax}, \sigma_0^2\, \mathbf{P}^{-1}\right) \tag{5.20}$$

As discussed in §3.2, for every null hypothesis, there are infinite choices of alternative hypotheses. In our case, if H_0 is not true, the first natural alternative hypothesis that could be posed is that the observations do not come from normally distributed population at all; i.e.,

$$H_a : \boldsymbol{l} \notin n\left(\mathbf{Ax}, \sigma_0^2\, \mathbf{P}^{-1}\right) \tag{5.21}$$

The test of H_0 against H_a above is done by the so-called χ^2 goodness of fit test that determines whether the histogram of the residuals (cf. §1.1) is compatible with a postulated probability density function that is the normal distribution (cf. Vanicek and Krakiwsky, 1986). Since experiences have shown that the typical measurement process in surveying follows the statistical law of normal distribution, here we concentrate on the case that the observations do come from some normally distributed population but with population means or variances or both that are different from those specified in H_0, resulting in the following four special alternative hypotheses:

$$H_{a1}: \boldsymbol{l} \in n\left(\mathbf{Ax}, \tilde{\sigma}_0^2\, \mathbf{P}^{-1}\right) \tag{5.22}$$

$$H_{a2}: \boldsymbol{l} \in n\left(\mathbf{Ax} - \nabla \boldsymbol{l}, \sigma_0^2 \mathbf{P}^{-1}\right) \tag{5.23}$$

$$H_{a3}: \boldsymbol{l} \in n\left(\mathbf{By}, \sigma_0^2 \mathbf{P}^{-1}\right) \tag{5.24}$$

$$H_{a4}: \boldsymbol{l} \in n\left(\mathbf{By}, \tilde{\sigma}_0^2 \mathbf{P}^{-1}\right) \tag{5.25}$$

First of all, H_{a1} means that an incorrect observational weighting scheme might have been adopted (i.e., $\tilde{\sigma}_0^2 \neq \sigma_0^2$). H_{a2} and H_{a3} postulate that blunders might exist in the observation data (i.e., $\nabla \boldsymbol{l} \neq 0$), and that the functional mathematical model might be wrong (i.e., $\mathbf{By} \neq \mathbf{Ax}$), respectively, and finally H_{a4} represents the case where both the functional model and the observational weighting might be incorrect. This discussion shows that when the null hypothesis H_0 is rejected there can be many choices of alternative hypothesis, and choosing the right one at the time is not a trivial thing. In the present example alone, any of the above four alternative hypotheses could be the reason for the rejection of H_0, and a proper selection of H_a requires good knowledge of the data collection procedures and analysis of the results.

Fortunately, in geodesy and surveying, the mathematical models relating observations to certain unknown parameters (e.g., coordinates) are very well defined, because they are based on geometrical and simple physical laws (Vanicek and Krakiwsky, 1986). This is especially true for the case of geodetic network positioning. Therefore, alternative hypotheses H_{a1} and H_{a2} are in fact the most frequently examined postulates when the null hypothesis fails the statistical test, and alternative hypotheses H_{a3} and H_{a4} are not of major concern in our context.

Since the residuals are the quantities to be examined after network adjustment, the probability density functions of the residual vector **v** under the null hypothesis H_0 and the alternative hypotheses H_{a1}, and H_{a2} can be derived as follows

$$\hat{\mathbf{v}} \big| H_0 \in n\left(\mathbf{0},\ \mathbf{C}_v\right) \tag{5.26}$$

$$\hat{\mathbf{v}} \big| H_{a1} \in n\left(\mathbf{0},\ \tilde{\mathbf{C}}_v\right) \tag{5.27}$$

$$\hat{\mathrm{v}}\big|\mathrm{H}_{a2} \in \mathrm{n}(\nabla\hat{\mathbf{v}},\ \mathbf{C}_v) \tag{5.28}$$

where **C**v, $\tilde{\mathrm{C}}_v$, and $\nabla\hat{\mathbf{v}}$ are calculated by

$$\mathbf{C}_v = \sigma_0^2\left[\mathbf{P}^{-1} - \mathbf{A}\left(\mathbf{A}^T\mathbf{P}\mathbf{A} + \mathbf{D}\mathbf{D}^T\right)^{-1}\mathbf{A}^T\right] \tag{5.29}$$

$$\tilde{\mathbf{C}}_v = \tilde{\sigma}_0^2\left[\mathbf{P}^{-1} - \mathbf{A}\left(\mathbf{A}^T\mathbf{P}\mathbf{A} + \mathbf{D}\mathbf{D}^T\right)^{-1}\mathbf{A}^T\right] \tag{5.30}$$

$$\nabla\hat{\mathbf{v}} = \left[\mathbf{A}\left(\mathbf{A}^T\mathbf{P}\mathbf{A} + \mathbf{D}\mathbf{D}^T\right)^{-1}\mathbf{A}^T - \mathbf{I}\right]\nabla l \tag{5.31}$$

Equation (5.26) shows that under the null hypothesis H_0 the observations $\boldsymbol{l}$ are burdened only with random errors with zero mean. In this case, the vector of observation residuals $\hat{\mathrm{v}}$ also consists of random variables with zero mean. The purpose of post-adjustment data screening is to examine the estimated residuals, using the tools of statistical testing, to determine whether or not anything has gone wrong with this basic postulate; that is, to determine whether the residuals give an indication that a certain observation(s) is erroneous or not. There are basically two well-known approaches to test the null hypothesis H_0 using the estimated residuals, depending on whether the variance factor σ_0^2 used for scaling the covariance matrix of observations is known or not. The first approach was credited to Baarda (1968) who proposed a global test for detection of outliers and data snooping for localization of gross errors in the case that σ_0^2 is known, whereas the second approach, called tau-test, was proposed by Pope (1976), dealing with the case when σ_0^2 is unknown. The principles of these two approaches and their applications in practice are discussed in the following §5.3 and §5.4, respectively.

5.3 Global Test and Data Snooping

Global test and data snooping are the most frequently used post-adjustment data screening techniques. Baarda (1968) proposed the global test for outlier detection and data snooping for the localization of gross errors. After network adjustment, the global test is applied first, which tests the compatibility of the estimated a posterior variance factor $\hat{\sigma}_0^2$ with the a priori selected variance factor σ_0^2. If the global test failed, it means something is wrong with the null hypothesis H_0, and further examination of the residuals is made by data snooping. Obviously, this approach requires that the a priori knowledge of the precision of the observations be available, that is, the a priori variance factor is known. Otherwise, a meaningful physical interpretation of the test results would be difficult. The theory and procedures discussed in Chapter 1 can be used to obtain an appropriate accuracy estimate of the observations.

5.3.1 Global Test on the a Posterior Variance Factor

From Equation (4.130), the estimated a posterior variance factor from a least squares network adjustment is as follows

$$\hat{\sigma}_0^2 = \frac{\hat{\mathbf{v}}^T \mathbf{P} \hat{\mathbf{v}}}{r}$$

Global test serves to examine the compatibility of the a posterior estimated variance factor $\hat{\sigma}_0^2$ with the a priori given variance factor σ_0^2, and it uses the following test statistic

$$y = \frac{r \cdot \hat{\sigma}_0^2}{\sigma_0^2} \tag{5.32}$$

It can be shown that under the null hypothesis H_0, the test statistic y belongs to the chi-square distribution with degrees of freedom being r; i.e.,

$$y|H_0 \in \chi^2(r) \tag{5.33}$$

Since the expectation of $\chi^2(r)$ is r, i.e.,

$$E\{y|H_0\} = r, \tag{5.34}$$

which leads to

$$E\left\{\frac{\hat{\sigma}_0^2}{\sigma_0^2}\middle| H_0\right\} = 1 \tag{5.35}$$

or

$$E\{\hat{\sigma}_0^2 | H_0\} = \sigma_0^2 \tag{5.36}$$

That is to say, under the null hypothesis H_0, the expected value of the estimated a posterior variance factor $\hat{\sigma}_0^2$ equals the a priori variance factor σ_0^2, any value smaller or larger than σ_0^2 means that they are not compatible.

Given the significance level α, then according to the principle of two-tailed statistical testing as discussed in §3.2, the null hypothesis H_0 will be rejected if either

$$y > \chi^2_{1-\alpha/2}(r) \tag{5.37}$$

or

$$y < \chi^2_{\alpha/2}(r) \tag{5.38}$$

holds. That is, H_0 will be accepted if

$$\chi^2_{\alpha/2}(r) \le y \le \chi^2_{1-\alpha/2}(r) \tag{5.39}$$

Recall that the weight matrix is defined by [cf. (4.116)]

$$\mathbf{P} = \sigma_0^2\, \mathbf{C}_l^{-1} \tag{5.40}$$

In practice, the a priori variance factor (also called variance of unit weight) is usually set to 1 (i.e., $\sigma_0^2 = 1$), and that leads to the weight matrix **P** being equated with the inverse of the variance-covariance matrix of observations, i.e.,

$$\mathbf{P} = \mathrm{C}_l^{-1} \tag{5.41}$$

In this case, the global test simplifies to examine whether or not the estimated a posterior variance factor is statistically equal to 1.0.

As discussed before, once H_0 is rejected, there will be an infinite choice of alternative hypotheses, and the two special cases we will concentrate on are

1. H_{a1} (incorrect observational weighting), and
2. H_{a2} (gross errors exist in the observation data).

Since at this stage one does not know which of the above factors caused the failure of the test, a step-by-step procedure should be followed.

First of all, H_{a1} should be examined first to check whether observations were not weighted properly. Note that the statistic y is calculated by

$$y = \frac{r \cdot \hat{\sigma}_0^2}{\sigma_0^2} = \hat{\mathbf{v}}^T \frac{\mathbf{P}}{\sigma_0^2} \hat{\mathbf{v}} = \hat{\mathbf{v}}^T \mathbf{C}_l^{-1} \hat{\mathbf{v}} \tag{5.42}$$

Equation (5.42) shows that computation of the statistic y has nothing to do with the choice of the a priori variance factor σ_0^2, and it depends on only two variables: $\hat{\mathbf{v}}$ and $\mathbf{C}_l$. Furthermore, from Equation (5.27) one can see that under alternative hypothesis H_{a1} the expected value of $\hat{\mathbf{v}}$ is zero. Thus, the value of y can be seen as inversely proportional to the variance-covariance matrix of observations $\mathbf{C}_l$. An overestimate of the observation random errors (i.e., assuming that the observations are less precise than the actual situation) will lead to a too small value of y, which falls under its lower boundary value calculated by Equation (5.37). On the other hand, an underestimate of the observation random errors (i.e., assuming that the observations are more precise than the actual situation) will lead to a too big value of y, which exceeds its upper boundary value calculated by Equation (5.38). Therefore, either an overestimate or an underestimate of observation random errors will produce a value of the statistic y that falls beyond its boundary values and leads to the rejection of the null hypothesis H_0.

From the relation between the statistic y and the estimated a posterior variance factor σ_0^2, we have

$$\hat{\sigma}_0^2 = \left(\frac{y}{r}\right)\sigma_0^2 \tag{5.43}$$

Thus, in the case that the observation errors are overestimated, the resulting a posterior variance factor $\hat{\sigma}_0^2$ will be much smaller than the a priori variance factor σ_0^2, and vice versa.

Rejection of H_0 caused by improper estimates of random observational errors is easier to recognize, since in this case the residuals are still normally distributed with zero means, no excessive residuals are likely to be produced. Therefore, if the global test shows that the estimated a posterior variance factor $\hat{\sigma}_0^2$ is not compatible with the a priori variance factor σ_0^2, and in the meantime the magnitudes of residuals are reasonably small as compared with the accuracy of the survey instruments being used, we have a reason to believe that the a priori given observation precision is not correct and the variance-covariance matrix $\mathbf{C}_l$ should be re-constructed. In practice, one usually modifies $\mathbf{C}_l$ by scaling it with the estimated a posterior variance-factor $\hat{\sigma}_0^2$, i.e.,

$$\tilde{\mathbf{C}}_l = \hat{\sigma}_0^2\, \mathbf{C}_l \tag{5.44}$$

A new network adjustment is then run using $\tilde{\mathrm{C}}_l$, which will lead to the acceptance of the null hypothesis H_0. On the other hand, however, if the global test failed and some residuals show excessive magnitudes, alternative hypothesis H_{a2} should be examined to test whether or not any gross errors exist in the observation data, and that is handled by the data snooping technique discussed below.

5.3.2 Data Snooping of Individual Residuals

Application of the data snooping technique is a combined process of *outlier detection* and *gross error localization and elimination*. The former searches for outliers, i.e., residuals that according to some statistical test rule exceed some boundary value, whereas the latter investigates which observation(s) contains a gross error, which caused the outlier(s), and then eliminates the erroneous observation(s) when justified.

Outlier Detection

As discussed before, although the observational residuals are known after a network adjustment, the gross errors $\nabla \boldsymbol{l}$ are not, leading to no conclusions as to which observations are the erroneous ones. Therefore, if there is a reason to believe that the null hypothesis H_0 was rejected because there are gross errors in the observations (i.e., H_{a2} is true), the crucial task is then to locate and eliminate the gross errors in the observation data. However, the alternative hypothesis H_{a2} is still too general since it does not give any information about the individual elements of the gross error vector $\nabla \boldsymbol{l}$, which refers to the entire model. Therefore, we need to propose a simpler and more specific alternative hypothesis H_{a2i} that will constrain the relationship between the estimated residuals $\hat{\mathrm{v}}$ and gross errors $\nabla \boldsymbol{l}$. Baarda's data snooping technique is to assume that only one observation at the time is erroneous. The specific alternative hypothesis H_{a2i} is therefore expressed as

$$H_{a2i}: \nabla \boldsymbol{l}_i = \mathbf{c}_i \; \nabla l \neq \mathbf{0} \quad (i = 1, \ldots, n) \tag{5.45}$$

where $\mathbf{c}_i = (0\,0 \ldots 0\,1\,0 \ldots 0)^T$, which is a vector of zeros except for identity at the i-th position, and ∇l is the magnitude of the gross error in the i-th observation l_i.

The statistic chosen to test the null hypothesis H_0 against the above alternative hypothesis H_{a2i} is constructed as follows (cf. Alberda, 1980)

$$w_i = \frac{-\mathbf{c}_i^T \, \mathbf{P} \, \hat{\mathbf{v}}}{\sigma_0 \sqrt{\mathbf{c}_i^T \, \mathbf{P} \, \mathbf{Q}_{\hat{v}} \, \mathbf{P} \, \mathbf{c}_i}} \tag{5.46}$$

which for uncorrelated observations reduces to the standardized residual

$$w_i = \frac{-\hat{v}_i}{\sqrt{r_i}\,\sigma_{li}} = \frac{\hat{v}_i}{\sigma_{\hat{v}i}} \tag{5.47}$$

where

$$\sigma_{\hat{v}i} = \sqrt{r_i}\, \sigma_{li} \tag{5.48}$$

It can be derived that under the null hypothesis H_0 the statistic w_i is normally distributed with zero mean and unit variance, i.e.,

$$w_i | H_0 \in n(0, 1) \tag{5.49}$$

According to the principle of two-tailed test approach, given the significance level α_0, H_0 is rejected, if either

$$w_i < n_{\alpha 0/2}(0, 1) \tag{5.50}$$

or

$$w_i > n_{1-\alpha 0/2}(0, 1) \tag{5.51}$$

holds. Otherwise, H_0 is accepted if

$$n_{\alpha 0/2}(0, 1) \leq w_i \leq n_{1-\alpha 0/2}(0, 1) \tag{5.52}$$

where $n_{\alpha_{0/2}}(0, 1)$ and $n_{1-\alpha_{0/2}}(0, 1)$ are the lower and upper boundary values of the statistic w_i calculated from the standard normal probability density function n(0, 1) under the given significance level α_0.

The term "data snooping" refers to the above one-dimensional test, examining only one standardized residual at the time, and the procedure is subsequently repeated if there is more than one gross error. Usually $\alpha_0 = 0.001$ is suggested, which leads to a boundary value of 3.29. Thus the null hypothesis is rejected and the i-th residual is flagged for rejection if

$$|w_i| > 3.29 \tag{5.53}$$

or

$$|\hat{v}_i| > 3.29\,\sigma_{v_i} \tag{5.54}$$

It should be noted that the value of test statistic w_i is calculated based on σ_{l_i} being known. Therefore, the data snooping procedure is reliable only when a good knowledge of stochastic properties of the observations are available. That is why if the global test fails, the alternative hypothesis H_{a1} should always be examined first.

Gross Error Localization and Elimination

After the residuals are flagged (i.e., outliers are detected), the next step is to localize the gross errors; that is, to determine whether or not the flagged residual $\hat{v}_i$ is caused by a gross error ∇l_i in its corresponding observation l_i. In general the success of gross error *localization* depends on the *geometry* of the network and the number and magnitudes of gross errors existing in the observation data. A general requirement for this is that a majority of observations should be statistically consistent, and obviously the maximum number of gross errors must be less than the number of redundancy of the system. Since we are dealing with gross errors of small or marginal magnitudes, the correct localization of gross errors from flagged residuals (i.e., outliers) are strongly dependent on the network geometry. In the case that only one gross error is assumed to exist at the time, the discussions in §5.1 made it clear that, because of the correlation among the estimated residuals, a flagged residual $\hat{v}_i$ may indicate a gross error ∇l_i in its corresponding observation l_i only if it has a dominant redundancy number r_i; that is,

$$r_i > |r_{ji}| \; (j \neq i,\ j = 1, \ldots, n) \qquad \text{[cf. Equation (5.16)]}$$

where r_i (i=1, ..., n) and r_{ij} (i, j=1, ..., n; i≠j) respectively refer to the diagonal and off-diagonal elements of the matrix **R** calculated by Equation (5.2). Otherwise, if the above equation does not hold true; that is, if

$$r_i \leq \max\left\{|r_{ji}| \; (j \neq i,\ j = 1, \ldots, n)\right\} \qquad \text{[cf. Equation (5.18)]}$$

due to the strong correlation among the residuals, it is difficult to determine which observation l_j is erroneous by a gross error ∇l_j (j≠i, j=1, ..., n) that has produced the excessive residual $\hat{v}_i$. That is to say, a gross error in any observation l_j (j≠i, j=1, ..., n) might have produced the outlying residual $\hat{v}_i$ of l_i. The localization of the gross error(s) is therefore difficult in this case.

Due to the fact that an outlying residual $\hat{v}_i$ may not necessarily be caused by a gross error in its corresponding observation l_i, or equivalently, a gross error in l_i may produce an outlying residual for any observation l_j (j≠i, j=1, ..., n), an automatic elimination of observations from the flagged residuals should never be done. Outlying residuals (i.e., outliers) can only be used to flag their corresponding observations for possible candidates of rejection, and the actual elimination of an observation must be based on a further analysis of data. Usually an iterative procedure is needed in order to accurately pinpoint all the possible gross errors existing in the observation data. Having a good knowledge of what kinds of gross errors might be expected and how observations are affected by them would certainly help the investigator (a surveyor or a surveying engineer) in making the right decision and therefore performing a detailed and meaningful analysis of the network.

It should be stressed again that post-adjustment data screening techniques deal with gross errors of small magnitude. Since least squares adjustment will smooth a large gross error(s) throughout its (their) neighboring observations, part or all of observations that are related to the erroneous observation may be flagged depending on the magnitude of the gross error. In this case, one should carefully examine the data to determine which observation is actually erroneous, and the isolation of gross errors by the statistical rejection of residuals is even more difficult, as a large error might have been distributed throughout the data set. The problem in this case is not that the data snooping technique does not pinpoint large errors, but that it points to many other non-existing large errors (Steeves, 1984), and the investigator has therefore to search through a large portion of the data set for non-existent errors. More than often, several least squares network adjustments are required to isolate all of the gross errors by "trial and error."

5.3.3 Undetectable Gross Errors

As discussed before, in post-adjustment data screening we concentrate on gross errors of small magnitudes. Due to the limited sensitivity of outlier detection techniques, however, no gross errors of arbitrarily small magnitude are detectable. This subsection discusses the lower bound values of both the global gross error vector and an individual gross error that are detectable under the selected significance level and power of the test. This also tells us that after the detection and elimination of gross errors by statistical methods there are still blunders or gross errors remaining in the data, and their influence on the estimated parameters should be considered, which will be treated in the next chapter introducing the concept of "reliability" of geodetic networks.

5.3.3.1 Lower Bound of the Detectable Global Gross Error Vector

Equation (5.35) shows that under the null hypothesis H_0, the expectation of $\hat{\sigma}_0^2/\sigma_0^2$ is equal to 1.0, but under the alternative hypothesis H_{a2}, it will not be equal to 1.0 anymore, and it can be derived as follows

$$
\begin{aligned}
E\left\{\frac{\hat{\sigma}_0^2}{\sigma_0^2}\middle| H_{a2}\right\} &= E\left\{\frac{(\hat{\mathbf{v}}+\nabla\hat{\mathbf{v}})^T\mathbf{P}(\hat{\mathbf{v}}+\nabla\hat{\mathbf{v}})}{r}\frac{1}{\sigma_0^2}\right\} \\
&= E\left\{\frac{\hat{\mathbf{v}}^T\mathbf{P}\hat{\mathbf{v}}}{r}\frac{1}{\sigma_0^2}\middle| H_0\right\} + \frac{\nabla\hat{\mathbf{v}}^T\mathbf{P}\nabla\hat{\mathbf{v}}}{r}\frac{1}{\sigma_0^2} \\
&= 1 + \frac{\lambda}{r}
\end{aligned}
\tag{5.55}
$$

where it is assumed that $\hat{\mathbf{v}}$ is calculated under H_0. $\nabla\hat{\mathbf{v}}$ is the influence of the gross error vector $\nabla\boldsymbol{l}$ on $\hat{\mathbf{v}}$, and

$$
\lambda = \frac{1}{\sigma_0^2}\nabla\mathbf{v}^T\mathbf{P}\nabla\mathbf{v}
\tag{5.56}
$$

Noting from Equation (5.5) that

$$
\nabla\hat{\mathbf{v}} = -\mathbf{R}\,\nabla\boldsymbol{l}
$$

we have

$$
\begin{aligned}
\lambda &= \frac{1}{\sigma_0^2}\nabla\boldsymbol{l}^T\mathbf{R}^T\mathbf{P}\mathbf{R}\nabla\boldsymbol{l} \\
&= \frac{1}{\sigma_0^2}\nabla\boldsymbol{l}^T\mathbf{P}\mathbf{Q}_v\mathbf{P}\nabla\boldsymbol{l}
\end{aligned}
\tag{5.57}
$$

Therefore, under the alternative hypothesis H_{a2}, the statistic y follows the non-central chi-square distribution $\chi^2(r, \lambda)$ with central parameter λ, i.e.,

$$
y\big|H_{a2} = \frac{r\hat{\sigma}_0^2}{\sigma_0^2}\bigg|H_{a2} \in \chi^2(r, \lambda)
\tag{5.58}
$$

Since the values of the gross error vector $\nabla\boldsymbol{l}$ is in general not known, the value of λ cannot be calculated, and therefore the probability β (see Figure 5.1) cannot

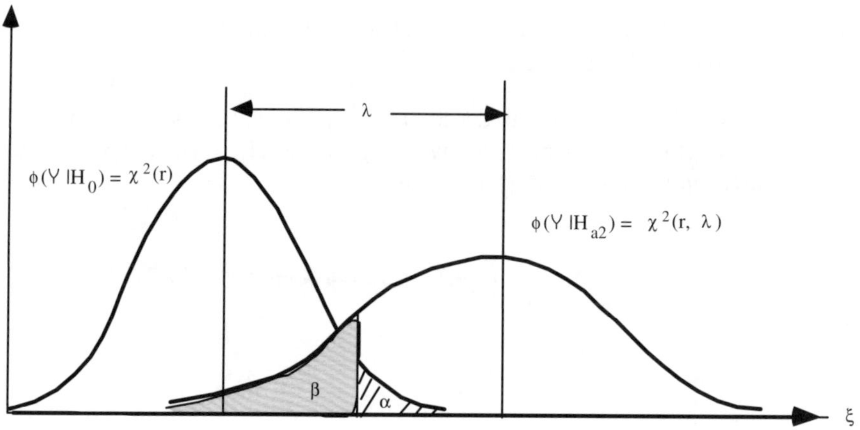

Figure 5.1. Detectable deviation (λ) of H_{a2} from H_0 given α and β

be computed in practice. Baarda (1968) proposed to invert the relation between the power of the test and the non-centrality parameter and use

$$\lambda = \lambda(\alpha, \beta, r) \tag{5.59}$$

to describe, in general form, the functional relationship between the probability levels α, β, and degrees of the freedom r. For a preset power of the test, Equation (5.59) gives the distance between H_0 and H_{a2}; i.e., the magnitude of the deviation of H_{a2} from H_0 that can be detected with that power. If we now require that an alternative, i.e., an error vector has to be detected with a probability of $1-\beta$ that is larger than a prescribed bound $1-\beta_0$ then the distance λ between H_0 and H_{a2} has to be larger than the bound λ_0 calculated by

$$\lambda_0 = \lambda(\alpha, \beta_0, r) \tag{5.60}$$

where λ_0 represents the minimum detectable deviation of H_{a2} from the null hypothesis H_0 given the significance level α and power of the test $1-\beta_0$. The values of λ_0 can be found in tables given in Pearson and Hartley (1951) or Pelzer (1971).

Let us normalize $\nabla \boldsymbol{l}$ by

$$\nabla \boldsymbol{l} = \overline{\nabla l} \cdot \mathbf{u} \tag{5.61}$$

where $\overline{\nabla l}$ is the length of the gross error vector $\nabla \boldsymbol{l}$, and $\mathbf{u}$ is a unit vector in the direction of $\nabla \boldsymbol{l}$ defined by

$$\mathbf{u} = \frac{\nabla \boldsymbol{l}}{\|\nabla \boldsymbol{l}\|} \tag{5.62}$$

with $\|\cdot\|$ denoting the norm of a vector. Considering the minimum detectable deviation of H_{a2} from H_0 (i.e., λ_0) and substituting Equation (5.61) into (5.57), one obtains the minimum detectable length of the gross error vector as follows

$$\overline{\nabla l_0} = \sigma_0 \sqrt{\frac{\lambda_0}{\mathbf{u}^T \mathbf{P} \mathbf{Q}_v \mathbf{P} \mathbf{u}}} \tag{5.63}$$

Under the given significance level α and power of the test $1-\beta_0$, if the length $\overline{\nabla l}$ of the global gross error vector ∇l is less than $\overline{\nabla l_0}$, it will not be detectable, otherwise, it is detectable. Assuming a power of the test of 80%, some typical values of λ_0 are listed in Table 5.1.

5.3.3.2 Lower Bound of the Detectable Magnitude of an Individual Gross Error

Equation (5.63) does not give any information about the minimum detectable magnitude for an individual gross error. This can be derived under the special alternative hypothesis H_{a2i} that assumes that there is only one gross error at the time [cf. Equation (5.45)]. In this case the minimum detectable magnitude $\nabla_0 l_i$ of a gross error ∇l_i in observation l_i can be calculated by simply substituting the vector $\mathbf{c}_i$ in Equation (5.45) into Equation (5.63) as follows:

$$\begin{aligned} \nabla_0 l_i &= \sigma_0 \sqrt{\frac{\lambda_0}{\mathbf{c}_i^T \mathbf{P} \mathbf{Q}_v \mathbf{P} \mathbf{c}_i}} \\ &= \sigma_{l_i} \sqrt{\frac{\lambda_0}{r_i}} \end{aligned} \tag{5.64}$$

Another way to derive $\nabla_0 l_i$ is to use the test statistic w_i defined by Equation (5.47). Note that the influence of an error ∇l_i in l_i onto the test statistic w_i is

$$\delta_i = \nabla w_i = -\frac{\nabla \hat{v}_i}{\sigma_{l_i} \sqrt{r_i}} = \frac{\nabla l_i \sqrt{r_i}}{\sigma_{l_i}} \tag{5.65}$$

which leads to a shift of the probability density function of w_i by δ_i. Thus, under the alternative hypothesis H_{a2i} the statistic w_i follows a non-central normal distribution with non-central parameter δ_i and unit variance, i.e.,

$$w_i | H_{a2i} \in n(\delta_i, 1) \tag{5.66}$$

Similar to the above discussions, since the actual value ∇l_i of a gross error and thereby the non-centrality parameter δ_i is not known, the probability β cannot be

Table 5.1. Some typical values of the lower bound λ_0 $(1-\beta_0 = 0.80)$

α	Degrees of Freedom (r)							
	2	5	10	20	30	40	50	85
0.05	9.6	13.4	16.5	21.0	25.3	28.5	32.0	40.0
0.01	14.0	18.3	22.7	29.0	34.5	39.0	42.0	50.0

computed in practice, and the inverse relation between the power of the test and the non-centrality parameter can be expressed by

$$\delta = \delta(\beta, \alpha) \tag{5.67}$$

For a preset significance level α_0 and power of the test $1-\beta_0$, the lower bound δ_0 of δ that is detectable by statistic testing reads

$$\delta_0 = \delta(\beta_0, \alpha_0) \tag{5.68}$$

The lower bound for just detectable gross errors $\nabla_0 l_i$ can be derived by substituting the actual influence δ_i of the gross error onto the test statistic by the lower bound δ_0, i.e.,

$$\nabla_0 l_i = \sigma_{l_i} \frac{\delta_0}{\sqrt{r_i}} \tag{5.69}$$

The calculation of δ_0 may be illustrated through Figure 5.2, from which one can see that (cf. also Förstner, 1986)

$$\beta = \Phi(n_{\alpha/2} - \delta_i) \tag{5.70}$$

where Φ represents the probability function of w_i under the alternative hypothesis H_{a2i}. According to the theory of mathematical statistics, the probability function Φ is calculated from the probability density function ϕ as follows

$$\Phi(\xi) = \int_{-\infty}^{\xi} \phi(\xi) d\xi \tag{5.71}$$

When α_0, β_0 are fixed, the noncentrality parameter is therefore computed by inverting Equation (5.71) as follows

$$\delta_0 = n_{\alpha_0/2} - \Phi^{-1}(\beta_0) \tag{5.72}$$

Since for $\beta < 0.5$, the function $\Phi^{-1}(\beta)$ is negative, the minimum bound δ_0 for the noncentrality parameter δ_i is always larger than the critical value $n_{\alpha/2}$ if one requires a power of at least 50%. This also means that the distance δ between H_0 and H_{a2i} must be much larger than the critical value $n_{\alpha/2}$ if one wants to correctly

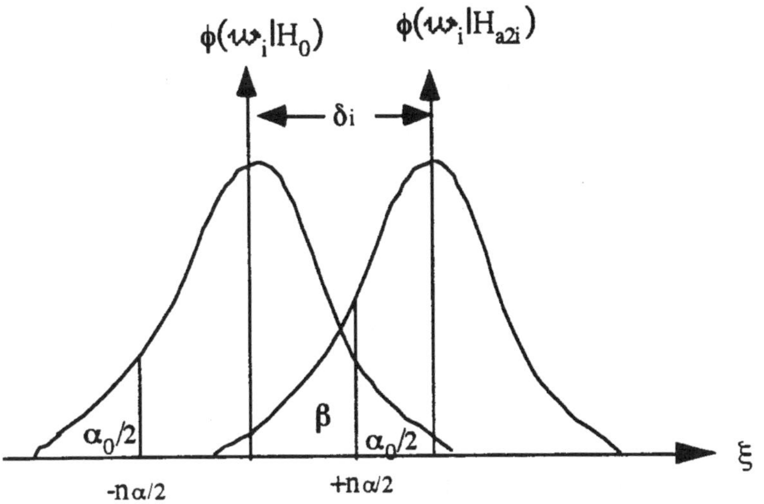

Figure 5.2. Density functions of test statistic w_i

reject H_0 with a high probability when it is actually false or when H_{a2i} is true. Values for δ_0 corresponding to some typical significance levels α_0 and powers of the test $(1-\beta_0)$ are listed in Table 5.2.

5.3.4 Choice of Significance Levels α, α_0, and Power of the Test $1-\beta_0$

Baarda (1968) proposed the use of global test for the detection of gross errors and "data snooping" for their localization. One may choose different significance levels α and α_0 as well as power of the test $(1-\beta)$ and $(1-\beta_0)$ for the global test and data snooping, respectively. The choice should be based on the philosophy that the decisions from both types of test should be consistent, i.e., the same boundary values should be found whether the global or the uni-variate test is performed. Referring to Equations (5.64) and (5.69), the above requirement leads to

$$\delta_0 = \sqrt{\lambda_0} \tag{5.73}$$

Since the noncentrality parameter δ_0 and λ_0 are the minimum deviation of the alternative hypothesis from the null hypothesis, which can just be detected with a given probability β_0, Baarda proposed to keep the power of the test constant for both tests but vary the levels of significance. For the global test it is "α" but for the uni-variate test (data snooping) it is α_0. If $\beta = \beta_0$ (constant), these levels of significance are interconnected through the non-centrality parameter λ_0. The following procedures can thus be followed (cf. also Kavouras, 1982)

Table 5.2. Lower bounds δ_0 for non-centrality parameter in dependency of the significance level α_0 and the required minimum power $1\text{-}\beta_0$ of the test

α_0 / $1\text{-}\beta_0$	0.01%	0.10%	1%	5%
50%	3.72	3.29	2.58	1.96
70%	4.41	3.82	3.10	2.48
80%	4.73	4.13	3.42	2.80
90%	5.17	4.57	3.86	3.24
95%	5.54	4.94	4.22	3.61
99%	6.22	5.62	4.90	4.29
99.90%	6.98	6.38	5.67	5.05

Step 1: choose α_0 and β_0, compute δ_0, then $\lambda_0 = \delta_0^2$
Step 2: compute α from $\lambda_0 = \lambda(\alpha, \beta_0, r)$

For instance, when the power of the test is chosen to be 90% and significance level α_0 0.001, we have from Table 5.2

$$\delta_0 = 4.57$$

which leads to

$$\lambda_0 = \delta_0^2 = 20.88$$

Assuming a degree of freedom of 20, we can search for the value of α from Table 5.1 as follows

$$\alpha = 0.05$$

The selection of β_0 is not as critical as the selection of α_0 for the outcome of the tests. The dependence between α and α_0 can also be found from nomograms (Baarda, 1968, 1976). Usually, the power of the test is chosen to be 80%, while $\alpha_0 = 0.001$.

5.4 Tau-Test

Baarda's global test and data snooping technique require that the observational precision or the variance factor σ_0^2 be known, i.e., all the variances are properly scaled. If, however, σ_0^2 is not adequately known or one does not want to rely on the a priori error estimates, the a posterior estimated variance factor $\hat{\sigma}_0^2$ is used instead. In this case, the global test on the variance is not performed, and the "data snooping" is modified by adopting the following new statistic proposed by Pope (1976); i.e.,

$$\tau_i = \frac{\hat{v}_i}{\hat{\sigma}_{\hat{v}_i}} = \frac{\hat{v}_i}{\hat{\sigma}_0 \sqrt{q_{\hat{v}_i}}} \tag{5.74}$$

which under the null hypothesis H_0 follows the τ-distribution with degrees of freedom r, that is,

$$\tau_i | H_0 \in \tau(r) \tag{5.75}$$

Mathematically, this is a multi-variate statistic since the residuals are used for the estimation of $\hat{\sigma}_{\hat{v}_i}$ through $\hat{\sigma}_0^2$. Given the significance level α, τ_i will be flagged as an outlier if

$$|\tau_i| > \tau_{\alpha/2} \tag{5.76}$$

where the critical value $\tau_{\alpha/2}$ can be calculated according to its relation to the student's t-distribution as follows [cf. (3.18)]

$$\tau_{\alpha/2}(r) = \frac{\sqrt{r}\, t_{\alpha/2}(r-1)}{\sqrt{r - 1 + t_{\alpha/2}^2(r-1)}}$$

The test is applied successively to all standardized residuals.

It should be mentioned that one used to wrongly assume that $\tau_i = \hat{v}_i / \hat{\sigma}_{\hat{v}_i}$ belongs to the student's t-distribution since "tau"-distribution is rarely discussed in the statistical textbooks and it is therefore not universally known under this name (Kavouras, 1982). Nevertheless, the two distributions give closer critical values as the degrees of freedom r increase. When r tends to infinite, they both approach the normal distribution.

The problem with the τ-test is that since the estimated variance factor is affected by the presence of gross errors in the observations, the larger the gross error among the data, the larger the estimated $\hat{\sigma}_0^2$ that tends to reduce the value of the statistic τ_i. Therefore, there is always a good chance for some gross errors to stay undetected by the tau-test, especially when they are of small magnitudes. This point should be kept in mind whenever the tau-test is used.

Similarly, after the residuals are flagged, localization and elimination of gross errors should be made using the same philosophy as discussed above for Baarda's data snooping technique.

Finally, the concept of maximum tau-test should be discussed that is defined as a test on the maximum τ calculated by (5.74) as follows

$$\begin{aligned} \alpha &= \Pr.\{\max[\tau_i, i = 1, \ldots, n] > \tau_\alpha\} \\ &= 1 - \Pr.\{(\tau_i \le \tau_\alpha) \text{ and } (\tau_2 \le \tau_\alpha) \text{ and} \ldots \text{and } (\tau_n \le \tau_\alpha)\} \end{aligned} \tag{5.77}$$

where Pr. means probability. Since the calculation of the exact joint probability in (5.77) is difficult, the so-called Bonferroni's inequality is used that reads

$$\text{Pr.}\left\{\bigcap_{i=1}^{n}\left(|\tau_i| \leq \tau_{\alpha/n}\right)\right\} \geq \left(1-\frac{\alpha}{n}\right)^n \cong 1-\alpha \tag{5.78}$$

This says that if all the individual tests of quantities, whether dependent or not, are executed at the significance level of $\alpha_0 = \alpha/n$, the joint probability (i.e., the confidence level) for the entire group will be at least $(1 - \alpha)$. Considering Equations (5.77) and (5.78), one can see that the maximum τ test can be performed simply by executing the individual test (5.76) with a reduced significance level of α/n, instead of α. That is, given the significance level α for the maximum τ test, an individual τ_i will be flagged as an outlier if

$$|\tau_i| > \tau_{\alpha/(2n)} \tag{5.79}$$

The maximum τ test is also called the in-context test, meaning that the existence of the other members of the data series are being considered simultaneously (cf. Vanicek and Krakiwsky, 1986). From (5.79), one can see that the maximum tau-test is more conservative. Since it uses a smaller α in testing τ_i, it is more reluctant to reject the tested hypothesis. Consequently, it protects us from committing the Type I error but not the Type II error.

5.5 A Proposed Strategy

5.5.1 Summary and Discussions

The concepts and pitfalls of outlier detection and localizing gross errors of small magnitude based on estimated observational residuals after a least squares network adjustment have been discussed in this chapter, assuming that gross errors of large magnitude have been detected and eliminated at an early stage by implementing a well designed observational procedure and/or applying the pre-adjustment data screening techniques. Two statistical techniques have been analyzed for post-adjustment data screening; that is, the Baarda's global test and data snooping technique and the tau-test. The former approach requires that one should have a good knowledge of the accuracy of the observations. Otherwise, the global test does not bear any practical meaning and the data snooping is substituted by the latter approach, i.e., the tau-test. Therefore, in order to perform a detailed and meaningful analysis of a network the investigator should have a good understanding of the survey procedure and the philosophy of the error-detection techniques. In addition, knowing the expected accuracy of the observations, what kinds of gross errors might be expected, and how observations are affected by gross errors (cf. Chapter 1) would certainly help the investigator to select the right alternative hypothesis and therefore to make the right decision.

When Baarda's approach is used, the global test on the estimated a posterior variance factor is performed first to check its compatibility with the a priori given variance factor. This test is very important since it gives general information about the behavior of both the underlying functional and the stochastic model of the network. If the global test fails, there is definitely something wrong with our basic assumptions (i.e., the null hypothesis H_0) and the error has to be discovered. Failure of the global test may in general be attributed either to the improper functional and/or stochastic model or to the presence of gross errors. Therefore, if the global test failed, efforts should be taken first to examine the stochastic properties of the observations to make sure that proper random error estimates of observations, i.e., estimates of observational precisions, have been achieved. After that, if the global test still failed we then have a reason to believe that possible gross errors exist in the observation data. In this case, since the global test is unable to pinpoint gross errors in the data set, the data snooping technique is employed to screen each individual residual for outliers. On the other hand, it is also quite possible that a significant error, detectable by data snooping, may stay hidden in the total sum of the squares of the residuals when the global test is performed, if the other errors turn out to be small. For this reason, it is, therefore, always recommended to apply the data snooping procedure even if the global test passed.

In the case that the observational accuracy is not sufficiently known, the global test on the estimated variance factor is not meaningful, and therefore Baarda's approach is replaced by the tau-test alone to screen the estimated observational residuals for outliers. As discussed above, when tau-test is used it should be kept in mind that since the estimated variance factor is affected by the presence of gross errors in the observations, the larger the gross error contained in the data, the larger the estimated variance factor, and therefore the smaller the value of the test statistic τ_i. In this case, there is always a good chance for some gross errors to stay undetected by the tau-test, especially when they are of small magnitudes.

As for the localization and elimination of gross errors, due to the possible correlation among the residuals, observations should never be rejected automatically in terms of outliers because an outlying residual does not necessarily mean that its corresponding observation contains a gross error. Therefore observations should only be flagged for possible candidates of rejection. A flagged observation can be deleted only if further analysis shows a good reason for the failure of the test. In addition to a good network geometry that produces dominant redundancy numbers [cf. (5.16)], to successfully pinpoint gross errors based on outliers also requires that an adequate number of observations are statistically consistent, forming a well-conditioned algebraic system (Kavouras, 1982). Obviously the number of gross errors must not exceed the redundancy of the system. Fortunately, in a well-conducted survey project the number of gross errors is always quite limited. In the case of multi-gross errors and strong correlation among the residuals, the localizing of gross errors is very difficult. This is also true for the case that gross errors of large magnitude exist. The best results may be achieved by adopting a systematic iterative testing procedure as discussed below, even though it still may not guarantee the complete success in every case.

Although post-adjustment data screening techniques concentrate on detection and elimination of small gross errors, one should always pay attention to the possibility of the existence of large gross errors in the case that the pre-adjustment data screening techniques failed to detect and eliminate all large gross errors or that the pre-adjustment data screening techniques were not applied at all. As discussed before, the existence of large gross errors in a network adjustment is easy to recognize since because of the nature of least squares minimization a large error tends to be distributed throughout the data set, and therefore outlying residuals (or outliers) may be produced for part or all of the observations related to the erroneous observation, pointing to many other non-existing gross errors, depending on the magnitude of the gross error in question. In this case, usually several least squares network adjustments are required in order to locate and eliminate the gross errors.

Finally, §5.3.3 tells us that, due to the limited sensitivity of the statistical testing procedures, no gross errors of arbitrarily small magnitude can be detected, and therefore even after all tests are passed, the system can only be considered as being free from significant gross errors. Some small gross errors may still remain in the data and affect the solution. Although one cannot detect these errors using standard techniques (e.g., data snooping), measures of the sensitivity of the detection can be given by considering the cost of making a wrong decision; that is, the cost of committing either Type I or Type II error (cf. §3.2.2). This sensitivity may be enhanced by performing an optimal network design before proceeding with the field work as discussed in Part II of this book.

5.5.2 A Proposed Strategy for Outlier Detection and Gross Error Elimination

According to the above discussions a proposed procedure for post-adjustment data screening is outlined here (see Figure 5.3).

First of all, the procedure starts with the pre-adjustment data screening to detect and eliminate gross errors of large magnitude. A least squares network adjustment with minimum datum constraints is then run that gives the estimates of the observational residuals $\hat{\mathbf{v}}$ and the a posterior variance factor $\hat{\sigma}_0^2$. With the estimated a posterior variance factor and the observational residuals, one can then proceed with the outlier detection and gross error localization and elimination.

If the observation precisions are adequately known, following the Baarda's approach for data screening, global test on the estimated variance factor $\hat{\sigma}_0^2$ is performed first. If it passes, we have a reason to believe that the observations have been properly weighted and no model errors exist. However, even in this case, there may still be some gross errors of small magnitudes that stay undetected by the global test. Further data screening should be performed using the data snooping technique. If no outliers are found, the procedure stops. Otherwise, if some outliers are found, further effort should be taken to localize and eliminate gross errors. After that, a minimum constrained network adjustment is run again, and the same procedure is repeated until no outliers are found.

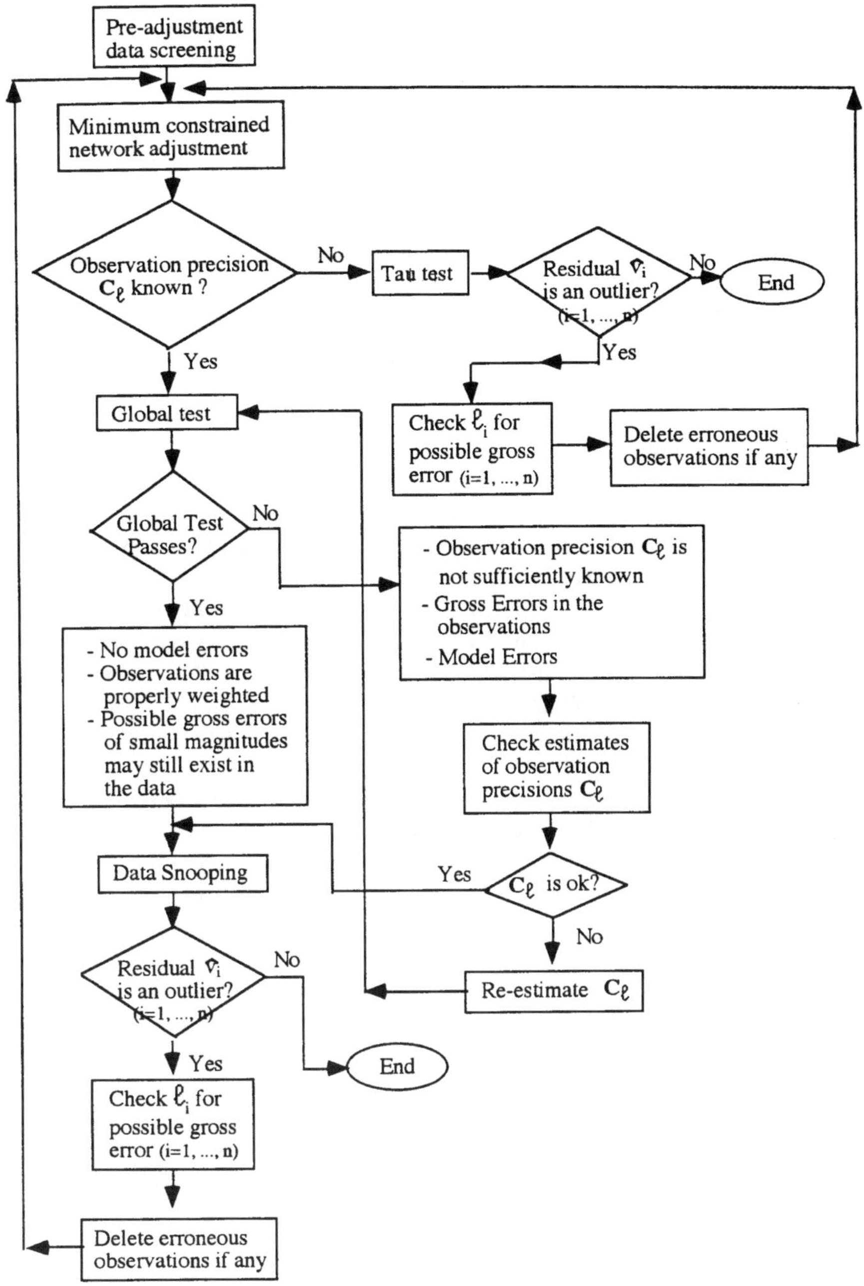

Figure 5.3. A proposed strategy for outlier detection and gross error elimination

If the global test failed, there are two main reasons; that is, either the observation precisions are not properly estimated or gross errors exist in the data. The precisions of observations should be examined first. If it is found that they have been properly estimated, we have a reason to believe that gross errors exist in the observation data, and the data snooping technique is employed to detect outliers and localize and eliminate gross errors. After deleting any erroneous observations, the network adjustment is performed again and the same data screening procedure is repeated starting with the global test until the global test passes and no outliers are discovered. Similarly, if the adopted observation precisions are found not appropriate (either overestimates or underestimates of observation random errors are possible), the observation precisions should be modified and the network adjustment is rerun. The data-screening procedure is repeated starting with the global test until the global test passes and no outliers found. After that the system is considered as being blunderless.

If the observation precisions are not adequately known, the global test is not meaningful and the tau-test should be followed to screen individual residuals for outliers. The localization and elimination of gross errors should follow the same philosophy discussed above.

As discussed before, care must be taken when rejecting observations based on outliers because, due to the possible correlations among the residuals, an observation that shows an outlying residual may not necessarily be the erroneous one. Therefore, very often an observation may be wrongly rejected and then a new network adjustment is performed. In this case, the new adjustment may produce the same outliers in the other observations as before because the gross errors are still there. When this happens, it is necessary to re-introduce a previously withdrawn observation and reject a new one. Therefore, a systematic iterative adjustment and testing procedure is needed in order to produce the best results. Such a procedure is in general very effective as long as the residuals are not highly correlated. In the case of multi-gross errors and highly correlated residuals, it is still difficult to guarantee complete success in every possible case (cf. also Kavouras, 1982), especially when the gross errors are of small magnitude. It should be noted that we should always try not to reject observations whenever possible, not only because of their cost, but also because with every new rejection, the network redundancy r decreases and the method becomes less effective, especially for small survey networks with fewer redundancy.

Finally, it should be noted that since statistical tests are based on limited experimental data they do not have the ability to protect us against wrong reasoning (cf. §3.2). The outcome of statistical tests should, therefore, be interpreted with care, and blind acceptance of the test results is never recommended. Actual decisions should be made in conjunction with the logic and the nature of the specific problem under study.

6 Network Quality Measures

After the observations have been screened and the mathematical model examined, the obtained network results may now be used, or forwarded to the client for use. In addition, the quality of the network results should also be measured and presented to the network user. Obviously, the quality of the network solutions depends on the quality of the input observations. The quality of a geodetic positioning network is in general described by precision, reliability, and economy. Conceptually, network precision describes how the precision of the observations affects the results through network geometry. It is a measure of the network's characteristics in propagating random errors, presuming the absence of gross errors and/or systematic effects in the observation data. On the contrary, network reliability describes how the network reacts to small biases in the observations. It refers to the robustness of the network; that is, its ability to resist undetectable gross errors in the observations. And finally, economy refers to the cost of the survey campaign, which could differ significantly from application to application, depending on the importance and specifications of the project. Since the cost has to be considered before the procedure to be followed has been designed (cf. §9.3), this chapter deals with the various measures for network precision and reliability.

6.1 Network Precision Measures

In general, precision measures of a geodetic network can be in the form of a variance-covariance matrix, a confidence region, or a scalar function of the elements of the variance-covariance matrix. It can be said that the global variance-covariance matrix $\mathbf{C}_x$ of the point coordinates contains complete information about the precision of a geodetic network since any other form of precision measures can be derived exclusively from it. The choice of a specific form of precision measures depends on the purpose that a network must serve. For instance, when defining a geodetic network for setting out an engineering structure, for controlling the breakthrough of tunneling, or for providing photogrammetric control, there may be special requirements, e.g., the precision of certain functions pertaining to specific points or group of points, and the network user wants to see whether the surveyed network has met these requirements, while for multi-purpose networks, such as national control networks, it could be quite difficult to define a proper precision measure.

Depending on the scope it covers, a precision measure can be distinguished as being of either global or local measure. The former refers to the whole network, while the latter some specific point or a group of points. Local measures can indicate weak zones in the network design and are an important tool for the comparison between different designs and methods. The various global and local precision measures for a geodetic network will be discussed in the following sections. They reflect different aspects of the network precision and have been widely used in the surveying community.

6.1.1 Global Precision Measures

6.1.1.1 The Global Variance-Covariance Matrix and Its Eigenvalue Decomposition

From Chapter 4, after a least squares network adjustment, the variance-covariance matrix of the estimated coordinates is expressed in either of the following forms [cf. (4.131) and (4.134)], depending on whether the a priori or the a posterior variance factor is used:

$$\mathbf{C}_x = \sigma_0^2 \mathbf{Q}_x \tag{6.1}$$

or

$$\hat{\mathbf{C}}_x = \hat{\sigma}_0^2 \mathbf{Q}_x \tag{6.2}$$

where $\mathbf{Q}_x$ is the cofactor matrix of the coordinate variates [cf. (4.137)], and $\mathbf{C}_x$ and $\hat{\mathbf{C}}_x$ represent the a priori and the a posterior variance-covariance matrix of the estimated coordinates, respectively.

Assuming that the dimension of the coordinate vector is u, an expanded form of the cofactor matrix $\mathbf{Q}_x$ and the variance-covariance matrix $\mathbf{C}_x$ and $\hat{\mathbf{C}}_x$ is by convention as follows

$$\mathbf{Q}_x = \begin{pmatrix} q_{11} & q_{12} & \cdots & q_{1i} & \cdots & q_{1u} \\ q_{21} & q_{22} & \cdots & q_{2i} & \cdots & q_{2u} \\ \vdots & \vdots & \vdots & \vdots & \vdots & \vdots \\ q_{i1} & q_{i2} & \cdots & q_{ii} & \cdots & q_{iu} \\ \vdots & \vdots & \vdots & \vdots & \vdots & \vdots \\ q_{u1} & q_{u2} & \cdots & q_{ui} & \cdots & q_{uu} \end{pmatrix} \tag{6.3}$$

$$\mathbf{C}_x = \begin{pmatrix} \sigma_1^2 & \sigma_{12} & \cdots & \sigma_{1i} & \cdots & \sigma_{1u} \\ \sigma_{21} & \sigma_2^2 & \cdots & \sigma_{2i} & \cdots & \sigma_{2u} \\ \vdots & \vdots & \vdots & \vdots & \vdots & \vdots \\ \sigma_{i1} & \sigma_{i2} & \cdots & \sigma_i^2 & \cdots & \sigma_{iu} \\ \vdots & \vdots & \vdots & \vdots & \vdots & \vdots \\ \sigma_{u1} & \sigma_{u2} & \cdots & \sigma_{ui} & \cdots & \sigma_u^2 \end{pmatrix} \quad (6.4)$$

$$\hat{\mathbf{C}}_x = \begin{pmatrix} \hat{\sigma}_1^2 & \hat{\sigma}_{12} & \cdots & \hat{\sigma}_{1i} & \cdots & \hat{\sigma}_{1u} \\ \hat{\sigma}_{21} & \hat{\sigma}_2^2 & \cdots & \hat{\sigma}_{2i} & \cdots & \hat{\sigma}_{2u} \\ \vdots & \vdots & \vdots & \vdots & \vdots & \vdots \\ \hat{\sigma}_{i1} & \hat{\sigma}_{i2} & \cdots & \hat{\sigma}_i^2 & \cdots & \hat{\sigma}_{iu} \\ \vdots & \vdots & \vdots & \vdots & \vdots & \vdots \\ \hat{\sigma}_{u1} & \hat{\sigma}_{u2} & \cdots & \hat{\sigma}_{ui} & \cdots & \hat{\sigma}_u^2 \end{pmatrix} \quad (6.5)$$

where

$$\sigma_i^2 = \sigma_0^2 q_{ii} \quad (i = 1, \ldots, u) \quad (6.6)$$

$$\sigma_{ij} = \sigma_0^2 q_{ij} \quad (i, j = 1, \ldots, u; i \neq j) \quad (6.7)$$

$$\hat{\sigma}_i^2 = \hat{\sigma}_0^2 q_{ii} \quad (i = 1, \ldots, u) \quad (6.8)$$

$$\hat{\sigma}_{ij} = \hat{\sigma}_0^2 q_{ij} \quad (i, j = 1, \ldots, u; i \neq j) \quad (6.9)$$

σ_i^2, and σ_{ij} ($j \neq i$) are called the a priori variances and covariances of coordinates, respectively, and $\hat{\sigma}_i^2$, and $\hat{\sigma}_{ij}$ ($j \neq i$) the a posterior variances and covariances of coordinates, respectively.

As discussed above, $\mathbf{C}_x$ or $\hat{\mathbf{C}}_x$ contains the complete information about the precision of the network. However, it should be noted that the coordinates **x** and the variance-covariance matrix $\mathbf{C}_x$ are datum dependent, i.e., they depend on the choice of the network datum. In the case that d datum parameters are needed to provide the network datum, the variance-covariance matrix $\mathbf{C}_x$ will have a rank defect of d; i.e.,

$$\text{rank}(\mathbf{C}_x) = u - d = h < u \quad (6.10)$$

where u is the total number of coordinate components of the network stations, and d the number of datum parameters. The transformation of variance-covariance matrix of coordinates referring to different network datums is accomplished by S-transformation as discussed in §4.3.5. Certainly, there exist some datum independent precision measures that will be discussed below.

Mathematically, since $\mathbf{Q}_x$ is a positive semi-definite matrix, it can be decomposed as follows:

$$\mathbf{Q}_x = \begin{pmatrix} \boldsymbol{V}_1 & \boldsymbol{V}_2 \end{pmatrix} \begin{pmatrix} \Lambda_1 & \mathbf{0} \\ \mathbf{0} & \mathbf{0} \end{pmatrix} \begin{pmatrix} \boldsymbol{V}_1^T \\ \boldsymbol{V}_2^T \end{pmatrix} = \boldsymbol{V}_1 \Lambda_1 \boldsymbol{V}_1^T \tag{6.11}$$

where

$$\Lambda_1 = \text{diag}(\lambda_1, \lambda_2, \ldots, \lambda_h) \tag{6.12}$$

with

$$\lambda_1 \geq \lambda_2 \geq \ldots \geq \lambda_h > 0 \tag{6.13}$$

$$\boldsymbol{V}_1 = \begin{pmatrix} \boldsymbol{\nu}_1 & \boldsymbol{\nu}_2 & \cdots & \boldsymbol{\nu}_h \end{pmatrix} \tag{6.14}$$

$$\boldsymbol{V}_2 = \begin{pmatrix} \boldsymbol{\nu}_{h+1} & \boldsymbol{\nu}_{h+2} & \cdots & \boldsymbol{\nu}_u \end{pmatrix} \tag{6.15}$$

λ_i (i=1, ..., h) represent the h (i.e., u-d) non-zero eigenvalues, and $\boldsymbol{V}_1$ and $\boldsymbol{V}_2$ are the submatrices of the normalized eigenvectors relating to the non-zero eigenvalues and zero eigenvalues, respectively.

An eigenvalue and its corresponding eigenvector are related through the following matrix equation

$$(\mathbf{Q}_x - \lambda_i \mathbf{I}) \boldsymbol{\nu}_i = \mathbf{0} \; (i = 1, \ldots, u) \tag{6.16}$$

where $\mathbf{I}$ is a u by u unit matrix. Expanding (6.16), we have

$$\begin{pmatrix} q_{11} - \lambda_i & q_{12} & \cdots & q_{1u} \\ q_{21} & q_{22} - \lambda_i & \cdots & q_{2u} \\ \vdots & \vdots & \vdots & \vdots \\ q_{u1} & q_{u2} & \cdots & q_{uu} - \lambda_i \end{pmatrix} \boldsymbol{\nu}_i = \mathbf{0} \quad (i = 1, \ldots, u) \tag{6.17}$$

where λ_i and $\boldsymbol{\nu}_i$ are the i-th eigenvalue and its associated eigenvector of $\mathbf{Q}_x$, respectively. Solution of the eigenvalue is made by equating the determinant of the coefficient matrix of $\boldsymbol{\nu}_i$ to zero, i.e.,

$$\det\begin{pmatrix} q_{11}-\lambda_i & q_{12} & \cdots & q_{1u} \\ q_{21} & q_{22}-\lambda_i & \cdots & q_{2u} \\ \vdots & \vdots & \vdots & \vdots \\ q_{u1} & q_{u2} & \cdots & q_{uu}-\lambda_i \end{pmatrix} = 0 \tag{6.18}$$

This leads to a polynomial of u-th order that is expressed as follows

$$f(\lambda) = \lambda^u + a_1\,\lambda^{u-1} + a_2\,\lambda^{u-2} + \cdots + a_{u-1}\,\lambda + a_u = 0 \tag{6.19}$$

where the index i is omitted since the equation holds for all λ_i (i=1, ..., u). $f(\lambda)$ is also called the characteristic polynomial of matrix $\mathbf{Q}_x$. After we obtained all the eigenvalues λ_i (i=1, ..., u) from Equation (6.19), they are substituted back into Equation (6.17) to solve for their corresponding eigenvectors $\boldsymbol{\nu}_i$ (i=1, ..., u). There exist well-designed algorithms and computer programs for the solution of eigenvalues and eigenvectors of any type of matrices (e.g., Hanson et al., 1990). For lower order polynomial, e.g., $u \leq 3$, they may be solved manually.

The eigenvectors can be normalized as follows

$$\tilde{\boldsymbol{\nu}}_i = \frac{\boldsymbol{\nu}_i}{\left\|\boldsymbol{\nu}_i\right\|} \tag{6.20}$$

where $\|\cdot\|$ represents the norm of a vector. When the L2 norm is used (cf. §10.5.1), we have:

$$\tilde{\boldsymbol{\nu}}_i^T \tilde{\boldsymbol{\nu}}_i = 1 \tag{6.21}$$

Although the eigenvalue decomposition of a cofactor matrix is purely a mathematical procedure, it possesses some geometrical meaning regarding the confidence region discussed below.

Finally, since the variance-covariance matrix ($\mathbf{C}_x$ or $\hat{\mathbf{C}}_x$) and the cofactor matrix ($\mathbf{Q}_x$) differ by only a scale, i.e., the a priori or the a posterior variance factor, they share the same eigenvalue decomposition as (6.11) with the same eigenvectors but with eigenvalues differing by the same scale.

6.1.1.2 The Global Confidence Region: The Confidence Hyper-Ellipsoid

A confidence region for the estimated coordinates represents the amount of trust one can place on the estimated positions, and it is established to test the compatibility of the estimated coordinates $\hat{\mathbf{x}}$ with their corresponding true values $\mathbf{x}$ as described by the following null hypothesis

$$H_0 : E\{\hat{\mathbf{x}}\} = \mathbf{x} \tag{6.22}$$

The statistic used for testing H_0 is constructed as follows

$$z = (\hat{\mathbf{x}} - \mathbf{x})^T \frac{\mathbf{Q}_x^-}{\sigma_0^2} (\hat{\mathbf{x}} - \mathbf{x}) \tag{6.23}$$

where σ_0^2 is the a priori variance factor, and $\mathbf{Q}_x^-$ a generalized inverse of the cofactor matrix $\mathbf{Q}_x$ of the estimated coordinates.

Under the null hypothesis H_0, the statistic z follows the chi-square distribution $\chi^2(h)$ with degrees of freedom being equal to the rank of matrix $\mathbf{Q}_x$ denoted by h, i.e.,

$$z|H_0 \in \chi^2(h) \tag{6.24}$$

Since the expected value (i.e., the true value) $\mathbf{x}$ of $\hat{\mathbf{x}}$ is in general unknown, given the significance level α, a tested value $\mathbf{x}$ is considered compatible with the estimated value $\hat{\mathbf{x}}$ at a probability of $(1-\alpha)$ if the following inequality holds

$$(\hat{\mathbf{x}} - \mathbf{x})^T \frac{\mathbf{Q}_x^-}{\sigma_0^2} (\hat{\mathbf{x}} - \mathbf{x}) \le \chi_{1-\alpha}^2(h) \tag{6.25}$$

where $\chi_{1-\alpha}^2(h)$ is the critical value.

If the a priori variance factor σ_0^2 is unknown, the a posterior variance factor $\hat{\sigma}_0^2$ will be used, and a new statistic z' is assembled as follows

$$z' = (\hat{\mathbf{x}} - \mathbf{x})^T \frac{\mathbf{Q}_x^-}{\hat{\sigma}_0^2 \cdot h} (\hat{\mathbf{x}} - \mathbf{x}) \tag{6.26}$$

which under H_0 follows the Fisher distribution F(h, r), i.e.,

$$z'|H_0 \in F(h, r) \tag{6.27}$$

Similarly, $\mathbf{x}$ is then considered to be compatible with $\hat{\mathbf{x}}$ at a probability of $(1-\alpha)$ if

$$(\hat{\mathbf{x}} - \mathbf{x})^T \frac{\mathbf{Q}_x^-}{\hat{\sigma}_0^2 \cdot h} (\hat{\mathbf{x}} - \mathbf{x}) \le F_{1-\alpha}(h, r) \tag{6.28}$$

where $F_{1-\alpha}(h, r)$ is the critical value.

Geometrically, either Equation (6.25) or (6.28) represents an h-dimensional confidence region or an h-dimensional hyper-ellipsoid centered at $\hat{\mathbf{x}}$, and any tested value $\mathbf{x}$ that falls within the hyper-ellipsoid is considered compatible with $\hat{\mathbf{x}}$ at the probability of $(1-\alpha)$.

From Equation (6.11), a generalized inverse of $\mathbf{Q}_x$ can be expressed by

$$\mathbf{Q}_{\mathrm{x}}^{-} = \begin{pmatrix} \boldsymbol{V}_1 & \boldsymbol{V}_2 \end{pmatrix} \begin{pmatrix} \Lambda_1^{-1} & \mathbf{0} \\ \mathbf{0} & \mathbf{0} \end{pmatrix} \begin{pmatrix} \boldsymbol{V}_1^{\mathrm{T}} \\ \boldsymbol{V}_2^{\mathrm{T}} \end{pmatrix} = \boldsymbol{V}_1 \Lambda_1^{-1} \boldsymbol{V}_1^{\mathrm{T}} \tag{6.29}$$

where

$$\Lambda_1^{-1} = \mathrm{diag}\begin{pmatrix} \lambda_1^{-1} & \lambda_2^{-1} & \cdots & \lambda_{\mathrm{h}}^{-1} \end{pmatrix} \tag{6.30}$$

Substituting Equation (6.29) into (6.25) and (6.28), respectively, and considering the following orthonormal coordinate transformation

$$\boldsymbol{u} = \boldsymbol{V}_1^{\mathrm{T}} (\hat{\mathbf{x}} - \mathbf{x}) \tag{6.31}$$

one obtains

$$\frac{1}{\sigma_0^2} \boldsymbol{u}^{\mathrm{T}} \Lambda_1 \boldsymbol{u} \le \chi_{1-\alpha}^2(\mathrm{h}) \tag{6.32}$$

and

$$\frac{1}{\hat{\sigma}_0^2 \cdot \mathrm{h}} \boldsymbol{u}^{\mathrm{T}} \Lambda_1 \boldsymbol{u} \le \mathrm{F}_{1-\alpha}(\mathrm{h}, \mathrm{r}) \tag{6.33}$$

Expanding the left-hand side of Equations (6.32) and (6.33), one obtains the following standard forms of the confidence hyper-ellipsoid

$$\sum_{\mathrm{i}=1}^{\mathrm{h}} \frac{u_{\mathrm{i}}^2}{\left[\sigma_0 \sqrt{\lambda_{\mathrm{i}} \cdot \chi_{1-\alpha}^2(\mathrm{h})}\right]^2} \le 1 \tag{6.34}$$

and

$$\sum_{\mathrm{i}=1}^{\mathrm{h}} \frac{u_{\mathrm{i}}^2}{\left[\hat{\sigma}_0 \sqrt{\lambda_{\mathrm{i}} \cdot \mathrm{h} \cdot \mathrm{F}_{1-\alpha}(\mathrm{h}, \mathrm{r})}\right]^2} \le 1 \tag{6.35}$$

From Equations (6.34) and (6.35), one can see that the lengths of the semi-axes of the hyper-ellipsoid are directly proportional to the square roots of the eigenvalues of $\mathbf{Q}_{\mathrm{x}}$, and the orientations of the semi-axes are defined by the eigenvectors of $\mathbf{Q}_{\mathrm{x}}$, respectively. When the a priori variance factor σ_0^2 is known, the lengths of the semi-axes a_{i} (i=1, ..., h) of the h-dimensional hyper-ellipsoid are calculated by

$$\mathrm{a}_{\mathrm{i}} = \sigma_0 \sqrt{\chi_{1-\alpha}^2(\mathrm{h})} \sqrt{\lambda_{\mathrm{i}}} \quad (\mathrm{i} = 1, \ldots, \mathrm{h}) \tag{6.36}$$

Otherwise, if the a posterior variance factor is used, they are

$$a_i = \hat{\sigma}_0 \sqrt{h \cdot F_{1-\alpha}(h, r)} \sqrt{\lambda_i} \quad (i = 1, \ldots, h) \tag{6.37}$$

Because of the relationship between the confidence region and the eigenvalues and the eigenvectors of $\mathbf{Q}_x$, the eigenvalue decomposition of $\mathbf{Q}_x$ plays a very important role in the geometrical interpretation of the network precision. Obviously, the smaller the eigenvalues of $\mathbf{Q}_x$, the smaller the confidence region, and therefore the better the network precision. As an eigenvalue approaches infinity, so is the respective axis of the confidence ellipsoid. It is this relationship that makes us choose the eigenvalues of $\mathbf{C}_x$ as small as possible through the appropriate design of the network in order to make as precise a statement about the zero hypothesis as possible.

Some other properties of special interest for the accuracy analysis about the eigenvalues of the variance-covariance matrix are given below (cf. also Caspary, 1987):

$$\sum_{i=1}^{u} \lambda_i = \operatorname{tr}(\mathbf{C}_x) \tag{6.38}$$

$$\prod_{i=1}^{u} \lambda_i = \det(\mathbf{C}_x) \tag{6.39}$$

$$\lambda_{min} \le \frac{\mathbf{f}^T \mathbf{C}_x \mathbf{f}}{\mathbf{f}^T \mathbf{f}} \le \lambda_{max} \tag{6.40}$$

Equations (6.38) and (6.39) mean that the sum and the product of all eigenvalues of a variance-covariance matrix equal the trace and determinant of the variance-covariance matrix, respectively. Equation (6.40) is the so-called Rayleigh relation that can be used to estimate the extreme values of the accuracy of certain functions of the parameter vector. Geometrically, the determinant of the variance-covariance matrix is proportional to the volume of the confidence hyper-ellipsoid. It should be noted that for a singular variance-covariance matrix $\mathbf{C}_x$, its determinant will be zero. In this case, the determinant is usually replaced by the product of the non-zero eigenvalues.

6.1.1.3 Scalar Precision Functions

When the network size becomes large, it is not practical to assess the precision of a network by inspecting or comparing each individual element of the global variance-covariance matrix of the estimated coordinates as the number of elements of the u by u matrix $\mathbf{C}_x$ is large. In this case, it may be necessary to compress the mass of information into one or a few representative measures that serve as an overall representation of the network precision and can be assessed easily. These representative measures are usually in the form of scalar functions of the elements of the full variance-covariance matrix, and they serve as an overall representation of the network precision. Listed below are some typical scalar precision functions (cf. Grafarend, 1974):

- Norm

$$f = \left\| \mathbf{C}_x \right\| \tag{6.41}$$

  with $\| \cdot \|$ denoting the norm of a matrix
- Trace

$$f = \text{Trace}\left(\mathbf{C}_x\right) = \lambda_1 + \lambda_2 + \ldots + \lambda_h \tag{6.42}$$

  with λ_1, λ_2, ..., λ_h being the non-zero eigenvalues of the matrix $\mathbf{C}_x$
- The maximum eigenvalue

$$f = \lambda_{max} \tag{6.43}$$

  with λ_{max} being the maximum eigenvalue of matrix $\mathbf{C}_x$
- Determinant

$$f = \text{Det}\left(\mathbf{C}_x\right) = \lambda_1 * \lambda_2 * \ldots * \lambda_h \tag{6.44}$$

  with λ_1, λ_2, ..., λ_h being the non-zero eigenvalues of matrix $\mathbf{C}_x$.

These are some typical examples of scalar precision functions. It should be noted that any scalar precision function may lead to loss of information. Therefore, the final precision measures must conform to the purpose of the network and especially the requirements of the user to ensure that the desired information has been presented.

6.1.2 Local Precision Measures

Any precision measures that do not cover the whole network are referred to here as local precision measures; for instance, the precision of a coordinate component, a point, or a group of points. A variety of measures have been in use that reflect the different aspects of the network precision. Nevertheless, they all can be derived from the previously introduced global variance-covariance matrix $\mathbf{C}_x$.

6.1.2.1 Station Variance-Covariance Matrix

Assume that a network consists of a total of m stations and the coordinate vector **x** is arranged sequentially in an ascending order of the point numbers as follows

$$\mathbf{x} = \begin{pmatrix} \mathbf{x}_1 & \mathbf{x}_2 & \cdots & \mathbf{x}_i & \cdots & \mathbf{x}_m \end{pmatrix}^T \tag{6.45}$$

where $\mathbf{x}_i$ (i=1, ..., m) represents a sub-vector of the coordinates of point i. In two- and three-dimensional space, $\mathbf{x}_i$ can be denoted respectively as follows

$$\mathbf{x}_i = \begin{pmatrix} x_i \\ y_i \end{pmatrix} \quad (i = 1, \ldots, m) \tag{6.46}$$

and

$$\mathbf{x}_i = \begin{pmatrix} x_i \\ y_i \\ z_i \end{pmatrix} \quad (i = 1, \ldots, m) \tag{6.47}$$

The global cofactor matrix $\mathbf{Q}_x$ of the coordinates is partitioned into block sub-matrices representing the cofactor matrix for each station and the cofactor matrix between pairs of stations as follows

$$\mathbf{Q}_x = \begin{pmatrix} \mathbf{Q}_{11} & \mathbf{Q}_{12} & \cdots & \mathbf{Q}_{1i} & \cdots & \mathbf{Q}_{1m} \\ \mathbf{Q}_{21} & \mathbf{Q}_{22} & \cdots & \mathbf{Q}_{2i} & \cdots & \mathbf{Q}_{2m} \\ \vdots & \vdots & \vdots & \vdots & \vdots & \vdots \\ \mathbf{Q}_{i1} & \mathbf{Q}_{i2} & \cdots & \mathbf{Q}_{ii} & \cdots & \mathbf{Q}_{im} \\ \vdots & \vdots & \vdots & \vdots & \vdots & \vdots \\ \mathbf{Q}_{m1} & \mathbf{Q}_{m2} & \cdots & \mathbf{Q}_{mi} & \cdots & \mathbf{Q}_{mm} \end{pmatrix} \tag{6.48}$$

where $\mathbf{Q}_{ii}$ (i=1, ..., m) represents the cofactor matrix for the estimated coordinates of stations i, and $\mathbf{Q}_{ij}$ (i, j=1, ..., m; i≠j) the cofactor matrix for the station pair i and j.

In two-dimensional space, $\mathbf{Q}_{ii}$ and $\mathbf{Q}_{ij}$ are 2 by 2 matrix as follows

$$\mathbf{Q}_{ii} = \begin{pmatrix} q_{x_i x_i} & q_{x_i y_i} \\ q_{y_i x_i} & q_{y_i y_i} \end{pmatrix} \quad (i = 1, \ldots, m) \tag{6.49}$$

$$\mathbf{Q}_{ij} = \begin{pmatrix} q_{x_i x_j} & q_{x_i y_j} \\ q_{y_i x_j} & q_{y_i y_j} \end{pmatrix} \quad (i, j = 1, \ldots, m; i \neq j) \tag{6.50}$$

Similarly, in three-dimensional space, $\mathbf{Q}_{ii}$ and $\mathbf{Q}_{ij}$ are 3 by 3 matrix as follows

$$\mathbf{Q}_{ii} = \begin{pmatrix} q_{x_i x_i} & q_{x_i y_i} & q_{x_i z_i} \\ q_{y_i x_i} & q_{y_i y_i} & q_{y_i z_i} \\ q_{z_i x_i} & q_{z_i y_i} & q_{z_i z_i} \end{pmatrix} \quad (i = 1, \quad ..., m) \tag{6.51}$$

$$\mathbf{Q}_{ij} = \begin{pmatrix} q_{x_i x_j} & q_{x_i y_j} & q_{x_i z_j} \\ q_{y_i x_j} & q_{y_i y_j} & q_{y_i z_j} \\ q_{z_i x_j} & q_{z_i y_j} & q_{z_i z_j} \end{pmatrix} \quad (i, j = 1, ..., m; i \neq j) \tag{6.52}$$

It can be shown that $\mathbf{Q}_{ii}$ is a symmetrical positive semi-definite matrix, whereas $\mathbf{Q}_{ij}$ is not. That is, in Equations (6.49) and (6.51), we have

$$\sigma_{y_i x_i} = \sigma_{x_i y_i} \tag{6.53}$$

$$\sigma_{y_i z_i} = \sigma_{z_i y_i} \tag{6.54}$$

$$\sigma_{x_i z_i} = \sigma_{z_i x_i} \tag{6.55}$$

The a priori and a posterior variance-covariance matrix $\mathbf{C}_{ii}$ and $\hat{\mathbf{C}}_{ii}$ for the estimated coordinates of station i is respectively as follows

$$\mathbf{C}_{ii} = \sigma_0^2 \mathbf{Q}_{ii} \quad (i = 1, ..., m) \tag{6.56}$$

$$\hat{\mathbf{C}}_{ii} = \hat{\sigma}_0^2 \mathbf{Q}_{ii} \quad (i = 1, ..., m) \tag{6.57}$$

Considering Equations (6.49) and (6.53), in two-dimensional space, $\mathbf{C}_{ii}$ and $\hat{\mathbf{C}}_{ii}$ can be expanded as follows:

$$\mathbf{C}_{ii} = \begin{pmatrix} \sigma_{x_i}^2 & \sigma_{x_i y_i} \\ \sigma_{x_i y_i} & \sigma_{y_i}^2 \end{pmatrix} \quad (i = 1, ..., m) \tag{6.58}$$

$$\hat{\mathbf{C}}_{ii} = \begin{pmatrix} \hat{\sigma}_{x_i}^2 & \hat{\sigma}_{x_i y_i} \\ \hat{\sigma}_{x_i y_i} & \hat{\sigma}_{y_i}^2 \end{pmatrix} \quad (i = 1, \ ..., \ m) \tag{6.59}$$

where

$$\sigma_{x_i}^2 = \sigma_0^2 \, q_{x_i x_i} \tag{6.60}$$

$$\sigma_{y_i}^2 = \sigma_0^2 \, q_{y_i y_i} \tag{6.61}$$

$$\sigma_{x_i y_i} = \sigma_0^2 \, q_{x_i y_i} \tag{6.62}$$

$$\hat{\sigma}_{x_i}^2 = \hat{\sigma}_0^2 \, q_{x_i x_i} \tag{6.63}$$

$$\hat{\sigma}_{y_i}^2 = \hat{\sigma}_0^2 \, q_{y_i y_i} \tag{6.64}$$

$$\hat{\sigma}_{x_i y_i} = \hat{\sigma}_0^2 \, q_{x_i y_i} \tag{6.65}$$

Similar expressions can be obtained for the three-dimensional case.

Finally, the a priori and a posterior covariance matrix referring to a pair of points i and j are as follows

$$\mathbf{C}_{ij} = \sigma_0^2 \mathbf{Q}_{ij} \quad (i, j = 1, \ldots, m; i \neq j) \tag{6.66}$$

$$\hat{\mathbf{C}}_{ij} = \hat{\sigma}_0^2 \mathbf{Q}_{ij} \quad (i, j = 1, \ldots, m; i \neq j) \tag{6.67}$$

The variance-covariance matrix $\mathbf{C}_{ii}$ or $\hat{\mathbf{C}}_{ii}$ (i=1, ..., m) contains full information about the precision of the estimated coordinates for station i. The smaller the variance of an estimated coordinate component, the better its precision, and the smaller the covariance between the estimated coordinate components, the weaker the correlation between them. The covariance matrix $\mathbf{C}_{ij}$ or $\hat{\mathbf{C}}_{ij}$ (i, j=1, ..., m; i≠j) represents only the correlation between the estimated coordinates of stations i and j, and apparently the smaller the covariance values, the weaker the correlation between stations i and j.

6.1.2.2 Standard Deviations and Confidence Intervals

In addition to the full variance-covariance matrix of the estimated coordinate components for each station, standard deviations and confidence intervals are some simplified forms of local precision measures for the estimated coordinate components. According to the above Equations (6.59), (6.60), (6.63) and (6.64), standard deviations of coordinates are defined as the square root of their corresponding variance as follows

$$\sigma_{x_i} = \sigma_0 \sqrt{q_{x_i x_i}} \tag{6.68}$$

$$\sigma_{y_i} = \sigma_0 \sqrt{q_{y_i y_i}} \tag{6.69}$$

$$\hat{\sigma}_{x_i} = \hat{\sigma}_0 \sqrt{q_{x_i x_i}} \tag{6.70}$$

$$\hat{\sigma}_{y_i} = \hat{\sigma}_0 \sqrt{q_{y_i y_i}} \tag{6.71}$$

Similar to the global confidence region as discussed in §6.1.1.2, confidence intervals of estimated coordinate components deal with assessing the compatibility of an estimated value of a coordinate component with a tested value. This is done by testing the following null hypothesis

$$H_0 \; E\{\hat{x}_i\} = \mu_{x_i} \tag{6.72}$$

against an alternative hypothesis

$$H_a \; E\{\hat{x}_i\} \neq \mu_{x_i} \tag{6.73}$$

using the following statistic x or $\tilde{x}$, depending on whether the a priori variance factor σ_0^2 is known or not, i.e.,

$$x = \frac{\left(\hat{x}_i - \mu_{x_i}\right)}{\sigma_0 \sqrt{q_{x_i x_i}}} \tag{6.74}$$

or

$$\tilde{x} = \frac{\left(\hat{x}_i - \mu_{x_i}\right)}{\hat{\sigma}_0 \sqrt{q_{x_i x_i}}} \tag{6.75}$$

It can be shown that, under the null hypothesis H_0, x and $\tilde{x}$ follow the standard normal distribution n(0, 1) and the student's t-distribution t(r) with r degrees of freedom, respectively. That is,

$$x | H_0 \in n(0,1) \tag{6.76}$$

$$\tilde{x} | H_0 \in t(r) \tag{6.77}$$

Therefore, given the significance level α, a tested value μ_{x_i} will be considered compatible with the estimated value $\hat{x}_i$ at a probability of $(1-\alpha)$ if

$$\frac{\left|\hat{x}_i - \mu_{x_i}\right|}{\sigma_0 \sqrt{q_{x_i x_i}}} \leq n_{\alpha/2}(0,1) \tag{6.78}$$

or

$$\frac{\left|\hat{x}_i - \mu_{x_i}\right|}{\hat{\sigma}_0 \sqrt{q_{x_i x_i}}} \leq t_{\alpha/2}(r) \qquad (6.79)$$

The above Equations (6.78) and (6.79) can be reformulated as follows

$$\hat{x}_i - n_{\alpha/2}(0,1)\sigma_0 \sqrt{q_{x_i x_i}} \leq \mu_{x_i} \leq \hat{x}_i + n_{\alpha/2}(0,1)\sigma_0 \sqrt{q_{x_i x_i}} \qquad (6.80)$$

$$\hat{x}_i - t_{\alpha/2}(r)\hat{\sigma}_0 \sqrt{q_{x_i x_i}} \leq \mu_{x_i} \leq \hat{x}_i + t_{\alpha/2}(r)\hat{\sigma}_0 \sqrt{q_{x_i x_i}} \qquad (6.81)$$

Here, the intervals

$$\left[\hat{x}_i - n_{\alpha/2}(0,1)\sigma_0 \sqrt{q_{x_i x_i}}, \hat{x}_i + n_{\alpha/2}(0,1)\sigma_0 \sqrt{q_{x_i x_i}}\right] \qquad (6.82)$$

and

$$\left[\hat{x}_i - t_{\alpha/2}(r)\hat{\sigma}_0 \sqrt{q_{x_i x_i}}, \hat{x}_i + t_{\alpha/2}(0,1)\hat{\sigma}_0 \sqrt{q_{x_i x_i}}\right] \qquad (6.83)$$

are called the (1–α) confidence intervals of the estimated coordinate component $\hat{x}_i$. Similarly, the confidence intervals for the estimated coordinate component $\hat{y}_i$ and $\hat{z}_i$ (in 3D case) can be derived as follows

$$\left[\hat{y}_i - n_{\alpha/2}(0,1)\sigma_0 \sqrt{q_{y_i y_i}}, \hat{y}_i + n_{\alpha/2}(0,1)\sigma_0 \sqrt{q_{y_i y_i}}\right] \qquad (6.84)$$

or

$$\left[\hat{y}_i - t_{\alpha/2}(r)\hat{\sigma}_0 \sqrt{q_{y_i y_i}}, \hat{y}_i + t_{\alpha/2}(r)\hat{\sigma}_0 \sqrt{q_{y_i y_i}}\right] \qquad (6.85)$$

and

$$\left[\hat{z}_i - n_{\alpha/2}(0,1)\sigma_0 \sqrt{q_{z_i z_i}}, \hat{z}_i + n_{\alpha/2}(0,1)\sigma_0 \sqrt{q_{z_i z_i}}\right] \qquad (6.86)$$

or

$$\left[\hat{z}_i - t_{\alpha/2}(r)\hat{\sigma}_0 \sqrt{q_{z_i z_i}}, \hat{z}_i + t_{\alpha/2}(r)\sigma_0 \sqrt{q_{z_i z_i}}\right] \qquad (6.87)$$

If a tested value μ_x (or μ_y, μ_z) falls within its corresponding confidence interval, it is considered compatible with the estimated value at a probability of (1–α).

6.1.2.3 Error Curve and Position Accuracy

In two-dimensional space, the standard deviations σ_x and σ_y discussed above represent only the precision of a positioned network point along the x and y axis of the coordinate system, respectively. Therefore, they do not give full information on the precision of a network position; for instance, we do not know the precision of the point in any direction other than the coordinate axes. The ***error curve***, also called ***pedal curve***, is then used for this purpose. It is defined by the locus of standard deviations in all directions.

From Figure 6.1, we can see that if the (x, y) coordinates are assumed to have an error (ε_x, ε_y) along the x and y axes, respectively, their projection ε_α on to a direction of azimuth α can be written as follows

$$\varepsilon_\alpha = \varepsilon_x \sin\alpha + \varepsilon_y \cos\alpha \tag{6.88}$$

According to the law of variance-covariance propagation, the standard deviation along the direction of azimuth α can be derived as follows

$$\sigma_\alpha = \sqrt{\sigma_x^2 \sin^2\alpha + \sigma_y^2 \cos^2\alpha + 2\sigma_{xy} \sin\alpha \cos\alpha} \tag{6.89}$$

By tracing the locus of the standard deviations in all directions ($\alpha = 0° \sim 360°$) an error curve can be produced as shown in Figure 6.2.

The value of α at which σ_α reaches its maximum can be obtained by setting the partial derivative of σ_α^2 with respect to α to zero; i.e.,

$$\frac{\partial\sigma_\alpha^2}{\partial\alpha} = 2\sin\alpha\cos\alpha\left(\sigma_x^2 - \sigma_y^2\right) + 2\sigma_{xy}\left(\cos^2\alpha - \sin^2\alpha\right) = 0 \tag{6.90}$$

Solving Equation (6.90), we obtain

$$\tan 2\alpha = \frac{2\sigma_{xy}}{\sigma_x^2 - \sigma_y^2} \tag{6.91}$$

where α is the azimuth of the direction of σ_{max}. The direction of σ_{min} is $\alpha + 90°$. The quadrant of 2α is determined by the algebraic signs of the numerator and the denominator.

The maximum and minimum standard deviations are respectively as follows

$$\sigma_{max}^2 = \frac{1}{2}\left[\sigma_x^2 + \sigma_y^2 + \sqrt{\left(\sigma_x^2 - \sigma_y^2\right)^2 + 4\sigma_{xy}^2}\right] \tag{6.92}$$

$$\sigma_{min}^2 = \frac{1}{2}\left[\sigma_x^2 + \sigma_y^2 - \sqrt{\left(\sigma_x^2 - \sigma_y^2\right)^2 + 4\sigma_{xy}^2}\right] \tag{6.93}$$

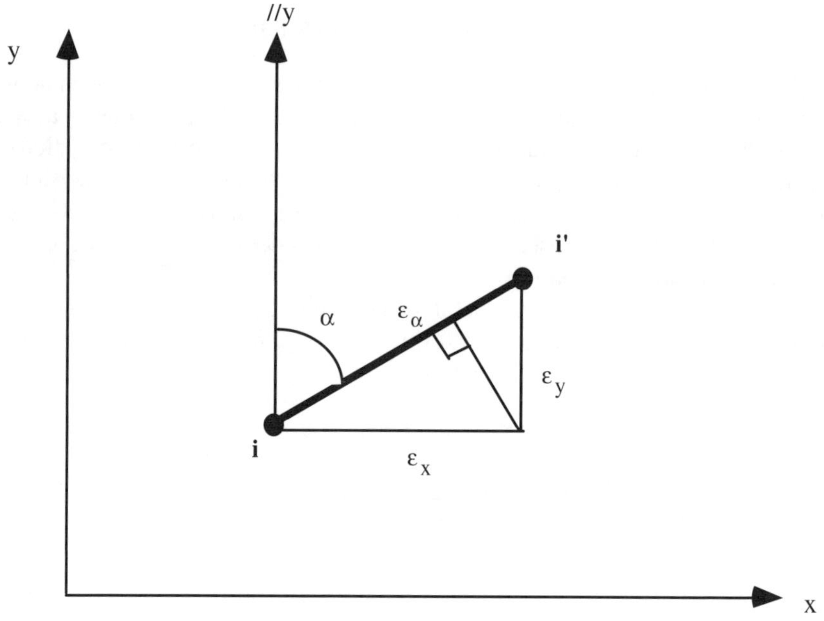

Figure 6.1. Projection of error components

Finally, the accuracy of an estimated position in two-dimensional space is defined by

$$\sigma_p = \sqrt{\sigma_x^2 + \sigma_y^2} \tag{6.94}$$

From Equations (6.92) and (6.93), we can see that

$$\sigma_p = \sqrt{\sigma_x^2 + \sigma_y^2} = \sqrt{\sigma_{max}^2 + \sigma_{min}^2} \tag{6.95}$$

It is, therefore, clear that σ_p represents the maximum expected standard deviation of the position of a point.

Similarly, in the three-dimensional space, the error curve becomes an error surface, and the position accuracy is calculated by

$$\sigma_p = \sqrt{\sigma_x^2 + \sigma_y^2 + \sigma_z^2} \tag{6.96}$$

6.1.2.4 Station Error Ellipses and Relative Error Ellipses

Station error ellipses are the special cases of the global confidence region (i.e., the hyper-ellipsoid) established by Equations (6.34) and (6.35) for the whole net-

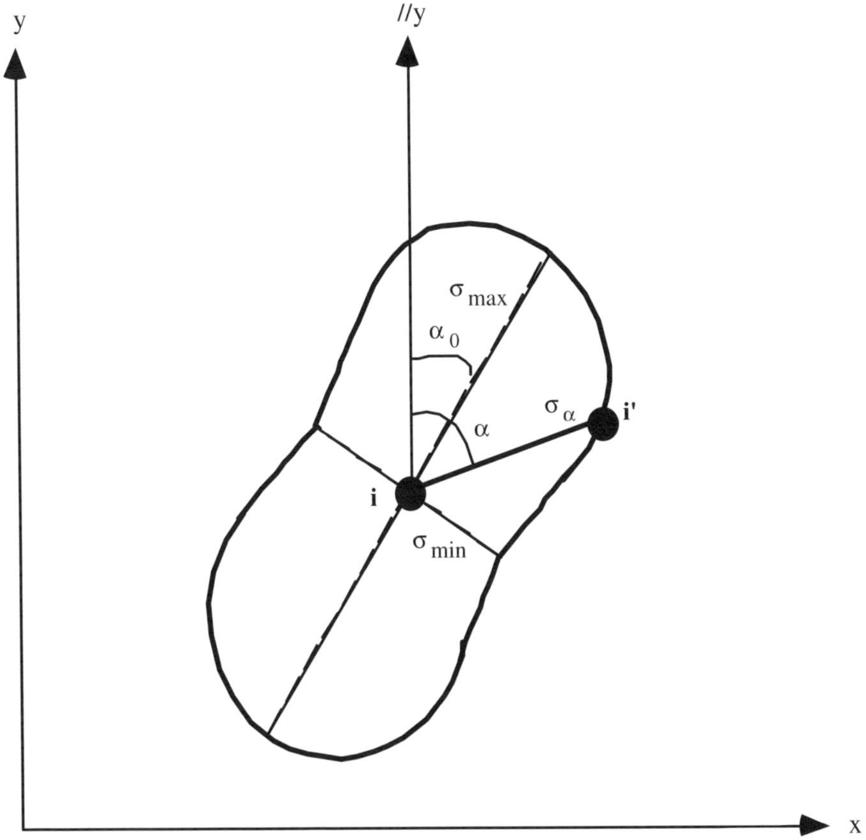

Figure 6.2. An error curve

work. Here, instead of assessing the compatibility of the estimated global coordinate vector $\hat{\mathbf{x}}$ with its true value $\mathbf{x}$, the estimated two coordinate components ($\hat{x}_i$, $\hat{y}_i$) for each station (in 2D case) are examined separately.

Denote

$$\hat{\mathbf{x}}_i = \begin{pmatrix} \hat{x}_i \\ \hat{y}_i \end{pmatrix} \quad (i = 1, \ldots, m) \tag{6.97}$$

$$\mathbf{x}_i = \begin{pmatrix} x_i \\ y_i \end{pmatrix} \quad (i = 1, \ldots, m) \tag{6.98}$$

In the case that the a priori variance factor σ_0^2 is known, testing of the null hypothesis

$$\mathrm{H}_0: E\{\hat{\mathbf{x}}_i\} = \mathbf{x}_i \tag{6.99}$$

against an alternative hypothesis

$$\mathrm{H}_a: E\{\hat{\mathbf{x}}_i\} \neq \mathbf{x}_i \tag{6.100}$$

is made using the following statistic

$$v = \frac{(\hat{\mathbf{x}}_i - \mathbf{x}_i)^T \mathbf{Q}_{ii}^{-} (\hat{\mathbf{x}}_i - \mathbf{x}_i)}{\sigma_0^2} \tag{6.101}$$

In the case that the cofactor matrix $\mathbf{Q}_{ii}$ has full rank, it can be shown that under the null hypothesis H_0, v is distributed as $\chi^2(2)$; that is,

$$v|\mathrm{H}_0 \in \chi^2(2) \tag{6.102}$$

Therefore, given the significance level α, the null hypothesis will be accepted at a probability of $(1-\alpha)$ if the following inequality holds

$$\frac{(\hat{\mathbf{x}}_i - \mathbf{x}_i)^T \mathbf{Q}_{ii}^{-} (\hat{\mathbf{x}}_i - \mathbf{x}_i)}{\sigma_0^2} \leq \chi_{1-\alpha}^2(2) \tag{6.103}$$

If we replace $\mathbf{Q}_{ii}$ by its eigenvalue decomposition, the above equation can be reformulated into the following standard form

$$\frac{u_1^2}{\left(\sigma_0\sqrt{\chi_{1-\alpha}^2(2)\lambda_1}\right)^2} + \frac{u_2^2}{\left(\sigma_0\sqrt{\chi_{1-\alpha}^2(2)\lambda_2}\right)^2} \leq 1 \tag{6.104}$$

where λ_1 and λ_2 are the eigenvalues of the cofactor matrix $\mathbf{Q}_{ii}$, and they can be obtained by solving the characteristic polynomial of the form of Equation (6.19) as follows

$$\lambda_1 = \frac{1}{2}\left[q_{x_ix_i} + q_{y_iy_i} + \sqrt{\left(q_{x_ix_i} - q_{y_iy_i}\right)^2 + 4\,q_{x_iy_i}^2}\right] \tag{6.105}$$

$$\lambda_2 = \frac{1}{2}\left[q_{x_ix_i} + q_{y_iy_i} - \sqrt{\left(q_{x_ix_i} - q_{y_iy_i}\right)^2 + 4\,q_{x_iy_i}^2}\right] \tag{6.106}$$

The orientations of λ_1 and λ_2 are defined by their corresponding eigenvectors that are solved using Equation (6.17), and it can be shown that λ_1 and λ_2 are respectively in the directions of azimuth α_1 and α_2 defined below

$$\tan 2\alpha_1 = \frac{2q_{x_i y_i}}{q_{x_i x_i} - q_{y_i y_i}} \tag{6.107}$$

$$\alpha_2 = \alpha_1 + 90° \tag{6.108}$$

Equation (6.104) shows that the semi-major and semi-minor axes of the confidence ellipse are respectively as follows

$$a = \sigma_0 \sqrt{\chi^2_{1-\alpha}(2)\lambda_1} \tag{6.109}$$

$$b = \sigma_0 \sqrt{\chi^2_{1-\alpha}(2)\lambda_2} \tag{6.110}$$

A standard station error ellipse is defined by setting

$$\chi^2_{1-\alpha}(2) = 1 \tag{6.111}$$

which holds for $1-\alpha = 39.4\%$. That is, the probability for the true point to lie within the standard ellipse is 39.4%. The semi-major and semi-minor axes of the standard error ellipses are therefore

$$a_0 = \sigma_0 \sqrt{\lambda_1} \tag{6.112}$$

$$b_0 = \sigma_0 \sqrt{\lambda_2} \tag{6.113}$$

At 95% confidence level, the semi-axes are expanded by a factor of 2.4484, i.e.,

$$a_0 = 2.4484\, \sigma_0 \sqrt{\lambda_1} \tag{6.114}$$

$$b_0 = 2.4484\, \sigma_0 \sqrt{\lambda_2} \tag{6.115}$$

Similarly, when the a priori variance factor σ_0^2 is not known, statistical assessment of the estimated coordinates for a station is done using the following statistic

$$\tilde{\nu} = \frac{(\hat{\mathbf{x}}_i - \mathbf{x}_i)^T \mathbf{Q}_{ii}^{-} (\hat{\mathbf{x}}_i - \mathbf{x}_i)}{2 \cdot \hat{\sigma}_0^2} \tag{6.116}$$

which, under the null hypothesis H_0, follows the Fisher distribution F(2, r); that is,

$$\tilde{\nu} | H_0 \in F_{1-\alpha}(2, r) \tag{6.117}$$

Given the significance level α, the error ellipse for the estimated station coordinates is defined by

$$\frac{u_1^2}{\left(\hat{\sigma}_0\sqrt{2F_{1-\alpha}(2,r)\lambda_1}\right)^2}+\frac{u_2^2}{\left(\hat{\sigma}_0\sqrt{2F_{1-\alpha}(2,r)\lambda_2}\right)^2}\leq 1 \tag{6.118}$$

where λ_1 and λ_2 are defined by the same Equations (6.105) and (6.106). The semi-major and semi-minor axes of the error ellipse are therefore

$$a=\hat{\sigma}_0\sqrt{2F_{1-\alpha}(2,r)\lambda_1} \tag{6.119}$$

$$b=\hat{\sigma}_0\sqrt{2F_{1-\alpha}(2,r)\lambda_2} \tag{6.120}$$

A standard error ellipse in this case is obtained by setting

$$2F_{1-\alpha}(2,r)=1 \tag{6.121}$$

The probability at which the above Equation (6.121) holds varies with the degrees of freedom r of the network, and it reaches its maximum of 39.4% when $r=\infty$. The semi-major and semi-minor axes of the standard error ellipses are therefore

$$a_0=\hat{\sigma}_0\sqrt{\lambda_1} \tag{6.122}$$

$$b_0=\hat{\sigma}_0\sqrt{\lambda_2} \tag{6.123}$$

By comparing Equations (6.105), (6.106), and (6.107) with Equations (6.92), (6.93), and (6.91), one can see that the semi-axes of both the standard error ellipses and the error curve share the same size and orientation. Figure 6.3 shows their relationship.

The standard error ellipse is not identical to the standard deviation curve. For narrow ellipses there is only a small segment for which the standard deviations are close to the length of the semi-minor axis. The standard deviation in an arbitrary direction α is equal to the projection of the standard ellipse onto that direction, and it is related to the elements of the standard error ellipse as follows:

$$\sigma_\alpha^2=a_0^2\cos^2(\alpha-\alpha_0)+b_0^2\sin^2(\alpha-\alpha_0) \tag{6.124}$$

where α is the azimuth of a direction, and α_0 as well as a_0 and b_0 are the azimuth of the semi-major axis, and the sizes of the semi-major and semi-minor axes of the standard error ellipse, respectively.

When a station error ellipse is used, the precision of the two coordinate components of the station is examined simultaneously. Station error ellipses are also

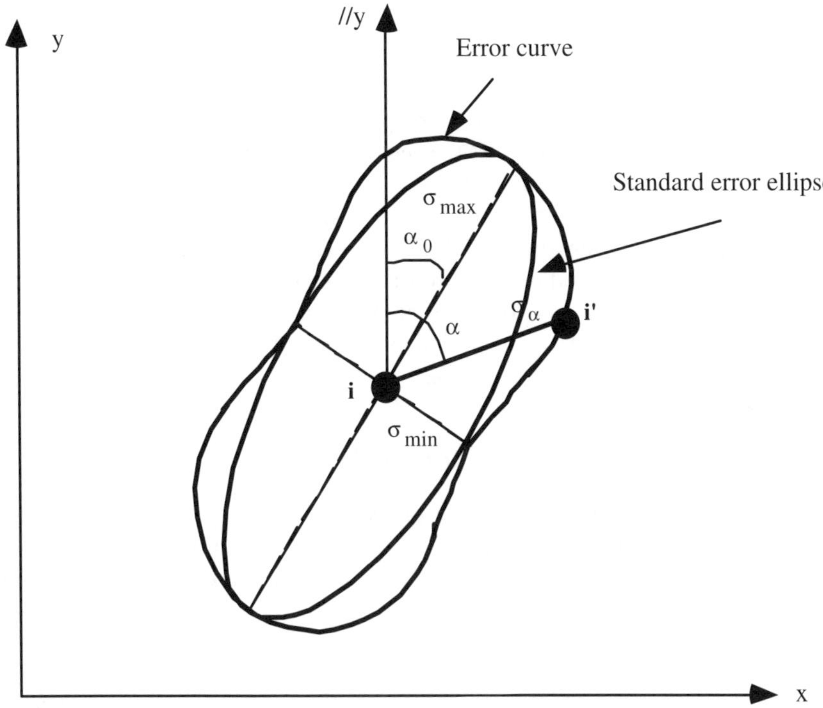

Figure 6.3. Error curve and standard station error ellipse

referred to as absolute error ellipses since they refer to the datum of the network. Relative error ellipses are confidence region established for the coordinate differences between any two network stations other than the network datum points. The elements of the relative error ellipses are computed using the formulae developed for the absolute error ellipses simply by replacing the elements of the cofactor matrix $\mathbf{Q}_{ii}$ for station coordinates $\hat{\mathbf{x}}_i$ by the cofactor matrix $\mathbf{Q}_{\Delta ij}$ that refers to the coordinate differences $(\hat{\mathbf{x}}_j - \hat{\mathbf{x}}_i)$ of the station pair i and j involved.

Denote

$$\Delta\hat{\mathbf{x}}_{ij} = \begin{pmatrix} \hat{x}_j \\ \hat{y}_j \end{pmatrix} - \begin{pmatrix} \hat{x}_i \\ \hat{y}_i \end{pmatrix} = \hat{\mathbf{x}}_j - \hat{\mathbf{x}}_i \quad (i, j = 1, \ldots, m; i \neq j) \tag{6.125}$$

According to the law of variance-covariance propagation, the cofactor matrix $\mathbf{Q}_{\Delta ij}$ of $\Delta\hat{\mathbf{x}}_{ij}$ can be derived as follows

$$\mathbf{Q}_{\Delta ij} = \mathbf{Q}_{jj} - \mathbf{Q}_{ji} - \mathbf{Q}_{ij} + \mathbf{Q}_{ii} \tag{6.126}$$

where $\mathbf{Q}_{jj}$, $\mathbf{Q}_{ji}$, $\mathbf{Q}_{ij}$, and $\mathbf{Q}_{ii}$ have been defined in Equations (6.48) to (6.50).

In network analysis, the station error ellipses and relative station error ellipses are usually superimposed on the network as shown below in Figure 6.4, in which the ellipses plotted using solid and dotted lines represent the station and relative error ellipses, respectively. This type of plot gives the investigator an intuitive view of the error distribution within the network, and therefore serves as a valuable tool for network quality assessment.

Finally, in the three-dimensional space, a station error ellipse and a relative error ellipse become a station error ellipsoid and a relative error ellipsoid, respectively, and their corresponding semi-axes and orientations can be calculated using the general formula (6.36) or (6.37) through the eigenvalue decomposition of the station cofactor matrix $\mathbf{Q}_{ii}$ and the relative station cofactor matrix $\mathbf{Q}_{\Delta ij}$ calculated from Equations (6.51), (6.52) and (6.126).

6.1.2.5 Scalar Precision Functions

For geodetic networks established for special purposes, one may require that a certain function $\mathbf{c}^T\mathbf{x}$ of the coordinates of one or a group of points be positioned with a certain pre-set precision. Let us assume a linear function of the coordinates as follows

$$f = \mathbf{c}^T\mathbf{x} = \sum_{i=1}^{u} c_i x_i \tag{6.127}$$

where $\mathbf{c} = (c_1\ c_2\ \dots\ c_u)^T$ is a vector of constants. The variance of the above function is thus calculated according to the law of error propagation as follows

$$\sigma_f^2 = \sigma_0^2\, \mathbf{c}^T \mathbf{Q}_x \mathbf{c} \tag{6.128}$$

In the case of a nonlinear function, its accuracy can be evaluated by using its corresponding Taylor series expansion of linear form. A practical example for precision measures of the type of Equation (6.128) can be seen in tunneling surveys where what is usually most concerned is the lateral horizontal breakthrough error that can be expressed as the function of the accuracies of the coordinates of the predesigned breakthrough points, and one usually requires that this error be less than a certain boundary value (cf. Kuang, 1993c). There may be many more other scalar precision measures, depending on one's purpose.

Finally, all precision measures presented so far suffer from being datum dependent. That is, the accuracy of an estimated position; for instance, the standard deviations, error ellipses, is only meaningful as an accuracy relative to the network datum; i.e., the zero-variance computational base (cf. §4.3.5). In the case that the conventional minimum constraints are used, the datum defining point is error-free; that is, the standard deviations, error curve, or error ellipses for this point disappear, whereas for all other points these measures increase with the distance from the datum point. This effect is less pronounced with the relative error ellipses, and can be completely avoided with invariant functions

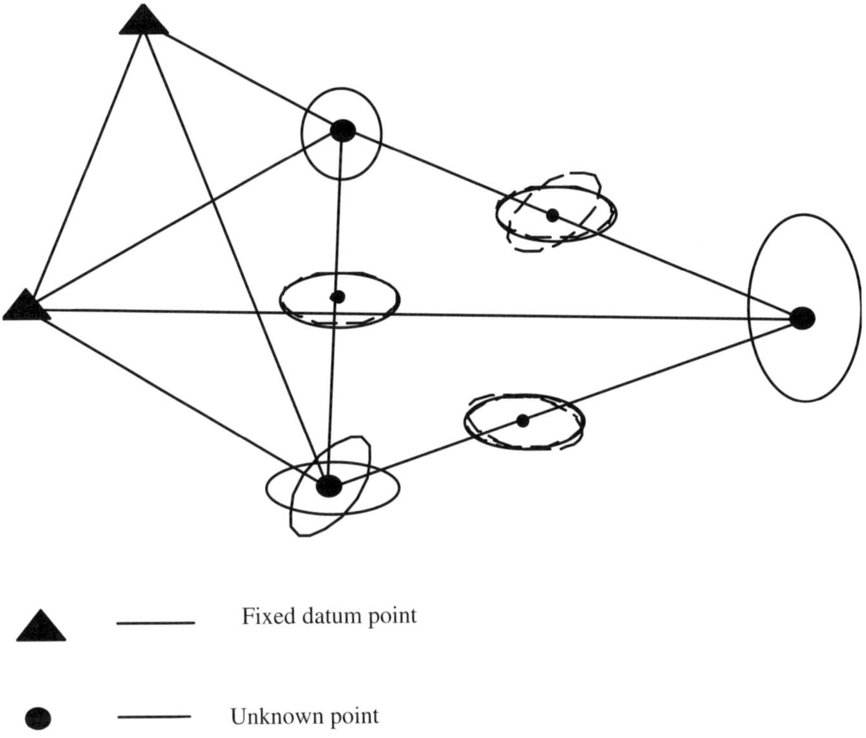

Figure 6.4. Station (or absolute) and relative error ellipses

(cf. Caspary, 1987, Koch, 1987). If the inner constraints are used, all point measures of accuracy will grow only slowly from the center of the network to the boundary.

6.2 Network Reliability Measures

The concept of reliability for geodetic networks originates from Baarda (1968). Generally, it refers to the ability of a network to resist gross errors in the observations. In this respect, usually the "***internal reliability***" and "***external reliability***" are distinguished. The former refers to the ability of a network to allow for the detection of gross errors by tests of hypothesis made with a specific confidence level (1–α) and power (1–β), while the latter is related to the effect of undetected gross errors on the estimated parameters and/or functions of them.

As discussed before, the method of least squares demands blunder and systematic error-free measurement data. Therefore, in order to obtain unbiased estimates of the unknown parameters under study, screening of the observations for gross errors and/or systematic effects plays a key role in network analysis, and that is why both the pre-adjustment and the post-adjustment data screening techniques have been developed and they are applied prior to and after the estimation of the parameters, respectively, to prevent them from smearing the results. Nevertheless, it also has been shown in §5.3.3 that all the present outlier detection techniques have their own limited sensitivity. That is, gross errors below a certain boundary value cannot be detected. The meaning of network reliability is twofold: the first is related to the controllability of the observations; that is, how sensitive a chosen outlier detection technique is when applied on the network, and the second mainly describes how sensitive the network itself is to these residual gross errors or small changes in the observations. In other words, it refers to the robustness of the network; namely, how well the network resists small inconsistencies of the observations, in contrast to the above discussed precision measures that relate mainly to random errors. High network reliability means it responds insignificantly to undetected gross errors.

6.2.1 Internal Reliability

As discussed above, internal reliability refers to the lower bounds for just detectable gross errors. Assuming one gross error at the time, the lower bound for just detectable gross errors $\nabla_0 l_i$ with Baarda's data snooping procedure has been derived in §5.3.3.2 as follows [cf. (5.70)]

$$\nabla_0 l_i = \sigma_{l_i} \frac{\delta_0}{\sqrt{r_i}} \quad (i = 1, \ldots, n) \tag{6.129}$$

where δ_0 represents minimum detectable distance between the null hypothesis H_0 and the alternative hypothesis H_{a2i} at the selected confidence level $(1-\alpha_0)$ and power of the test $(1-\beta_0)$. r_i and σ_{l_i} are the redundancy number and the standard deviation of observation l_i, respectively. Thus, a gross error larger than the above boundary value can be detected. Otherwise, it is not detectable. That is to say, a gross error as large as $\nabla_0 l_i$ or less may still remain in the data and affect the final results.

$\nabla_0 l_i$ (i=1, ..., n) is considered the local measures of internal reliability of a geodetic network (Baarda, 1968). It represents the minimum gross error detectable with selected α_0 and β_0. The average boundary value

$$\overline{\nabla_0 l} = \frac{1}{n} \sum_{i=1}^{n} \nabla_0 l_i \tag{6.130}$$

can be used as a global measure of internal reliability for a geodetic network.

The internal reliability can also be seen as a measure for the controllability of the observations. It is clear from the above equations that once the precision of the observation, the significance level α_0, and the required minimum power $(1-\beta_0)$ are fixed, the boundary value $\nabla_0 l_i$ is inversely proportional to the redundancy number r_i which represents the strength of the network geometry. The smaller the redundancy number r_i of the observation, the larger a gross error has to be in order to be detectable, and vice versa. An observation is best controlled in the case of r_i being equal to one (i.e., its maximum possible value). This occurs, however, only if the true value of the observation is known. If r_i is equal to zero (i.e., its minimum possible value), a test of the observation is not possible since the boundary value is infinite. In this case, the observation cannot be controlled at all.

6.2.2 External Reliability

External reliability relates to the effect of possibly undiscovered observational gross errors ∇l on the estimates of unknown coordinates. From the parameter estimation Equation (4.126), the influence of ∇l on the estimated coordinates, i.e., $\nabla \hat{\mathbf{x}}$ is given by

$$\nabla \hat{\mathbf{x}} = \left(\mathbf{A}^T \mathbf{P}\ \mathbf{A} + \mathbf{D}\mathbf{D}^T\right)^{-1} \mathbf{A}^T \mathbf{P} \nabla \boldsymbol{l} \tag{6.131}$$

Assuming one gross error at the time, the effect of the maximum undetectable blunder $\nabla_0 l_i$ in observation l_i is therefore

$$\nabla_{0,i}\ \hat{\mathbf{x}} = \left(\mathbf{A}^T \mathbf{P}\ \mathbf{A} + \mathbf{D}\mathbf{D}^T\right)^{-1} \mathbf{A}^T \mathbf{P} \mathbf{c}_i \nabla_0 l_i \quad (i = 1, \ldots, n) \tag{6.132}$$

where $\mathbf{c}_i$ is a vector of zeros except for identity at the i-th position.

Since $\nabla_{0,i}\ \hat{\mathbf{x}}$ is datum dependent, Baarda (1976) proposed a new standardized variate

$$\begin{aligned}
\lambda_{i,0} &= \frac{1}{\sigma_0^2}\left(\nabla_{0,i}\ \hat{\mathbf{x}}\right)^T \mathbf{Q}_x^{-}\left(\nabla_{0,i}\ \hat{\mathbf{x}}\right) \\
&= \frac{\nabla_0 l_i^2}{\sigma_0^2} \mathbf{c}_i^T \mathbf{P}\mathbf{A}\left(\mathbf{A}^T \mathbf{P}\ \mathbf{A} + \mathbf{D}\mathbf{D}^T\right)^{-1} (\mathbf{Q}_x)^{-} \left(\mathbf{A}^T \mathbf{P}\ \mathbf{A} + \mathbf{D}\mathbf{D}^T\right)^{-1} \mathbf{A}^T \mathbf{P} \mathbf{c}_i \\
&= \frac{\nabla_0 l_i^2}{\sigma_0^2} \mathbf{c}_i^T \mathbf{P}\mathbf{A}\left(\mathbf{A}^T \mathbf{P}\ \mathbf{A} + \mathbf{D}\mathbf{D}^T\right)^{-1} \mathbf{A}^T \mathbf{P} \mathbf{c}_i \quad (i = 1, \ldots, n)
\end{aligned} \tag{6.133}$$

which is invariant with respect to the coordinates defined. However, one is interested mainly in the maximum value of this variate that is related to the minimum deviation from the null hypothesis that can be detected with a significance level α_0, and power of the test $1-\beta_0$. So the following variate (Baarda, 1976)

$$\lambda_0 = \max\left(\lambda_{i,0}\right) \quad (i = 1, \ldots, n) \tag{6.134}$$

is considered as a measure of the external reliability of the estimated coordinates.

Substituting σ_0^2 from (6.129) and using the vector $\mathbf{c}_i$, Equation (6.133) becomes

$$\lambda_{i,0} = \delta_0^2\left(\frac{1}{r_i} - 1\right) \tag{6.135}$$

Equation (6.135) shows again that the larger the redundancy number r_i, the smaller is the effect, and vice versa.

Although $\lambda_{i,0}$ is considered as a datum independent global measure of external reliability, it is difficult to perceive physically. For geodetic network established for special engineering purposes, more useful measures are the local external reliability measures that refer to the effect of the maximum undetectable blunder $\nabla_0 l_i$ in observation l_i on the estimated station coordinates $\hat{\mathbf{x}}_j$ (j=1, ..., m), or on a specific coordinate component $\hat{\mathbf{x}}_k$ (k=1, ..., u), or on a specific function $\mathbf{c}^T\hat{\mathbf{x}}$ of the coordinate vector. For instance, in tunneling networks we require that the breakthrough error caused by gross errors in observations be under a certain bound in order to ensure a successful breakthrough of the tunnel (cf. Kuang, 1993c).

In two-dimensional space, the coordinates $\hat{\mathbf{x}}_j$ (j=1, ..., m) of station j are related to the global coordinate vector $\hat{\mathbf{x}}$ as follows

$$\begin{aligned}\hat{\mathbf{x}}_j &= \begin{pmatrix}\hat{x}_j\\ \hat{y}_j\end{pmatrix} = \begin{pmatrix}0 & \cdots & 0 & 1 & 0 & 0 & \cdots & 0\\ 0 & \cdots & 0 & 0 & 1 & 0 & \cdots & 0\end{pmatrix}\hat{\mathbf{x}}\\ &= \mathbf{H}_j\hat{\mathbf{x}}\end{aligned} \tag{6.136}$$

where $\mathbf{H}_j$ is a 2 by u matrix with zeros except for ones at the (1, 2j–1) and (2, 2j) positions. The effect of a gross error on $\hat{\mathbf{x}}_j$ is therefore

$$\begin{aligned}\nabla_{0,\,i}\,\hat{\mathbf{x}}_j &= \mathbf{H}_j\nabla_{0,\,i}\,\hat{\mathbf{x}}\\ &= \mathbf{H}_j\left(\mathbf{A}^T\mathbf{P}\,\mathbf{A} + \mathbf{D}\mathbf{D}^T\right)^{-1}\mathbf{A}^T\mathbf{P}\mathbf{c}_i\nabla_0 l_i \quad (i = 1, \ldots, n;\, j = 1, \ldots, m)\end{aligned} \tag{6.137}$$

Similarly, the effect of $\nabla_0 l_i$ on a coordinate component $\hat{x}_k$ is

$$\begin{aligned}\nabla_{0,\,i}\,\hat{\mathbf{x}}_k &= \mathbf{h}_k^T\nabla_{0,\,i}\,\hat{\mathbf{x}}\\ &= \mathbf{h}_k^T\left(\mathbf{A}^T\mathbf{P}\,\mathbf{A} + \mathbf{D}\mathbf{D}^T\right)^{-1}\mathbf{A}^T\mathbf{P}\mathbf{c}_i\nabla_0 l_i \quad (i = 1, \ldots, n;\, k = 1, \ldots, u)\end{aligned} \tag{6.138}$$

where $\mathbf{h}_k$ is a u by 1 vector of zeros except for identity at the k-th position.

Datum independent measures for $\nabla_{0,i}\,\hat{\mathbf{x}}_j$ and $\nabla_{0,i}\,\hat{\mathbf{x}}_k$, when needed, can also be derived as follows

$$\begin{aligned}(\lambda_{i,0})_j &= \frac{1}{\sigma_0^2}(\nabla_{0,i}\,\hat{\mathbf{x}}_j)^T \mathbf{Q}_{jj}^{-}(\nabla_{0,i}\,\hat{\mathbf{x}}_j) \\ &= \frac{\nabla_0 l_i^2}{\sigma_0^2}\mathbf{c}_i^T\mathbf{PA}\left(\mathbf{A}^T\mathbf{P}\,\mathbf{A}+\mathbf{DD}^T\right)^{-1}\mathbf{H}_j^T * \\ &\quad *\left(\mathbf{H}_j\,\mathbf{Q}_x\,\mathbf{H}_j^T\right)^{-}\mathbf{H}_j\left(\mathbf{A}^T\mathbf{P}\,\mathbf{A}+\mathbf{DD}^T\right)^{-1}\mathbf{A}^T\mathbf{Pc}_i\end{aligned} \tag{6.139}$$

$$(\lambda_{i,0})_k = \frac{1}{\sigma_0^2}(\nabla_{0,i}\,\hat{x}_k)^2\,\frac{1}{q_{kk}} \tag{6.140}$$

Finally, the influence of $\nabla_0 l_i$ on a function $f = \mathbf{c}^T\hat{\mathbf{x}}$ of $\hat{\mathbf{x}}$ is as follows

$$\nabla_{0,i}\,f = \mathbf{c}^T\nabla_{0,i}\,\hat{\mathbf{x}} = \mathbf{c}^T\left(\mathbf{A}^T\mathbf{P}\,\mathbf{A}+\mathbf{DD}^T\right)^{-1}\mathbf{A}^T\mathbf{Pc}_i\nabla_0 l_i \tag{6.141}$$

Similar expressions can be derived for the three-dimensional case. One can see that the reliability of a geodetic network depends on the geometry of the network, i.e., the configuration matrix and on the weight matrix of observations, not on the actual observations. Thus the problem of reliability should be considered at the design stage to ensure the detection of gross errors as small as possible and to minimize the effects of the undetected ones on the estimated parameters (cf. §9.2 and §10.2.2, etc.).

7 Deformation Network Analysis

7.1 Purpose of Deformation Monitoring Networks

Deformation refers to the changes a deformable body undergoes in its shape, dimension, and position. It can be said that any object, natural or man-made, undergoes changes in space and time. The determination and interpretation of the changes are the main goal of deformation surveys. Deformation surveys are one of the most important activities in surveying, especially in engineering surveying. Their results are directly relevant to the safety of human life and engineering structures. Deformation surveys can provide not only the geometric status of the deformed object, but also information on its response to loading stress. This provides a better understanding of the mechanics of deformations and the checking of various theoretical hypotheses on the behavior of a deformable body. Examples of deformation surveys include the monitoring of ground deformations due to mining exploitation, withdrawal of oil or underground water, or construction of large reservoirs; the monitoring of accumulation of stress near active tectonic plate boundaries; and the checking of the stability of large or complex structures (e.g., hydro-electric dams).

As with conventional geodetic positioning surveys, deformation measurements are also undertaken in three phases:

- Design of the surveying scheme
- The field observation campaign, and
- Post-analysis of the data.

Nevertheless, a distinguished feature with deformation networks is that since they are established to detect position changes, e.g., displacements or deformation parameters, depending on the attributes of the specific problem, they are considered changeable, either dynamically or kinematically, in space and time, in contrast to the conventional geodetic positioning networks that are considered in a static sense with determination of relative positions being the sole objective. Over the last two decades, considerable efforts have been made by geodesists to develop new methodologies and instrumentation for deformation monitoring as well as new theories and algorithms for the optimal network design and post-analysis of the data (cf. Chrzanowski, 1986; Chrzanowski et al., 1986; Pelzer, 1971; Chen, 1983; Caspary, 1987; Kuang, 1991, etc.). The optimal design of deformation monitoring networks will be discussed in Chapter 11. The purpose of this chapter is to address the concerns and principles of geometrical deformation analysis. It should be noted that theories of geodetic network analysis discussed in Chapters 1 to 6 also form the basis for deformation network analysis. For instance, accuracy analysis of observations, observation pre-processing, pre- and post-adjustment data screening techniques, and the principle of least squares estimation all are the issues encountered in the establishment and analysis of a geodetic deformation network. In addition, since the end products of a deformation network are deformation parameters, instead of station coordinates, estimation and statistical diagnostic checking of a postulated deformation model become a distinguished issue in deformation network analysis.

This chapter continues with a review of the concepts and definitions of basic deformation parameters and deformation models in §7.2. The geodetic and non-geodetic monitoring techniques are then discussed in §7.3, and finally, the concepts of geometrical analysis of deformations are addressed in §7.4. For more detailed discussions on this subject, the interested reader is referred to the given literature.

7.2 Basic Deformation Parameters and the Deformation Model

If acted upon by external forces (loads), any real material deforms, i.e., changes its dimensions, shape, and position. According to Sokolnikoff (1956) the basic deformation parameters are rigid body translation, rigid body rotation (or relative translation and rotation of one "block" with respect to another), strain tensor, and differential rotation components. If a time factor is involved, the derivative of the above quantities with respect to time is used instead. According to Chen (1983) and Chrzanowski et al. (1983), the above deformation parameters in three-dimensional space can be obtained if the displacement field $d(x, y, z; t\text{-}t_0)$ is known. The displacement field can be approximated by fitting a selected deformation model to displacements determined at discrete points as follows

$$\mathbf{d}(x, y, z; t - t_0) = \mathbf{B}(x, y, z; t - t_0)\,\mathbf{e} \tag{7.1}$$

where $\mathbf{d}$ is the vector of displacement components of point (x, y, z) at time t with respect to t_0, $\mathbf{B}$ a matrix of base functions, and $\mathbf{e}$ the vector of unknown deformation parameters.

The mathematical model (7.1) can be explicitly written as

$$\mathbf{d} = \begin{pmatrix} u(x, y, z; t - t_0) \\ v(x, y, z; t - t_0) \\ w(x, y, z; t - t_0) \end{pmatrix} = \begin{pmatrix} \mathbf{B}_u(x, y, z; t - t_0)\mathbf{e} \\ \mathbf{B}_v(x, y, z; t - t_0)\mathbf{e} \\ \mathbf{B}_w(x, y, z; t - t_0)\mathbf{e} \end{pmatrix} \tag{7.2}$$

where u, v, and w represent displacement components in the x, y, and z directions respectively, and they are functions of both position and time. From (7.2), the non-translational deformation tensor can be calculated by

$$\mathbf{E} = \begin{pmatrix} \frac{\partial u}{\partial x} & \frac{\partial u}{\partial y} & \frac{\partial u}{\partial z} \\ \frac{\partial v}{\partial x} & \frac{\partial v}{\partial y} & \frac{\partial v}{\partial z} \\ \frac{\partial w}{\partial x} & \frac{\partial w}{\partial y} & \frac{\partial w}{\partial z} \end{pmatrix} \tag{7.3}$$

and the normal strains, shear strains and the differential rotations around x, y, z axes are respectively

$$\varepsilon_x = \frac{\partial u}{\partial x} \tag{7.4a}$$

$$\varepsilon_y = \frac{\partial v}{\partial y} \tag{7.4b}$$

$$\varepsilon_z = \frac{\partial w}{\partial z} \tag{7.4c}$$

$$\varepsilon_{xy} = \left(\frac{\partial u}{\partial y} + \frac{\partial v}{\partial x}\right) / 2 \tag{7.4d}$$

$$\varepsilon_{xz} = \left(\frac{\partial u}{\partial z} + \frac{\partial w}{\partial x}\right) / 2 \tag{7.4e}$$

$$\varepsilon_{yz} = \left(\frac{\partial v}{\partial z} + \frac{\partial w}{\partial y}\right) / 2 \tag{7.4f}$$

$$\omega_x = \left(\frac{\partial v}{\partial z} - \frac{\partial w}{\partial y}\right)/2 \tag{7.4g}$$

$$\omega_y = \left(\frac{\partial u}{\partial z} - \frac{\partial w}{\partial x}\right)/2 \tag{7.4h}$$

$$\omega_z = \left(\frac{\partial u}{\partial y} - \frac{\partial v}{\partial x}\right)/2 \tag{7.4i}$$

In addition, certain functions of these strain parameters, e.g., maximum strain (ε), dilatation (Δ), pure shear (r_1), simple shear (r_2), and total shear (r) may also be of interest and they are defined as follows (cf. Frank, 1966)

$$\varepsilon = \sqrt{\varepsilon_x^2 + \varepsilon_y^2 + \varepsilon_z^2} \tag{7.5}$$

$$\Delta = \varepsilon_x + \varepsilon_y + \varepsilon_z \tag{7.6}$$

$$r_1 = \left(\varepsilon_x - \varepsilon_y\right) \tag{7.7}$$

$$r_2 = 2\,\varepsilon_{xy} \tag{7.8}$$

$$r = \sqrt{r_1^2 + r_2^2} \tag{7.9}$$

As for the selection of a deformation model, it depends on the a priori information that is available and, especially, from whatever trend or change is exhibited by the measurements or by the location of the stations. In deformation analysis, the whole area covered by the deformation surveys is treated as a non-continuous deformable body consisting of separate continuous deformable blocks. The blocks may undergo relative rigid body displacements and rotation, and each block may change its shape and dimensions. In the case of single point movement, the given point is treated as a separate block being displaced as a rigid body in relation to the undeformed block composed of the remaining points in the network. Examples of typical deformation models in two-dimensional space are given below (cf. Chrzanowski et al., 1983; Chen, 1983; Chrzanowski et al., 1986):

1. Single point displacement or a rigid body displacement of a group of points, say, block B (Figure 7.1a) with respect to block A. The deformation model is expressed as

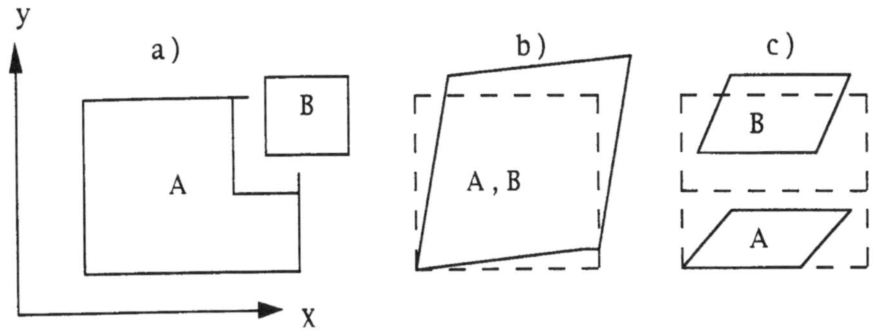

Figure 7.1. Typical deformation models (after Chrzanowski et al., 1983)

$$u_A = 0 \tag{7.10a}$$

$$v_A = 0 \tag{7.10b}$$

$$u_B = a_0 \tag{7.10c}$$

$$v_B = b_0 \tag{7.10d}$$

where the subscripts represent all the points in the indicated blocks.

2. Homogeneous strain in the whole body and differential rotation (Figure 7.1b), the deformation model is

$$u = \varepsilon_x\, x + \varepsilon_{xy}\, y - \omega\, y \tag{7.11a}$$

$$v = \varepsilon_{xy}\, x + \varepsilon_y\, y + \omega\, x \tag{7.11b}$$

where the physical meaning of the coefficients is defined in (7.4a,b,d,i) with ω_z in (7.4i) being replaced by ω.

3. A deformable body with one discontinuity (Figure 7.1c), say, between blocks A and B, and with different linear deformations in each block plus a rigid body displacement of B with respect to A. The deformation model is written as

$$\begin{aligned} u_A &= \varepsilon_{xA}\, x + \varepsilon_{xyA}\, y - \omega_A\, y \\ v_A &= \varepsilon_{xyA}\, x + \varepsilon_{yA}\, y + \omega_A\, x \end{aligned} \tag{7.12}$$

and

$$\begin{aligned} u_B &= a_0 + \varepsilon_{xB}(x - x_0) + \varepsilon_{xyB}(y - y_0) - \omega_B(y - y_0) \\ v_B &= b_0 + \varepsilon_{xyB}(x - x_0) + \varepsilon_{yB}(y - y_0) + \omega_B(x - x_0) \end{aligned} \tag{7.13}$$

where x_0, y_0 are the coordinates of any point in block B.

The components Δu_i and Δv_i of a total relative dislocation at any point i located on the discontinuity line between blocks A and B can be calculated by

$$\begin{aligned} \Delta u_i &= u_B(x_i, y_i) - u_A(x_i, y_i) \\ \Delta v_i &= v_B(x_i, y_i) - v_A(x_i, y_i) \end{aligned} \tag{7.14}$$

Usually, the actual deformation model is a combination of the above simple models or, if more complicated, it is expressed by nonlinear displacement functions that require the fitting of higher order polynomials or other suitable functions. If time dependent deformation parameters are sought, the above deformation models will contain time variables. For instance, in the model of homogeneous strain, if a linear time dependence is assumed, the model becomes:

$$\begin{aligned} u(x, y, t) &= \dot{\varepsilon}_x\, x\, t + \dot{\varepsilon}_{xy}\, y\, t - \dot{\omega}\, y\, t \\ v(x, y, t) &= \dot{\varepsilon}_{xy}\, x\, t + \dot{\varepsilon}_y\, y\, t + \dot{\omega}\, x\, t \end{aligned} \tag{7.15}$$

where $\dot{\varepsilon}_x$, $\dot{\varepsilon}_y$, $\dot{\varepsilon}_{xy}$, and $\dot{\omega}$ represent time derivatives of the strains and differential rotations.

7.3 Geodetic and Non-Geodetic Methods for Deformation Monitoring

Acquisition of deformation parameters is one of the main goals of deformation monitoring. Different methodologies and techniques have been used for this purpose. As compared with other types of surveys, deformation measurements have the following characteristics (cf. Chen, 1983):

- Higher accuracy requirement
  For example, in engineering projects, an accuracy of ±1 mm or higher might be a typical requirement
- Repeatability of observations
  The periods of resurveys range from seconds to years, depending on the rate of deformation
- Integration of different types of observations
  Here not only geodetic methods should be considered but also non-geodetic instrumentation, e.g., pendulum, tiltmeters, strainmeters, mechanical and laser alignment, hydrostatic levels, and others in order to get more complete information
- Network may be incomplete, scattered in space and time
- Sophisticated analysis of the acquired data in order to avoid the misinterpretation of measuring errors as deformation and local phenomena as a global status

Deformation measurements may be categorized as being of local, regional, continental or global scale, depending on their extent. Examples of local deformation studies include investigation of deformations that occur in man-made structures or in localities of coal mining, petroleum production, extensive water pumping for industry, etc. A good example for regional deformation studies is deformation monitoring of the earth-crust near the plate boundaries, and this is usually done using a fault-crossing network to monitor the creep between two tectonic blocks along the fault (cf. Chen, 1983). Finally, examples of deformations of global nature are the polar motion, variation of the earth's rotation, and relative motion between tectonic plates, etc. Here in this book, we concentrate on deformation monitoring of local scale, with a possible extension for regional applications. Usually, such a monitoring scheme consists of a geodetic network, plus some isolated non-geodetic observables that may not be geometrically connected with the geodetic network as discussed below in §7.3.2.

Geodetic methods, which include terrestrial geodetic methods, photogrammetric methods, and space techniques, are used to monitor the magnitude and rate of horizontal and vertical deformations of structures, the ground surface, and accessible parts of subsurface instruments in a wide variety of construction situations. Frequently, these methods are entirely adequate for deformation monitoring. In *non-geodetic* methods, we have geotechnical and specialized monitoring devices. They are required only if greater accuracy is sought or if measuring points are inaccessible to *geodetic* methods. However, in general, whenever non-geodetic instruments are used to monitor deformation, geodetic methods are also used to relate measurements to a reference datum.

7.3.1 The Geodetic Methods

According to Chrzanowski (1981), in deformation measurements by geodetic methods, whether they are performed for monitoring engineering structures or ground subsidence in mining areas or tectonic movements, the monitoring networks can be divided into relative networks and reference networks.

In relative networks (see Figure 7.2), all the survey points are assumed to be located on the deformable body, the purpose in this case is to identify the deformation model, i.e., to distinguish, on the basis of repeated geodetic observations, between the deformations caused by the extension and shearing strains, by the relative rigid body displacements, and by the single point displacements. In reference networks, however, some of the points are, or are assumed to be, outside the deformable body (object), thus serving as reference points for the determination of absolute displacements of the object points (see Figure 7.3).

A list of the most commonly used geodetic methods for deformation monitoring and their associated approximate accuracies is given in Table 7.1. Over the last two decades the availability of increasingly reliable and accurate EDM equipment has radically changed conventional surveying practices. EDM devices require fewer personnel than conventional optical instruments, are faster to use, and are more accurate. Therefore, horizontal triangulation monitoring networks have been gradu-

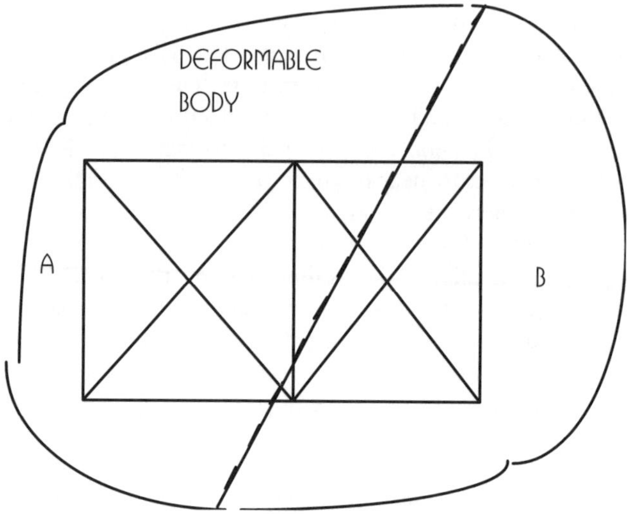

Figure 7.2. A relative monitoring network

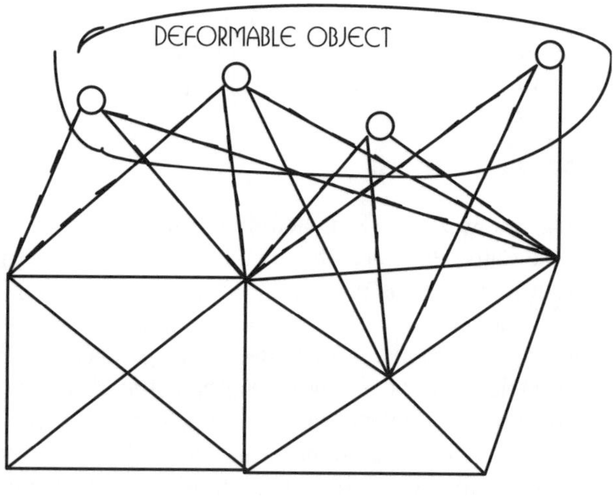

Figure 7.3. A reference monitoring network

ally replaced by trilateration or triangulateration networks. Depending on the models, an EDM instrument can have a range of a few meters to several tens of kilometers. High-precision EDM instruments are those with multiple wavelengths to reduce the effect of tropospheric refraction internally, e.g., Terrameter (Hugget, 1982), and those with high-modulation frequencies, e.g., Kern ME 5000. The former can

Table 7.1. The geodetic methods for deformation monitoring

Methods	Achievable Accuracy (s)
• Elevations by optical leveling	0.1 mm over a few tens of meters to about 1 mm over long distance
• Distance measurements with tapes or wires	0.1 mm over a few meters to about 2 ppm over a few hundred meters
• Offsets from a baseline using theodolite and scale	0.3–2 mm
• Traversing	1/30,000–1/150,000
• Triangulation	1/30,000–1/1,000,000
• Electromagnetic Distance Measurement (EDM)	0.2 mm or 0.1 ppm to 5 ppm
• Trigonometric leveling	2 mm √km
• Photogrammetric methods	1/5000–1/100,000
• Space techniques	
VLBI	0.01 ppm
SLR	0.01 ppm
GPS	0.1ppm–2 ppm

achieve an accuracy in the order of 0.1 ppm over distance of several kilometers, while the latter can give 0.3 mm standard deviation over a few hundred meters.

Trigonometric leveling is much more economical than conventional geodetic leveling when Third-order accuracy is adequate and measuring points are physically inaccessible. According to Chrzanowski (1983), even the First-order and Second-order accuracy can be achieved by a modified trigonometric leveling, i.e., the leap-frog or reciprocal trigonometric leveling. This method is especially advantageous in mountainous areas.

Both terrestrial and aerial photogrammetry have been extensively used in the determination of deformations of large structures (e.g., Faig, 1978), and ground subsidence (Faig and Armenakis, 1982). In terrestrial photogrammetry, an example is to use phototheodolites to take successive photographs from a fixed station along a fixed baseline. Movements are identified in a stereocomparator by stereoscopic advance or recession of pairs of photographic plates in relation to stable background elements. The procedure defines the components of movements taking place in the plane of the photograph. The photogrammetric method has the advantage that hundreds of potential movements are recorded on a single stereo photographic pair, allowing an appraisal of the overall displacement pattern in a minimum time.

Geodetic space techniques, e.g., VLBI (Very Long Base Line Interferometry), SLR (Satellite Laser Ranging), and GPS (Global Positioning System) can provide deformation data of global extent, such as polar motion, variation of the earth's rotation, and relative motion between tectonic plates. GPS has some advantages with which the conventional terrestrial surveys are not able to compete. For instance, it has high accuracy (0.1–2 ppm) with either short or long range, and it does not require intervisibility between network stations, etc. Especially with the full coverage and economic operations available today, GPS has been widely accepted in practice by surveyors, and it has been increasingly used to replace the conventional terrestrial surveys to establish geodetic control networks. In this regard, GPS is also a valuable tool for local or regional deformation monitoring surveys.

7.3.2 The Non-Geodetic Methods

The geodetic methods, through interconnections among the monitoring stations, can provide very useful information on the global deformation status of the monitored object and, in most cases, can also provide information on its rigid body translations and rotations with respect to reference points located outside the deformation area. However, as mentioned above, geodetic methods are limited only to open areas, and they require intervisibility between the survey stations, or between the monitoring stations and the satellites, or between the object points and the cameras. The deformations inside the deformable body, e.g., in foundations or foundation rocks of large engineering structures and relative movements of different layers of soil or rock formations in slope stability studies, etc., can only be approached by non-geodetic methods.

Non-geodetic methods include geotechnical instrumentation and other specialized monitoring devices. For instance, borehole inclinometers, and extensometers, are examples of geotechnical instruments, while inverted pendulum, hydrostatic levels, laser interferometers, and diffraction aligning equipment are made for some specialized monitoring purposes. Geotechnical instrumentation does not require intervisibility between the stations and can be easily adapted for continuous and telemetric data acquisition with an instantaneous display of the deformations, which is very advantageous in comparison with slow, labor intensive geodetic surveys. Non-geodetic methods are usually used to measure three types of deformations, i.e., extensions and strains; tilts and inclinations; and alignment. Typical examples of non-geodetic instruments with their associated accuracies are listed in Table 7.2 (cf. also Chrzanowski, 1986).

One has to remember that the non-geodetic methods, despite their indisputable advantages, also have weak points: a) the measurements are very localized and they may be affected by local disturbances that do not represent the actual deformations; b) since the local observables are not geometrically connected with observables at other monitoring stations, and global trend analysis of the deformations is much more difficult than in the case of geodetic surveys unless the observing stations are very densely spaced (Chrzanowski, 1986).

7.4 Geometrical Analysis of Deformations

The purpose of geometrical analysis of deformations is to determine the geometrical status of a deformable body, i.e., the change of its shape and dimensions. Since the deformations are usually small, careful analysis and statistical testing of

Table 7.2. The non-geodetic methods for deformation monitoring

Types of Deformation	Methods and Sample Instruments	Typical Accuracy
Extensions and strains	• Wire and Tape extensometers	
	* ISETH Distometer	0.05 mm
	* CERN Distinvar	0.05 mm
	* Rock Spy	0.02–0.2 mm
	• Rod and Tube extensometers	
	* Single-point extensometers	0.01–0.02 mm
	* Multi-point extensometers	0.01–0.02 mm
	* Torpedo-type extensometers	0.1 mm
	• Michelson type laser interferometers	
	* Laser strain meters	0.0004 ppm
Tilts and inclinations	• Precision tiltmeters	
	* High-precision mercury tiltmeter	0.0002"
	* Electrolevel	0.25"
	* Talyvel	0.5"
	• Hydrostatic leveling	
	* Elwaag001	0.03 mm/40 m
	* Nivomatic telenivelling system	0.1 mm/24 m
	• Suspended and Inverted Pendulum	0.1 mm
Alignment	• Mechanical Methods	
	* Steel wire alignment	0.1 mm
	* Nylon line alignment	0.035 mm–0.070 mm
	• Direct optical alignment	1–10 ppm
	• Alignment with laser diffraction gratings	0.1–1 ppm

the results are required. It has been shown that the approach proposed in, among others, Chen (1983) and Chrzanowski et al. (1986) is efficient for this purpose. The approach involves the following four phases:

i. evaluation of the observation data
ii. preliminary identification of the deformation models
iii. estimation of the deformation parameters; and
iv. diagnostic checking of the models and the final selection of the "best" model.

The task at phase i is to perform separate network adjustment for each epoch of observations to detect and eliminate any possible gross errors and systematical effects in the survey data. The theories described in Chapters 1 to 5 can be used for this purpose. The task at phases ii to iv deals with the identification and estimation of the deformations, and the mathematical tools used are discussed in the following subsections.

7.4.1 The Functional Relationship Between the Deformation Models and the Observed Quantities

Any observation, geodetic, or non-geodetic measurement made in deformation surveys, will contribute to the determination of deformation parameters and should be fully utilized in the analysis (Chrzanowski et al., 1986; Teskey, 1987; Kuang, 1991). In the RLG coordinate system, the functional relationships between different geodetic observable types and the selected deformation model can be obtained according to the observation modeling given in §4.1.1. It should be noted that raw measurements must be pre-processed according to procedures described in Chapter 2 before they can be used.

Coordinates

Observed coordinates of a point i; for instance, the coordinates derived from photogrammetric measurements or obtained from space techniques, are related to displacements of the point as follows

$$\begin{pmatrix} x_i(t) \\ y_i(t) \\ z_i(t) \end{pmatrix} = \begin{pmatrix} x_i(t_0) \\ y_i(t_0) \\ z_i(t_0) \end{pmatrix} + \begin{pmatrix} u_i \\ v_i \\ w_i \end{pmatrix} \tag{7.16}$$

where $[x_i(t_0), y_i(t_0), z_i(t_0)]$ and $[x_i(t), y_i(t), z_i(t)]$ are point coordinates observed at time epoch t_0 and t, respectively, and $[u_i, v_i, w_i]$ are the displacements of the point.

Coordinate Differences

Observed coordinate differences between points i and j, e.g., height difference (leveling) observation, pendulum (displacement) measurement, and alignment survey are related to differences of displacements between the two points as follows

$$\begin{pmatrix} x_j(t)-x_i(t) \\ y_j(t)-y_i(t) \\ z_j(t)-z_i(t) \end{pmatrix} = \begin{pmatrix} x_j(t_0)-x_i(t_0) \\ y_j(t_0)-yi(t_0) \\ z_j(t_0)-z_i(t_0) \end{pmatrix} + \begin{pmatrix} u_j-u_i \\ v_j-v_i \\ w_j-w_i \end{pmatrix} \tag{7.17}$$

where $[x_i(t_0), y_i(t_0), z_i(t_0)]$ and $[x_j(t_0), y_j(t_0), z_j(t_0)]$ as well as $[x_i(t), y_i(t), z_i(t)]$ and $[x_j(t), y_j(t), z_j(t)]$ are coordinates for points i and j observed at time epoch t_0 and t, respectively, and $[u_i, v_i, w_i]$ and $[u_j, v_j, w_j]$ are the displacements of the two points.

If the components of the displacement obtained from a pendulum observation do not coincide with the coordinate axes, a transformation to the common coordinate system has to be performed. Similarly, a coordinate transformation may be required in alignment surveys that provides a transverse displacement of a point with respect to a straight line defined by two base points.

Azimuth

A geodetic azimuth from point i to point j is related to the horizontal displacements of the two points as follows [cf. (4.34)]:

$$\alpha_{ij}(t)=\alpha_{ij}(t_0)+\left[\frac{\Delta y_{ij}^0}{\left(L_{ij}^0\right)^2} \quad -\frac{\Delta x_{ij}^0}{\left(L_{ij}^0\right)^2}\right]\begin{bmatrix} u_j-u_i \\ v_j-v_i \end{bmatrix} \tag{7.18}$$

where $\alpha_{ij}(t_0)$ and $\alpha_{ij}(t)$ are the azimuths observed at time epoch t_0 and t, respectively. The observation of a horizontal angle is expressed as the difference of two azimuths.

Spatial Distances

An observed spatial distance between points i and j is related to the differences of displacements of the two points as follows [cf. (4.33)]

$$s_{ij}(t)=s_{ij}(t_0)+\left(\frac{\Delta x_{ij}^0}{s_{ij}^0}, \quad \frac{\Delta y_{ij}^0}{s_{ij}^0}, \quad \frac{\Delta z_{ij}^0}{s_{ij}^0}\right)\begin{pmatrix} u_j-u_i \\ v_j-v_i \\ w_j-w_i \end{pmatrix} \tag{7.19}$$

where $s_{ij}(t_0)$ and $s_{ij}(t)$ are the spatial distances observed at time epoch t_0 and t, respectively.

Strains

Observation of a strain along the azimuth α_{ij} and zenith distance Z_{ij} at a point i is related to the deformation tensor by

$$\varepsilon(t)=\varepsilon(t_0)+\mathbf{c}^T\,\mathbf{E}\,\mathbf{c} \tag{7.20}$$

where $\varepsilon(t_0)$ and $\varepsilon(t)$ are the strains observed at time epoch t_0 and t, respectively. Matrix **E** has been defined by Equation (7.3) before, and vector **c** is defined by

$$\mathbf{c}^T=\left(\sin Z_{ij}\,\sin\alpha_{ij}, \sin Z_{ij}\,\cos\alpha_{ij}, \cos Z_{ij}\right) \tag{7.21}$$

with α_{ij} and Z_{ij} meaning the same as above.

Zenith Distances

An observed zenith distance Z_{ij} at point i to point j is related to the differences of displacements of the two points as follows [cf. (4.36)]

$$Z_{ij}(t) = Z_{ij}(t_0) + \begin{pmatrix} \frac{\Delta x_{ij}^0 \Delta z_{ij}^0}{L_{ij}^0 \left(s_{ij}^0\right)^2}, & \frac{\Delta y_{ij}^0 \Delta z_{ij}^0}{L_{ij}^0 \left(s_{ij}^0\right)^2} & -\frac{L_{ij}^0}{\left(s_{ij}^0\right)^2} \end{pmatrix} \begin{pmatrix} u_j - u_i \\ v_j - v_i \\ w_j - w_i \end{pmatrix} \tag{7.22}$$

where $Z_{ij}(t_0)$ and $Z_{ij}(t)$ are the zenith distances at time epoch t_0 and t, respectively.

Tilts

Observation of a horizontal tiltmeter is related to the deformation by

$$\tau(t) = \tau(t_0) + (\partial w / \partial x)\sin\alpha + (\partial w / \partial y)\cos\alpha \tag{7.23}$$

where α is the orientation of the tiltmeter, and $\tau(t_0)$ and $\tau(t)$ are the measured tilts at time epoch t_0 and t, respectively.

Once a deformation model has been postulated, the displacements u, v, w in Equations (7.16) to (7.23) and their derivatives are then replaced by the deformation model that is explicitly expressed in (7.2). Thus, all the observations can be expressed as functions of the unknown deformation parameters **e**.

7.4.2 Estimation of Deformation Parameters

Let $\boldsymbol{l}_i$ (i=0, 1, 2, ..., k) be the n_i-vector of observations with weight matrix $\mathbf{P}_i$ in epoch i, which includes both geodetic and non-geodetic observables. Considering (7.1), estimation of the deformation parameters is based on the following Gauss-Markov model:

$$H_0: E\{l_i\} = E\{l_0\} + \mathbf{A}_i \mathbf{B}_i \mathbf{e} \tag{7.24a}$$

$$D\{l_i\} = \sigma_0^2 \, \mathbf{P}_i \tag{7.24b}$$

where $\mathbf{A}_i$ is the configuration matrix that relates the observables to deformation model; $\mathbf{B}_i$ is the coefficient matrix of deformation model, it is a function of position and time; and **e** is the vector of deformation parameters. The parameters **e** may be estimated from the following mathematical model

$$\begin{pmatrix} l_0 \\ l_1 \\ \vdots \\ l_k \end{pmatrix} + \begin{pmatrix} \mathbf{v}_0 \\ \mathbf{v}_1 \\ \vdots \\ \mathbf{v}_k \end{pmatrix} = \begin{pmatrix} \mathbf{I} & \mathbf{0} \\ \mathbf{I} & \mathbf{A}_1\mathbf{B}_1 \\ \vdots & \vdots \\ \mathbf{I} & \mathbf{A}_k\mathbf{B}_k \end{pmatrix} \begin{pmatrix} \xi \\ \mathbf{e} \end{pmatrix} \tag{7.25}$$

with ξ being a vector of nuisance parameters and the weight matrix (assuming there is no correlation between epochs)

$$\mathbf{P} = \begin{pmatrix} \mathbf{P}_0 & \mathbf{0} & \mathbf{0} & \cdots & \mathbf{0} \\ \mathbf{0} & \mathbf{P}_1 & \mathbf{0} & \cdots & \mathbf{0} \\ \vdots & \vdots & \vdots & \vdots & \vdots \\ \mathbf{0} & \cdots & \mathbf{0} & \mathbf{P}_{k-1} & \mathbf{0} \\ \mathbf{0} & \cdots & \mathbf{0} & \mathbf{0} & \mathbf{P}_k \end{pmatrix} \tag{7.26}$$

Applying the least squares criterion to the above model and eliminating ξ allow the vector of deformation parameters $\mathbf{e}$ and its cofactor matrix to be calculated from

$$\begin{aligned} \hat{\mathbf{e}} = &\left(\sum_1^k \mathbf{B}_i^T \mathbf{A}_i^T \mathbf{P}_i \mathbf{A}_i \mathbf{B}_i - \sum_1^k \mathbf{B}_i^T \mathbf{A}_i^T \mathbf{P}_i \left(\sum_0^k \mathbf{P}_i \right)^{-1} \sum_1^k \mathbf{P}_i \mathbf{A}_i \mathbf{B}_i \right)^{-1} * \\ &* \left(\sum_1^k \mathbf{B}_i^T \mathbf{A}_i^T \mathbf{P}_i l_i - \sum_1^k \mathbf{B}_i^T \mathbf{A}_i^T \mathbf{P}_i \left(\sum_0^k \mathbf{P}_i \right)^{-1} \sum_1^k \mathbf{P}_i l_i \right) \end{aligned} \tag{7.27}$$

$$\mathbf{Q}_e = \left(\sum_1^k \mathbf{B}_i^T \mathbf{A}_i^T \mathbf{P}_i \mathbf{A}_i \mathbf{B}_i - \sum_1^k \mathbf{B}_i^T \mathbf{A}_i^T \mathbf{P}_i \left(\sum_0^k \mathbf{P}_i \right)^{-1} \sum_1^k \mathbf{P}_i \mathbf{A}_i \mathbf{B}_i \right)^{-1} \tag{7.28}$$

7.4.3 Statistical Testing of Deformation Models and Choice of the Best Model

In general, the deformation model is not fully understood or it may even be completely unknown. The main objective of statistical testing of deformation models is therefore to check the appropriateness of a postulated deformation model. This involves a global test on the goodness of fit of the model, and test on the significance of individual deformation parameters.

Goodness of Fit Test

Test on the goodness of fit of a postulated deformation model **Be** can be done using the following statistic that, under the null hypothesis H_0 [Equation (7.24)], follows the Fisher distribution $F(r_e, r)$, i.e.,

$$w_e = \frac{\hat{\sigma}_e^2}{\hat{\sigma}_0^2} \in F(r_e, r) \tag{7.29}$$

where $\hat{\sigma}_e^2$ and r_e are respectively the a posterior estimated variance factor and degrees of freedom associated with the estimation of deformation parameters described above. $\hat{\sigma}_0^2$ and r are respectively the weighted variance factor and degrees of freedom associated with the separate network adjustment of different epochs of observations as given below

$$\hat{\sigma}_0^2 = \frac{r_0 \hat{\sigma}_{00}^2 + r_1 \hat{\sigma}_{01}^2 + \cdots + r_k \hat{\sigma}_{0k}^2}{r_0 + r_1 + \cdots + r_k} \tag{7.30}$$

$$r = r_0 + r_1 + \cdots + r_k$$

where $\hat{\sigma}_{0i}^2$ and r_i are the estimated a posterior variance factor and degrees of freedom for the network adjustment of epoch i (i=0, 1, ..., k), respectively.

Given the significance level α, the null hypothesis is accepted if

$$w_e \le F_{1-\alpha}(r_e, r) \tag{7.31}$$

Otherwise, the postulated deformation model is considered as being inappropriate, and a different deformation model should be postulated and re-evaluated.

Test on the Significance of Individual Deformation Parameters

The significance of an individual deformation parameter $\hat{e}_i$ or a set of u_i parameters $\hat{\mathbf{e}}_i$, which is a subset of $\hat{\mathbf{e}}$, is done using the following test statistics:

$$y_i = \frac{\hat{e}_i^2}{q_{ii}\, \hat{\sigma}_0^2} \tag{7.32}$$

$$y_e = \frac{\hat{\mathbf{e}}_i^T\, \mathbf{Q}_{ii}^{-1}\, \hat{\mathbf{e}}}{u_i\, \hat{\sigma}_0^2} \tag{7.33}$$

where q_{ii} and $\mathbf{Q}_{ii}$ can be extracted from matrix $\mathbf{Q}_e$ as given in Equation (7.28).

Under the following null hypothesis

$$H_0: E\{\hat{e}_i\} = 0 \tag{7.34a}$$

or

$$E\{\hat{\mathbf{e}}_i\} = \mathbf{0} \tag{7.34b}$$

statistic y_i and y_e belong to the Fisher F(1, r) and F(u_i, r) distribution, respectively. Therefore, given the significance level α, the above null hypothesis will be rejected if

$$y_i > F_{1-\alpha}(1, r) \tag{7.35}$$

or

$$y_e > F_{1-\alpha}(u_i, r) \tag{7.36}$$

In this case, the tested parameters are considered significant. If insignificant parameters are detected, these parameters are then deleted from the deformation model, and the new model is re-estimated and re-evaluated. This process continues until the global test passes and all the parameters involved are statistically significant.

Choice of the Best Deformation Model

A postulated deformation model is acceptable if it passes the global test with an acceptable probability and all the parameters involved are significant. If more than one postulated deformation model passes the global test, then the model with the fewest significant parameters is selected. For some other details on the preliminary identification of deformation models, the interested reader is referred to the given references.

Part II

Network Optimal Design

8 Geodetic Network Optimal Design: An Overview

8.1 Statement of the Problem

As discussed in the Introduction, network design is the first step towards establishing a geodetic network. In order to prevent the whole operation from failing, the surveyors/engineers in charge should know about the result of their work according to the pre-set objectives before any measurement campaign is started. At the design stage of a geodetic network, the fundamental problem that a surveying engineer faces is how to decide on its configuration, i.e., the point location, how to choose the types of geodetic observations, and how to measure the network. Conceptually, the purpose of ***network optimal design*** or ***network optimization*** is in general understood to design an optimum network configuration and an optimum observing plan that will satisfy the pre-set network quality with the minimum effort. In other words, after the network quality requirements, i.e., the quality of the final products (for instance, precision and reliability), are given, the technique of network optimization allows for the finding of such an optimum network configuration and an optimum set of observations that will satisfy these requirements with the minimum cost (cf. Grafarend, 1974; Cross, 1985; Schmitt, 1985; Schaffrin, 1985; Kuang, 1991). Practically, the technique of network optimization serves to help us make decisions on which instruments should be selected from the hundreds of available models of various geodetic instruments and where they should be located and how the network should be measured in order to estimate the unknown parameters and achieve the desired network quality criteria derived from and determined by the purpose of the network. An optimized surveying scheme will, therefore, ensure the most economic field campaign, and it will help in identifying and eliminating gross errors in observations as well as in minimizing the effects of undetectable gross errors existing in the observation. Of interest to surveying engineers is that it enables surveyors to avoid unnecessary observations, and therefore may result in saving considerable time and effort in the field.

In general, the main task of optimal design of a geodetic network is comprised of:

- The determination of the optimal distribution of network points
- The selection of measurement technique(s), and
- The computation of the optimal distribution of required observational precisions among heterogeneous observables.

Due to the complexity of the problem, in the past it was very difficult, if not impossible, to solve for all aspects of network optimization in a single mathematical procedure. Instead, the problem of network design was divided into subproblems in which some progress could be made. The accepted classification was (Grafarend, 1974):

- Zero-Order Design (ZOD)—choosing an optimum reference system
- First-Order Design (FOD)—choosing the optimum locations for the stations
- Second-Order Design (SOD)—choosing what observations to make and with what precision to make them, and
- Third-Order Design (THOD)—choosing how to improve an existing network.

To the above design problems a fifth one can be added, called the Combined Design (COMD) (Grafarend et al., 1979) problem, where both the First- and Second-Order Design problems have to be optimally solved simultaneously. A network should be designed to satisfy the pre-set precision, reliability, and cost criteria in such a way that (cf. Schmitt, 1985):

- The postulated precision of the network elements, and of arbitrary estimable quantities, if any, can be realized
- It is as sensitive as possible to statistical testing procedures, which allow, for example, the detection of gross errors, in the measurements, as small as possible in magnitude, and
- The marking of the points and the performance of the measurements are satisfying some cost criteria.

It can be said that surveyors have been designing geodetic networks ever since their inception. With the development of computer technology the design approaches have evolved from the most primitive method based on intuition and/or empirical formulae (e.g., Asplund, 1963, etc.), through computer simulation (e.g., Ashkenazi, 1965; Nickerson, 1979, etc.), to analytical methods (e.g., Grafarend, 1974; Cross, 1985; Schaffrin, 1985; Schmitt, 1985; Wimmer, 1982; Jäger and Vogel, 1990; etc.; Kuang, 1991), and an abundance of research papers and articles have appeared concerning these subjects. Over the past two decades geodesists have done extensive theoretical and practical work to find the most appropriate analytical methods for the solution of optimal design problems, and great developments have been made recently. The recently developed multi-objective optimization methodology offers the possibility of solving all the aspects of network design problems in a single mathematical procedure in either one-, two-, or three-dimensional space (cf. Kuang, 1991, 1992f). Applications of the new methodology to solving different network design problems in one-, two-, and three-dimensional space have been elaborated by the author in a series of publications (cf. Kuang et al., 1991; Kuang 1992a-f, 1993a-h, 1994; Kuang and Chrzanowski 1992a,b, 1994). A brief review of the development of the network design methods, including both the criterion setup and the solution methodology, is given below, and this review should help the reader(s) unfamiliar with the subjects better understand the concerns and problems involved.

8.2 The "Trial and Error" Method and Computer Simulation

A primitive method that is suited for FOD, SOD, and THOD is the "trial and error" method, coupled with intuition and empirical formula. In this method, a solution to the design problem is postulated and the design and cost criteria computed. Should either of these criteria not be fulfilled, a new solution is postulated (usually by slightly altering the original postulate) and the criteria are recomputed. The procedure is repeated until a satisfactory (unlikely to be the optimum) network is found. With the development of modern computers, this "trial and error" method usually bears the name of the computer simulation that may greatly facilitate the computation and interpretation as summarized by the following steps (cf. Cross, 1985):

Step 1 Specify precision and reliability criteria
Step 2 Select an observation scheme (stations, observations, and weights)
Step 3 Compute the covariance matrices of the desired least squares estimates and derive the values of the quantities specified as precision and reliability criteria
Step 4 If these values are close to those specified in Step 1 then go to the next stage; otherwise alter the observation scheme (by removing observations or decreasing weights if the selected network is too good, or by adding observations or increasing weights if it is not good enough) and return to Step 3, and
Step 5 Compute the cost of the network and consider the possibility of returning to Step 2 and restarting the process with a completely different type of network (e.g. a traverse instead of triangulation). Stop when it is believed that the optimum (minimum cost) network has been found.

The method has been used for decades now and is well established. Some descriptions of software include Mepham and Krakiwsky (1983), Cross (1981), and Frank and Misslin (1980). Some efficient interactive graphics systems were described, for instance, in Nickerson (1979) and Conzett et al. (1980). With an interactive computer system with graphic terminals, the design can be criticized and directly improved in a dialogue mode.

The advantage of the simulation method is that arbitrary decision criteria can be used and compared together, in order to find the required design. There is no need to bring these criteria into a strong mathematical form that is indispensable if one uses purely analytical solutions with discrete risk functions. The obvious disadvantages of the method are that the optimum network may never be found and also a very large amount of work may be involved.

8.3 Analytical Methods

The starting point of analytical optimization techniques with regard to geodetic measurements was due to the dissertation of Helmert (1868), entitled "Studien über rationelle Vermessungen im Gebiet der höheren Geodäsie." Since that time several of the most exceptional geodesists have contributed to this subject, e.g., Schreiber (1882), Jung (1924), and Wolf (1961). They all attempted to minimize some objective function that describes the cost, precision, or reliability within a geodetic project by a scalar value. Baarda (1962) proposed a completely different concept that dealt with a so-called criterion matrix to be best approximated by the actual covariance matrix of the estimated parameters. These criterion matrices possess "ideal" structure (in a certain sense) that has to be specified in each case. Grafarend (1972) introduced the Taylor-Karman structured idealized variance-covariance matrix of Cartesian coordinates in two- and three-dimensional geodetic networks based on the theory of turbulence. In Grafarend and Schaffrin (1979), TK-structures were studied generally with variance matrices for azimuths, angles, and distances derived from general and special dispersion matrices of Cartesian coordinates and coordinate differences. In a subsequent publication, Schaffrin and Grafarend (1982) dealt with the problem of allocated criterion matrices that were computed using generalized inverses from the idealized variance-covariance matrix of azimuths, angles, or distances constructed under the postulate of homogeneity and isotropy. Molenaar (1981) extended the two-dimensional concept of Baarda into three dimensions by using quaternion algebra and spherical coordinates.

The solution strategies of the optimization problems in the different orders of the design are dependent both on the mathematical form to which the problem has been brought, and on the shape of the objective function that is representing the aim of the design. In contrast to the computer simulation method, the so-called "analytical" methods offer specific algorithms for the solution of particular design problems. Once set in motion, such an algorithm will automatically produce a network that will satisfy the user quality requirements and that will, in some mathematical sense, be optimum. Unfortunately, for decades almost all of the advances in purely analytical methods have been in finding solutions only for the Second-Order Design (SOD) problem. A comprehensive review of the methods used were given in Cross and Fagir (1982), among others, and they are summarized below to help the reader understand the various problems and concerns involved.

Firstly, the equations for SOD start with the following expression for the cofactor matrix $\mathbf{Q}_x$ of the estimated coordinates of the network points:

$$\mathbf{Q}_X = \left(\mathbf{A}^T \mathbf{P}\, \mathbf{A}\right)^{-1} \tag{8.1}$$

Inverting both sides of the above equation, we obtain

$$\mathbf{A}^T \mathbf{P}\, \mathbf{A} = \mathbf{Q}_X^{-1} = \mathbf{P}_X \quad \text{(Grafarend, 1974)} \tag{8.2}$$

where the network configuration matrix **A** has been assumed to be of full rank; that is, no datum defects and configuration defects exist. **P** is the observation weight matrix, and $\mathbf{P}_x$ the inverse of the criterion matrix.

8.3.1 Solution of the SOD by Generalized Inverse

Bossler et al. (1973) proposed a solution for (8.2) using the Moore-Penrose inverse, i.e.,

$$\mathbf{P} = \left(\mathbf{A}^+\right)^T \mathbf{P}_X\, \mathbf{A}^+ \tag{8.3}$$

The solution results in a positive-definite weight matrix. The problem with this approach is that since a fully populated observation weight matrix is not realizable by practical measurements, the solution is useless for practical applications.

In order to solve for a diagonal observation weight matrix **P**, i.e., for uncorrelated observations, the basic Equation (8.2) is reformulated using the Khatri-Rao product as follows (cf. Grafarend 1974; Schaffrin 1977)

$$\left(\mathbf{A}^T \Theta\, \mathbf{A}^T\right)\mathbf{p} = \text{vec}\left(\mathbf{P}_X\right) \tag{8.4}$$

where **p** is a vector containing the diagonal elements of **P**, Θ represents the Khatri-Rao product, and the operator "vec" produces a vector by stacking the columns of a matrix one under another (see Appendix A). Note that in a design problem where m new stations are connected by n observations for two-dimensional networks, (8.4) will be a set of m(2m+1) independent equations with n unknowns. One solution for **p** using the Moore-Penrose inverse is therefore

$$\mathbf{p} = \left(\mathbf{A}^T \Theta\, \mathbf{A}^T\right)^+ \text{vec}\left(\mathbf{P}_X\right) \tag{8.5}$$

Theoretically, Equation (8.5) is a minimum norm solution of (8.4); that is, it tries to best fit the criterion matrix with the variance-covariance matrix of the station coordinates in the sense that the sum of squares of the observational weights, i.e., $\mathbf{p}^T\mathbf{p}$, will be a minimum.

Due to the potential difficulties with inverting a poorly conditioned matrix $\mathbf{A}^T\Theta\mathbf{A}^T$ in Equation (8.5), Schaffrin et al. (1977), Schmitt (1977) suggested to rewrite (8.4) through the eigenvalue decomposition of the matrix $\mathbf{P}_x$ in order to improve the results. The eigenvalue decomposition of $\mathbf{P}_x$ reads

$$\mathbf{P}_x = \mathbf{E}\,\Lambda\,\mathbf{E}^T \tag{8.6}$$

where $\mathbf{E}$ is a matrix consisting of the normalized eigenvectors of $\mathbf{P}_x$, and Λ a diagonal matrix with diagonal elements being the eigenvalues of $\mathbf{P}_x$. Equation (8.6) is also called the canonical form of $\mathbf{P}_x$.

Substituting Equation (8.6) into Equation (8.4), one obtains a solution of the observational weights through Equation (8.5) as follows

$$\mathbf{p} = \left(\mathbf{Z}^T \ominus \mathbf{Z}^T\right)^+ \text{vec}\,(\Lambda) \tag{8.7}$$

where matrix $\mathbf{Z}$ is calculated by

$$\mathbf{Z} = \mathbf{A}\,\mathbf{E} \tag{8.8}$$

(8.5) and (8.7) should certainly produce identical solutions. Unfortunately, both solutions have proved to be impractical, since according to Cross and Whiting (1981), among others, they would regularly produce negative observation weights. These clearly have no physical meaning and are therefore difficult to interpret. For practical applications, one may simply discard observations with negative weights but this leads to disconnected networks, i.e., networks split into several independent sections. Alternatively, they also tried to discard only the observation with the least negative weight and the solution process repeated, but tests have shown that the observation with the least negative weight is rarely the least valuable.

8.3.2 Linear Programming

In order to avoid negative weights, the technique of linear programming is used. Boedecker (1977) solved (8.4) for the case of gravity networks by linear programming. Following his suggestion, Cross and Thapa (1979) attempted to find a solution for $\mathbf{p}$ in (8.4) such that the resulting network would have a covariance matrix that would, in some sense, be better than the criterion matrix. Therefore, a network is bound to satisfy this design criteria if the variances in the covariance matrix are forced to be smaller than those in the criterion matrix and, conversely, the covariances larger (cf. also Cross, 1985). But, since (8.4) involves an inversion of the criterion matrix, these inequalities were reversed and the linear programming problem was posed as

Minimize

$$\sum_{i=1}^{n} p_i$$

Subject to

$$\left(\mathbf{A}^{\mathrm{T}} \Theta\, \mathbf{A}^{\mathrm{T}}\right)\mathbf{p} \geq \mathrm{vec}\left(\mathbf{P}_{\mathrm{x}}\right) \quad \text{(diagonal elements)} \tag{8.9}$$

$$\left(\mathbf{A}^{\mathrm{T}} \Theta\, \mathbf{A}^{\mathrm{T}}\right)\mathbf{p} \leq \mathrm{vec}\left(\mathbf{P}_{\mathrm{x}}\right) \quad \text{(off - diagonal elements)} \tag{8.10}$$

$$\mathrm{p_i} \geq 0 \quad \text{for all i} \tag{8.11}$$

Unfortunately, the method usually yields networks that do not satisfy the design criteria. The reason is that the simple reversal of inequality signs due to the inversion of the criterion matrix is not valid. It seems impossible to predict, for a given choice of method, the correct inequality signs for (8.9) and (8.10). In order to avoid the inversion of the criterion matrix, Cross and Whiting (1980) suggested to expand the left-hand side of (8.2) using an unspecified generalized inverse, i.e.,

$$\left(\mathbf{A}^{\mathrm{T}}\right)^{-} \mathbf{P}^{-}\, \mathbf{A}^{-} = \mathbf{Q}_{\mathrm{x}} \tag{8.12}$$

which, after application of the Khatri-Rao product, becomes

$$\left[\left(\mathbf{A}^{-}\right)^{\mathrm{T}} \Theta \left(\mathbf{A}^{-}\right)^{\mathrm{T}}\right]\mathbf{w} = \mathrm{vec}\left(\mathbf{Q}_{\mathrm{x}}\right) \tag{8.13}$$

where **w** is a vector containing the reciprocals of required weights of the observations, i.e., the diagonal elements of **P**. The above linear programming problem is then restated as follows:

Maximize

$$\sum_{\mathrm{i=1}}^{\mathrm{n}} \mathrm{w_i}$$

Subject to

$$\left[\left(\mathbf{A}^{-}\right)^{\mathrm{T}} \Theta \left(\mathbf{A}^{-}\right)^{\mathrm{T}}\right]\mathbf{w} \leq \mathrm{vec}\left(\mathbf{Q}_{\mathrm{x}}\right) \quad \text{(diagonal elements)} \tag{8.14}$$

$$\left[\left(\mathbf{A}^{-}\right)^{\mathrm{T}} \Theta \left(\mathbf{A}^{-}\right)^{\mathrm{T}}\right]\mathbf{w} \geq \mathrm{vec}\left(\mathbf{Q}_{\mathrm{x}}\right) \quad \text{(off - diagonal elements)} \tag{8.15}$$

$$\mathrm{w_i} \geq 0 \quad \text{for all i} \tag{8.16}$$

The objective function for this setup must be maximized in order to reduce the total work. The difficulty with this approach is that a suitable generalized inverse could not be found. Cross and Whiting (1980) have tried to use the Moore-Penrose inverse even though they showed that theoretically it was not valid.

8.3.3 Quadratic Programming

Schaffrin et al. (1980) suggested the use of the linear complementary algorithm to overcome the negative weight problem. This involves determining a best-fit solution of (8.4) in the sense of least squares, subject to a number of linear constraints that in addition to describing the required precision and cost of the network, also ensure that **P** is non-negative. The mathematical setup is essentially equivalent to a Quadratic Programming (QP) problem and can be written as

Minimize

$$\left[\left(\mathbf{A}^{\mathrm{T}} \ominus \mathbf{A}^{\mathrm{T}}\right)\mathbf{p} - \mathrm{vec}\left(\mathbf{P}_{\mathrm{X}}\right)\right]^{\mathrm{T}}\left[\left(\mathbf{A}^{\mathrm{T}} \ominus \mathbf{A}^{\mathrm{T}}\right)\mathbf{p} - \mathrm{vec}\left(\mathbf{P}_{\mathrm{X}}\right)\right] \tag{8.17}$$

Subject to

$$\left(\mathbf{A}^{\mathrm{T}} \ominus \mathbf{A}^{\mathrm{T}}\right)\mathbf{p}\ (\geq; =; \leq)\ \mathrm{vec}\left(\mathbf{P}_{\mathrm{X}}\right) \tag{8.18}$$

$$\mathbf{c}^{\mathrm{T}}\mathbf{p} \leq \mathrm{d} \tag{8.19}$$

$$\mathrm{p_i} \geq 0 \quad \text{for all i} \tag{8.20}$$

where **c** is a vector of coefficients relating observation weight to cost and d the total allowable cost. Note that the difficulty regarding the inequality signs for use in (8.18) arises again but Schaffrin et al. (1980) states that it may be avoided by reforming (8.17) and (8.18) using the canonical formulation (8.6) and restricting (8.18) to the rows that correspond to the eigenvalues of $\mathbf{P}_x$ within vec Λ. Equation (8.18) then becomes

$$\left(\mathbf{Z}^{\mathrm{T}} \ominus \mathbf{Z}^{\mathrm{T}}\right)\mathbf{p} \geq \mathrm{vec}\,(\Lambda) \tag{8.21}$$

Schaffrin et al. (1980) reported a successful application of the above model to a geodetic network with Taylor-Karman structured criterion matrices.

8.3.4 Iterative Generalized Inverse

Wimmer (1982b) proposed a method by reforming the basic mathematical statement of the Second-Order Design as follows

$$\mathbf{P} = \mathbf{P}\,\mathbf{P}^{-1}\,\mathbf{P} \tag{8.22}$$

Post-multiplying both sides of (8.22) by $\mathbf{A}\,(\mathbf{A}^T\,\mathbf{P}\,\mathbf{A})^{-1}$ and pre-multiplying by $(\mathbf{A}^T\,\mathbf{P}\,\mathbf{A})^{-1}\mathbf{A}^T$ yield

$$\left(\mathbf{A}^T\,\mathbf{P}\,\mathbf{A}\right)^{-1} = \left[\left(\mathbf{A}^T\,\mathbf{P}\,\mathbf{A}\right)^{-1}\mathbf{A}^T\,\mathbf{P}\right]\mathbf{P}^{-1}\left[\mathbf{P}\,\mathbf{A}\left(\mathbf{A}^T\,\mathbf{P}\,\mathbf{A}\right)^{-1}\right] \tag{8.23}$$

Denote

$$\mathbf{G} = \mathbf{P}\,\mathbf{A}\left(\mathbf{A}^T\,\mathbf{P}\,\mathbf{A}\right)^{-1} \tag{8.24}$$

Substituting (8.1) and (8.24) into (8.23), one obtains

$$\mathbf{Q}_x = \mathbf{G}^T\mathbf{P}^{-1}\mathbf{G} \tag{8.25}$$

Applying the Khatri-Rao product to (8.25) and rearranging them yields

$$\left(\mathbf{G}^T \odot \mathbf{G}^T\right)\mathbf{w} = \mathrm{vec}\left(\mathbf{Q}_x\right) \tag{8.26}$$

Denote

$$\mathbf{H} = \left(\mathbf{G}^T \odot \mathbf{G}^T\right) \tag{8.27}$$

Substituting (8.27) into (8.26) yields

$$\mathbf{H}\,\mathbf{w} = \mathrm{vec}\left(\mathbf{Q}_x\right) \tag{8.28}$$

where **w** contains the reciprocals of the diagonal elements of **P**.

This formulation is of a structure similar to (8.4) but has the considerable advantage of being in terms of the criterion matrix itself rather than its inverse. All solutions to (8.28) must, however, be iterative because, according to (8.24), the matrix **G** is itself in terms of **P**. Hence we must first assume a set of values for **P**, solve (8.28) for **w** (and hence **P**) and use this value to recompute **G**. The process is repeated until **P** ceases to change. Cross and Fagir (1982) reported a successful application of the approach to designing a leveling network.

From the above discussion one can see that the problem common to most of the previous formulations for SOD is that in order to produce a system of linear equations for the solution of the optimal weights the optimization equations are formulated in terms of the fit of the inverse of the criterion matrix, instead of the criterion matrix itself. That makes it difficult to achieve the design criteria, since a good fit of the inverse of the criterion matrix does not necessarily mean the good fit of the criterion matrix itself. The success of SOD depends on both the setup of the criterion matrix and proper formulation of the mathematical model. Practically, no universally applicable and efficient analytical method existed for network optimization, and therefore the method of computer simulation or “trial and error” was still the production method in practice.

The justification of the First-Order Design has been called in question for a time for large scale networks of general purpose since there is practically no margin because of topographical realities (Schmitt, 1982). Koch (1982, 1985) proposed an innovative algorithm for the First-Order Design (FOD) of engineering or deformation networks of local scale by introducing differential position changes. This algorithm is, nevertheless, semi-analytical since in his mathematical modeling the derivatives needed were still provided by numerical methods.

8.3.5 A New Methodology

Having recognized the above problems, geodesists continue to search for the most appropriate analytical methods for network design. The present author started to develop a new methodology for the optimal design of integrated deformation monitoring schemes with geodetic and non-geodetic observables in 1988 and successfully completed the research in a form of a Ph.D. dissertation titled *Optimization and Design of Deformation Monitoring Schemes*, in 1991. This newly developed methodology allows for a fully analytical multi-objective optimization of a monitoring scheme. In the meantime, in the works of Jäger and Vogel (1990a,b), Koch's concept was extended by a fully analytical differentiation of the variance-covariance matrix of the estimated coordinates with respect to both the coordinates and observational weights. The main feature with the multi-objective optimization methodology developed by the author is that the Taylor series technique widely used in linearizing nonlinear scalar functions has been introduced to linearize nonlinear matrix functions related to network design, and that makes it possible to perform either separate or simultaneous fully analytical optimal solution of the network configuration and the observational plan using the methodology of operations research. Fully analytical optimal solution for the problems of First-Order (FOD), Second-Order (SOD), Third-Order (THOD), or the Combined First-Order and Second-Order Design can now be easily performed. In addition, all the quality aspects of precision, reliability, sensitivity, and cost of a network can be expressed fully analytically in terms of the unknown parameters to be optimized and considered simultaneously in constructing a global objective function based on the theory of multi-objective optimization. Hence, all the aspects of the conventional network design problems are now solved in a single mathematical procedure. The developed methodology is also universally applicable to the optimal design of networks in either one-, two-, or three-dimensional space. The author has written a FORTRAN-77 software package implementing all the features of the developed methodology, and excellent results were achieved when it was applied to the planning of both deformation monitoring and geodetic control surveys (cf. Kuang et al. 1991; Kuang 1992a-f, 1993a-h, 1994; Kuang and Chrzanowski, 1992a,b, 1994). Full details of the new methodology for the optimal design of geodetic positioning and deformation networks are given in Chapters 10 and 11, respectively, and a few application examples are included in Chapter 12 to demonstrate in detail how the methodology is applied in actual surveying projects.

9 Formulation of Optimality Criteria

Mathematically, optimization of geodetic networks means minimizing or maximizing an objective function that represents the "quality of the network" (Schmitt, 1985). As discussed in Chapter 6, three general criteria are used to evaluate this quality; that is, precision, reliability, and economy. Precision is the measure of how the network propagates random observational errors; reliability describes the ability of the redundant observations to check observational errors and the network's ability to resist residual gross errors in the network; and finally, economy is expressed in terms of the cost of the observation program. These criteria must be fulfilled for the different design problems of a geodetic network. Thus, conceptually, network optimization means to design a precise and reliable network that can also be realized in an economical way. This chapter discusses the mathematical setup of the various network quality criteria, i.e., the objective functions, which will be used in the mathematical modeling of network optimization problems as discussed in the following Chapters 10 and 11.

9.1 Optimality Criteria for Precision

In general, the optimality criteria for precision are dictated by the precision measures used to describe the network. As discussed in §6.1, full information about the precision of a geodetic network is contained in the global variance-covariance matrix $\mathbf{C}_x$ of point coordinates. The purpose for which a network must serve is decisive for the determination of the precision required. For instance, when defining a geodetic network for setting out an engineering structure, for controlling the breakthrough of tunneling, or for providing photogrammetric control, there may be special requirements, e.g., the accuracy of certain functions pertaining to specific points or group of points should satisfy certain pre-set values. In the case of the general purpose networks, however, the precision requirements cannot be specified so easily. In such cases some concepts of "ideal" networks are proposed. These ideal precision criteria can either be based on theoretic results, such as those of Grafarend (1972), on homogeneous and isotropic networks (Taylor-Karman Structure), or they can be derived from empirical studies with real networks. A great deal of work has been carried out in the field of user precision requirements for geodetic networks. In general, they can be classified into two types of precision criteria. One consists of the scalar functions of the elements of the variance-covariance matrix of the coordinates, and the other is the criterion matrices. It should be kept in mind that any precision criterion, whether in the form of scalar functions, or criterion matrices, must satisfy the desires of the network users.

9.1.1 Scalar Risk Function

Scalar precision criteria are made of global or local scalar precision measures (cf. §6.1.1.3 and §6.1.2.5) that serve as an overall representation of the precision of a network. Different scalar precision criteria for geodetic networks were proposed by Grafarend (1974). Typical examples are the N-, A-, E-, S-, and D-optimality criteria, which have the following meaning:

- N-Optimality

$$f = \| \mathbf{C}_x \| = \min \tag{9.1}$$

with $\| \cdot \|$ denoting the norm of a matrix

- A-Optimality

$$f = \mathrm{Trace}\left(\mathbf{C}_x\right) = \lambda_1 + \lambda_2 + \ldots + \lambda_h = \min \tag{9.2}$$

with $\lambda_1, \lambda_2, \ldots, \lambda_h$ being the non-zero eigenvalues of the matrix C_x

- E-Optimality

$$f = \lambda_{max} = \min \tag{9.3}$$

with λ_{max} being the maximum eigenvalue of matrix $\mathbf{C}_x$

- S-Optimality

$$f = \left(\lambda_{max} - \lambda_{min}\right) = \min \tag{9.4}$$

with $(\lambda_{max} - \lambda_{min})$ being the spectral width of matrix $\mathbf{C}_x$

- D-Optimality

$$f = \mathrm{Det}\left(\mathbf{C}_x\right) = \lambda_1 * \lambda_2 * \ldots * \lambda_h = \min \tag{9.5}$$

with $\lambda_1, \lambda_2, \ldots, \lambda_h$ being the non-zero eigenvalues of the variance-covariance matrix $\mathbf{C}_x$, and Det the determinant of a matrix.

Finally, if we demand some certain function f of coordinates **x**

$$f = \mathbf{c}^T \mathbf{x} \tag{9.6}$$

have the highest precision, an optimality criterion may be written as

$$c_f = \mathbf{c}^T \mathbf{C}_x \mathbf{c} = \min \tag{9.7}$$

where **c** is a vector of constants. A practical example for the constraints of the type of Equation (9.7) can be seen in tunneling surveys where what is usually of most concern is the lateral horizontal breakthrough error that can be expressed as the function of the accuracies of the coordinates of the predesigned breakthrough points, and one usually requires that this error be minimum or be less than a certain boundary value (cf. Kuang, 1993c). There may be many more other scalar precision criteria, depending on one's purpose.

Since global scalar precision measures of a network are rather coarse characteristics of the variance-covariance matrix $\mathbf{C}_x$, scalar precision criteria based on them therefore do not control individual values of the elements of the variance-covariance matrix. They can be used as criteria for the comparison of different designs. However, it is difficult to use a scalar precision criterion as an absolute criterion whose numerical value must, for instance, not surpass a certain pre-assigned value related to the purpose of the network (Alberda, 1974). Therefore, the application of scalar precision criteria in practice is limited. In the case that the precision criteria are set up for all the individual elements of the variance-covariance matrix of point coordinates, the use of a criterion matrix is recommended, through which a much more detailed control of precision is provided, as discussed below (cf. also Grafarend et al., 1979). Whenever a criterion matrix is used in the optimization procedure, the optimal design of a network is sought such that the error of the fit of the variance-covariance matrix of the estimated coordinates to the criterion matrix is a minimum.

9.1.2 Criterion Matrices

A criterion matrix is an artificial variance-covariance matrix possessing an ideal structure, where "ideal" means that it represents the optimal accuracy situation in the planned network. The structure of the criterion matrix also depends on the purpose of the network. For general purpose networks, such as the control networks for regional or national mapping, it is necessary to have some concepts of "ideal" networks. These ideal precision criteria can be based on theoretical results, such as those of Grafarend (1972) and Baarda (1973), on homogeneous and isotropic networks, i.e., the Taylor-Karman structure (abbreviated as TK-structure). They can also be derived from empirical studies with real networks. A special case of the Taylor-Karman structured criterion matrix is the identity matrix. For special purpose networks, the elements of the criterion matrix can be computed from the user's requirements, such as the shape of error ellipses or the accuracy of the derived quantities (Schmitt, 1985). For instance, in establishing networks for setting out an engineering structure, for controlling the breakthrough of tunneling, etc., there may be special accuracy requirements for certain functions pertaining to specific points or group of points (cf. Cross, 1985). Computation procedures for Taylor-Karman structured criterion matrices and examples of criterion matrices for special purpose networks are discussed below. Finally, a simple but efficient criterion matrix can be a diagonal matrix whose diagonal elements are the allowable variance of the estimated coordinates. If, instead of a scalar risk function, a criterion matrix is introduced in an optimal design procedure, the solution to the optimization problem must approximate it as closely as possible as given by Equation (9.33).

9.1.2.1 The Taylor-Karman Structure

The structure of the criterion matrix for general purpose networks, such as the control networks for regional or national mapping, has been extensively studied by Grafarend (1972) and Baarda (1973). The results are in the form of the general Taylor-Karman structured criterion matrix or its chaotic structure. The theory of Taylor-Karman structured criterion matrix is quite involved, a comprehensive review is given below (cf. also Schmitt, 1977; Schaffrin 1985).

Grafarend (1972) introduced the Taylor-Karman structured idealized variance-covariance matrix of Cartesian coordinates in two- and three-dimensional geodetic networks based on the theory of turbulence. Suppose we are given a two-point tensor function

$$\Phi = \Phi_{ij}(\mathbf{r}, \mathbf{r}), \mathrm{i}, \mathrm{j} \in \{1, 2, 3\} \tag{9.8}$$

where $\mathbf{r}$ is the position vector OP, $\mathbf{r}'$ the position vector OP′, and O the origin of the coordinate system.

Let $\mathbf{e}_i(\mathbf{r})$ be a system of three orthonormal basis vectors at point P, $\mathbf{e}_j(\mathbf{r}')$ has the same meaning at point P′. Then the Second-order form associated to the tensor function of second rank is written by

$$\Phi = \Phi_{ij}(\mathbf{r}, \mathbf{r})\mathbf{e}_i(\mathbf{r})\mathbf{e}_j(\mathbf{r}) \tag{9.9}$$

The summation has to be carried out over two-by-two identical subscripts.

- Φ is called homogeneous if it is translation invariant, i.e.,

$$\Phi_{ij}(\mathbf{r} + \mathbf{t}, \mathbf{r}' + \mathbf{t}) = \Phi_{ij}(\mathbf{r}, \mathbf{r}') \tag{9.10}$$

  with $\mathbf{t}$ a displacement vector. The tensor function $\Phi_{ij}(\mathbf{r}, \mathbf{r})$ is strictly homogeneous if it is a function of the difference between the position vectors $(\mathbf{r} - \mathbf{r})$.
- Φ is called isotropic if it is rotation invariant, i.e.,

$$\Phi_{ij}(\mathbf{Rr},\ \mathbf{Rr}') = \mathbf{R}\Phi_{ij}(\mathbf{r},\ \mathbf{r}')\mathbf{R}^T \tag{9.11}$$

where **R** is a rotation matrix. Provided that the Second-order form Φ is rotation invariant, the tensor function $\Phi_{ij}(\mathbf{r},\ \mathbf{r})$ is written as

$$\Phi_{ij}(\mathbf{r},\ \mathbf{r}') = \delta_{ij}\ \Phi(\mathbf{r},\ \mathbf{r}') \tag{9.12}$$

where $\Phi(\mathbf{r},\ \mathbf{r}')$ is a scalar function, and δ_{ij} is the Kronecker identity matrix.

- Φ is called homogeneous and isotropic if it is translation and rotation invariant. Provided that the Second-order form Φ is translation and rotation invariant, the tensor function $\Phi_{ij}(\mathbf{r},\ \mathbf{r})$ is written as

$$\Phi_{ij}(\mathbf{r},\ \mathbf{r}') = \Phi_m(|\mathbf{r}-\mathbf{r}'|) + \left[\Phi_l(|\mathbf{r}-\mathbf{r}'|) - \Phi_m(|\mathbf{r}-\mathbf{r}'|)\right]\Delta x_i \Delta x_j / |\mathbf{r}-\mathbf{r}'|^2 \tag{9.13}$$

where the characteristic functions Φ_l and Φ_m are called the longitudinal, and lateral functions, respectively. They are functions of the distance $|\mathbf{r} - \mathbf{r}'|$ between the points **P** and **P′**. (Δx_1 Δx_2 Δx_3) is the vector of the differences of Cartesian coordinates between the points **P** and **P′**.

Equation (9.13) has been named after Taylor (1935) and Karman (1937) and was introduced to geodesy by Grafarend (1972). By applying the two-point tensor function on geodetic networks, for the special case of homogeneity and isotropy, an expression for the submatrix of cross-covariance between points P_i and P_j on the plane within the TK-structured criterion matrix equals to (Grafarend, 1972):

$$(\mathbf{C}_{ij}) = \begin{pmatrix} \Phi_m(s) & 0 \\ 0 & \Phi_m(s) \end{pmatrix} + [\Phi_l(s) - \Phi_m(s)] \begin{pmatrix} (x_i - x_j)^2 / s^2 & (x_i - x_j)(y_i - y_j)/s^2 \\ (x_i - x_j)(y_i - y_j)/s^2 & (y_i - y_j)2/s^2 \end{pmatrix} \tag{9.14}$$

where

$$\Phi_m(s) = \frac{4\,d^2}{s^2} - 2\ k_0(s/d) - \frac{4\,d}{s} k_1(s/d) \tag{9.15}$$

$$\Phi_l(s) = -\frac{4\,d^2}{s^2} + 2\ k_0(s/d) + \frac{4\,d}{s} k_1(s/d) + \frac{2\,s}{d} k_1(s/d) \tag{9.16}$$

$$s = |r_i - r_j| = \sqrt{(x_i - x_j)^2 + (y_i - y_j)^2} \tag{9.17}$$

in which k_0 and k_1 are the modified Bessel functions of the second degree and of zero and First-order, d is the characteristic length. The magnitude of the characteristic distance d of a network still remains a problem. For example, according to Schmitt (1980), d should be chosen smaller than the minimum distance between any arbitrary two points of the network, while Wimmer (1982a) recommends the maximum distance (or diameter) of the network as an upper bound for 10 d (cf. also Schaffrin, 1985). Equation (9.14) leads to an error situation characterized by identical error circles at the netpoints as shown by Figure 9.1 where the solid and dotted ellipses represent the station and relative error ellipses, respectively.

9.1.2.2 The Chaotic Structure

Baarda (1973) and Alberda (1974) proposed to use a special case of this structure called the "**chaotic structure**." In that case, the relative error ellipses are also circles with a radius that depends on the distance between the points to which they refer (see Figure 9.2). They adopted the following relation (cf. also Molenaar, 1986b).

$$\Phi_m(s) = \Phi_1(s) = d^2 - r^2 s,\ r^2 > 0 \tag{9.18}$$

for a suitable constant r^2. The submatrix of cross-covariance between points P_i and P_j on the plane is

$$(\mathbf{C}_{ij}) = \begin{pmatrix} d^2 & 0 & d^2 - d_{ij}^2 & 0 \\ & d^2 & 0 & d^2 - d_{ij}^2 \\ \text{symmetric} & & d^2 & 0 \\ & & & d^2 \end{pmatrix} \tag{9.19}$$

where d^2 = constant; and $d_{ij}^2 = d_{ji}^2 = f(l_{ij})$ is a positive monotonic non-decreasing function of the side length l_{ij}. The d^2 and d_{ij}^2 should be chosen so that **C** is positive definite. However, the variance-covariances of all kinds of coordinate differences do not have anything to do with d^2. For example, for i, j≠1:

$$\sigma^2_{(x_i - x_1)} = 2\ d_{i1}^2 \tag{9.20}$$

$$\sigma^2_{(x_i - x_1)(x_j - x_1)} = d_{1i}^2 + d_{1j}^2 - d_{ij}^2 \tag{9.21}$$

From (9.20) and (9.21), we can see that by subtracting x_1 and y_1 from all other x_i and y_i we obtain new coordinate variates whose covariance matrix does not

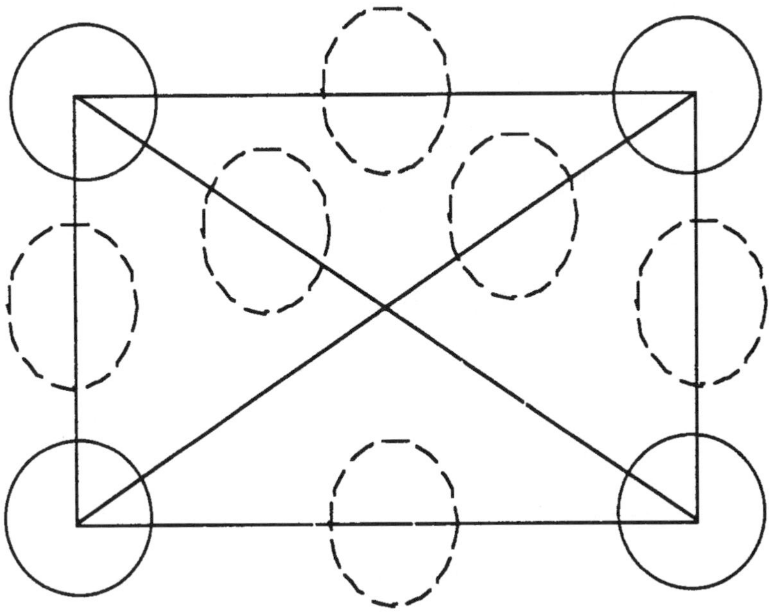

Figure 9.1. Station and relative error ellipses from a TK structured criterion matrix

contain d^2 any more. So it plays no role in the formulation of the criterion matrix for precision, only the function d_{ij}^2 is important. It is called the "**Choice function**" because one can choose this function to set a criterion. This function should be defined so that the transformed criterion matrix referred to a S-base $\mathbf{C}_s$ is positive definite. Baarda (1973) proposed that

$$d_{ij}^2 = c_0 + c_1\, l_{ij} \,,\ c_0 > 0 \,,\ c_1 > 0 \tag{9.22}$$

This function has been used in Netherlands for network reconnaissance. Alternative choices are (cf. Karadaidis, 1984; Meissl, 1976)

$$d_{ij}^2 = c_1^2\, c_2\, \ln\left\{1 + \frac{l_{ij}}{c_2}\right\} + c_0^2 \tag{9.23}$$

or

$$d_{ij}^2 = c_1^2 \left[1 - \exp\left(-c_2^2\, l_{ij}\right)\right] + c_0^2 \tag{9.24}$$

Other choices are possible too. The actual choice of the function depends on the type of the network. Once a choice has been made, the parameters c_0, c_1, and c_2

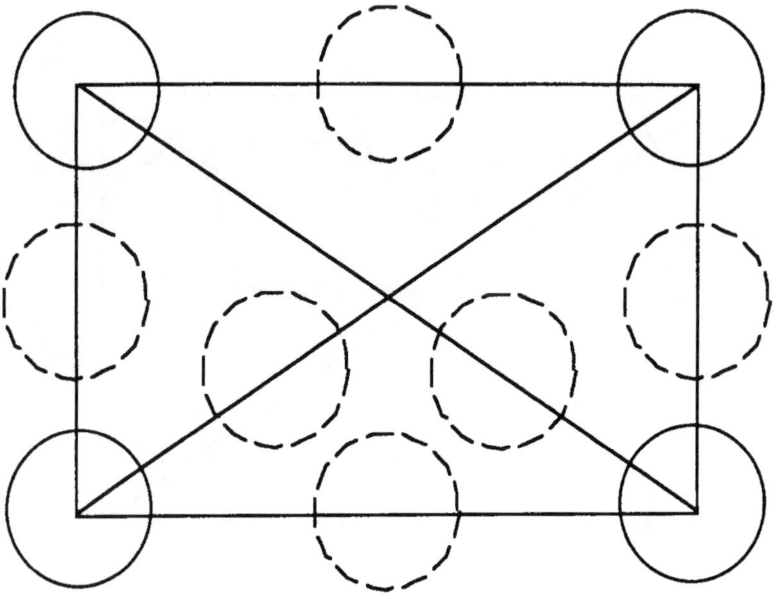

Figure 9.2. Station and relative error ellipses from a criterion matrix of chaotic structure

can be used to set the actual criterion for precision. This may depend on the class of network, e.g.: densification level, terrain type, etc. The error distribution of a chaotic structured criterion matrix is shown by Figure 9.2 in which the solid and dotted ellipses represent the station and relative error ellipses, respectively, and they all are of circular shape.

9.1.2.3 Modification of the Present Covariance Matrix

The potential problem with the Taylor-Karman structured criterion matrices is that the requirement for statistical homogeneity and isotropy is too strict for real networks, and it may be, therefore, very difficult, if not entirely possible, to realize in practice. For special purpose networks such as met in civil engineering and deformation measurements, the elements of the criterion matrix can be computed from user requirements, such as the shape of error ellipses or the accuracy of derived quantities. In this respect, an example was given by Koch (1982). He suggested constructing a criterion matrix by modifying the present covariance matrix of coordinates of geodetic netpoints in the following way:

Let $(\hat{x}_i, \hat{y}_i)$ and $(\hat{x}_j, \hat{y}_j)$ be the estimates of the projected coordinates of two points of a two-dimensional network and let their variance-covariances be given by

$$C\begin{pmatrix}\hat{x}_i\\ \hat{y}_i\\ \hat{x}_j\\ \hat{y}_j\end{pmatrix} = \begin{pmatrix}\sigma^2_{x_i} & \sigma_{x_iy_i} & \sigma_{x_ix_j} & \sigma_{x_iy_j}\\ & \sigma^2_{y_i} & \sigma_{y_ix_j} & \sigma_{y_iy_j}\\ \text{symmetric} & & \sigma^2_{x_j} & \sigma_{x_jy_j}\\ & & & \sigma^2_{y_j}\end{pmatrix} \tag{9.25}$$

By the law of error propagation the covariance matrix of the coordinates differences $(\hat{x}_i, \hat{y}_i)$ and $(\hat{x}_j, \hat{y}_j)$ of the two points follow

$$C\begin{pmatrix}\hat{x}_j - \hat{x}_i\\ \hat{y}_j - \hat{y}_i\end{pmatrix} = \begin{pmatrix}\sigma^2_{x_i} + \sigma^2_{x_j} - 2\sigma_{x_ix_j} & \sigma_{x_jy_j} - \sigma_{x_jy_i} - \sigma_{x_iy_j} + \sigma_{x_iy_i}\\ \text{symmetric} & \sigma^2_{y_i} + \sigma^2_{y_j} - 2\sigma_{y_iy_j}\end{pmatrix} \tag{9.26}$$

According to Equations (6.105) and (6.106), one can see that a confidence ellipse of circular shape follows for the first point from the equality

$$\left(\sigma^2_{x_i} - \sigma^2_{y_i}\right)^2 + \left(2\sigma_{x_iy_i}\right)^2 = 0 \tag{9.27}$$

which is satisfied by

$$\sigma^2_{x_i} = \sigma^2_{y_i} \text{ and } \sigma_{x_iy_i} = 0 \tag{9.28}$$

and for the second point by

$$\sigma^2_{x_j} = \sigma^2_{y_j} \text{ and } \sigma_{x_jy_j} = 0 \tag{9.29}$$

If, in addition

$$\sigma_{x_ix_j} = \sigma_{y_iy_j} \text{ and } \sigma_{x_jy_i} = -\sigma_{x_iy_j} \tag{9.30}$$

a relative confidence ellipse of circular shape is also obtained for the coordinate differences of the two points.

Let $\mathbf{C}_x$ be the covariance matrix of the coordinates of the network to be optimized, the elements of the covariance matrix for the points to be optimized are then changed according to Equations (9.28) to (9.30) so that smaller confidence ellipses with circular shapes for the coordinates and the coordinate differences of these points are obtained. This changed covariance matrix is then used as the criterion matrix for the optimization.

Finally, a simple but efficient criterion matrix suitable for either one- or multi-dimensional networks can be a diagonal matrix whose diagonal elements are established according to the required accuracies of the estimated coordinates (cf. Kuang, 1991).

9.1.2.4 The Datum Problem for Criterion Matrix

The Taylor-Karman structured criterion matrix (9.14) and (9.19) refer to a point field where all coordinates are stochastic. In principle, the criterion matrix should be created independently of any linear model connecting the sought parameters to certain observations. However, that is not possible in real point fields where a network datum for coordinates should be defined. In order to make the criterion matrix comparable with a "real" covariance matrix that has a defined datum, the criterion matrix should be transformed with the same datum parameters. This transformation may be accomplished by executing the so-called S-transformation [cf. Equation (4.174)]

$$\mathbf{C}_{\mathrm{S}}^{\mathrm{c}} = \mathbf{S}\,\mathbf{C}_{\mathrm{TK}}^{\mathrm{c}}\,\mathbf{S}^{\mathrm{T}} \tag{9.31}$$

where $\mathrm{C}_{\mathrm{TK}}^{\mathrm{c}}$ is the Taylor-Karman structured criterion matrix, and the matrix $\mathbf{S}$ is expressed as

$$\mathbf{S} = \left[\mathbf{I} - \mathbf{H}\left(\mathbf{D}^{\mathrm{T}}\mathbf{H}\right)^{-1}\mathbf{D}^{\mathrm{T}}\right] \tag{9.32}$$

with $\mathbf{I}$ being an identity matrix, matrix $\mathbf{D}$ and $\mathbf{H}$ describe the minimum constrained and its corresponding inner constrained network datum matrix, respectively (cf. §4.2.3 and §4.2.4).

If a criterion matrix is used in the optimization procedure, then the optimal design of a network is sought such that the error of the fit of the variance-covariance matrix of the estimated coordinates to the criterion matrix referred to the same datum is a minimum, i.e., the objective function can be written as:

$$\left\|\mathbf{C}_{\mathrm{X}} - \mathbf{C}_{\mathrm{S}}^{\mathrm{c}}\right\| = \text{minimum} \tag{9.33}$$

9.2 Optimality Criteria for Reliability

The reliability of a network depends on the geometry of the network, i.e., the configuration matrix and on the weight matrix of observations, not on the actual observations. Thus the problem of reliability should be considered at the design stage to ensure the detection of gross errors as small as possible and to minimize the effects of the undetected ones on the estimated parameters. At the design stage, the formulation of objective functions for network reliability should therefore be based on the following considerations:

- Gross errors should be detected and eliminated as completely as possible. An undetected gross error $\nabla_0 l_i$ in an observation l_i should be small in comparison with the standard deviation σ_{l_i} of l_i
- The effect of an undetected error $\nabla_0 l_i$ on estimated unknown parameters and/or functions of them should be as small as possible

From (5.69) and (6.134), we can see that the larger the redundancy number r_i the smaller is the size of the undetectable gross errors as well as its influence on the estimated coordinates. Thus a general criterion suitable for both the internal and external reliability of a geodetic network can be posed as (cf. Kuang, 1991).

$$\| \mathbf{r} \| = \text{maximum} \tag{9.34}$$

where $\mathbf{r}$ is a vector consisting of the redundancy numbers of observations, i.e.,

$$\mathbf{r} = \left(r_1,\ r_2,\ \ldots,\ r_n\right)^T \tag{9.35}$$

and $\| \cdot \|$ represents the norm of a vector.

It has been argued that it is desirable to have approximately a constant value for all r_i (i=1, ..., n) so that the ability of detecting gross errors is the same in every part of the network. If certain observation variates have larger boundary values, they are insufficiently checked and the network should be improved locally. With this in mind, a special reliability criterion can be of the type (cf. Baarda 1968; Van Mierlo 1981; Alberda 1980; Kuang, 1991):

$$\min \left(r_i\right) = \max \tag{9.36}$$

9.3 Optimality Criteria for Economy

A geodetic network should be designed in order that it satisfies the user precision and reliability requirements with the least cost. The cost of a survey campaign depends on many factors, and it may vary considerably from project to project. When a fully analytical optimization approach is used, it is then essential to write down a general mathematical cost function involving the observations to be considered during the design process. For instance, Cross and Whiting (1981) proposed to evaluate the cost of a leveling network by

$$\text{Cost} = \text{constant} \times \text{length of leveling} \tag{9.37}$$

thus minimizing cost amounts to minimizing the total length of leveling. However, cost functions for two- or three-dimensional networks can be much more complex. Since the analytical design techniques require the cost function to be primarily in terms of unknown parameters to be optimized, that is, the network configuration and observing plan, this, in practice, is not usually the case. Also, the methods may require cost functions to be continuous, which again is unlikely to be true.

According to Schmitt (1982, 1985), among others, an approximate approach to bring cost criteria into a mathematical form is to split the costs for the measurements into constant terms, e.g., cost of driving to the stations, cost of setting up the instruments, and cost of signaling the points. The remaining free parameters are then the repetition numbers of the observations. Different instrumentations can be considered in this concept by the introduction of special efficiency numbers for these instruments or error formulas for a single measurement. By making the assumption that the smaller the weight of an observation the less expensive it is to observe, a simplified formulation for minimum cost criteria can be (cf. also Schaffrin, 1985)

$$\| \mathbf{P} \| = \text{minimum} \tag{9.38}$$

where $\| \cdot \|$ represents the norm of the weight matrix $\mathbf{P}$.

10 Formulation and Solution of Optimization Problems

This chapter deals with the formulation and solution of mathematical models for geodetic network optimization. It starts with a discussion on the conventional "trial and error" approach for network design in §10.1. In order to develop a fully analytical approach for network optimal design, the various network quality criteria are first brought into a strong mathematical form in §10.2 and §10.3 in terms of the unknown parameters to be optimized. General mathematical models for network optimization are then formulated in §10.4 based on the theory of operations research, and some special cases of the models are established in §10.5 that are solvable by standard linear or quadratic programming techniques. Finally, the concept of multi-objective optimization is reviewed in §10.6, and based on the theory of multi-objective optimization a universally applicable Multi-Objective Optimization Model (MOOM) for geodetic network design is established that may guarantee a feasible solution in any case as long as the design problem is properly defined.

10.1 The "Trial and Error" Versus the Analytical Approach for Network Design

The conventional approach to designing a geodetic control network may be summarized below and illustrated by Figure 10.1:

Step 1. The first step is to select preliminary locations of control stations, which are represented by shaded triangles in Figure 10.1, on a map or on photographs based on the specific needs of survey project control required by the agency that requests the control.

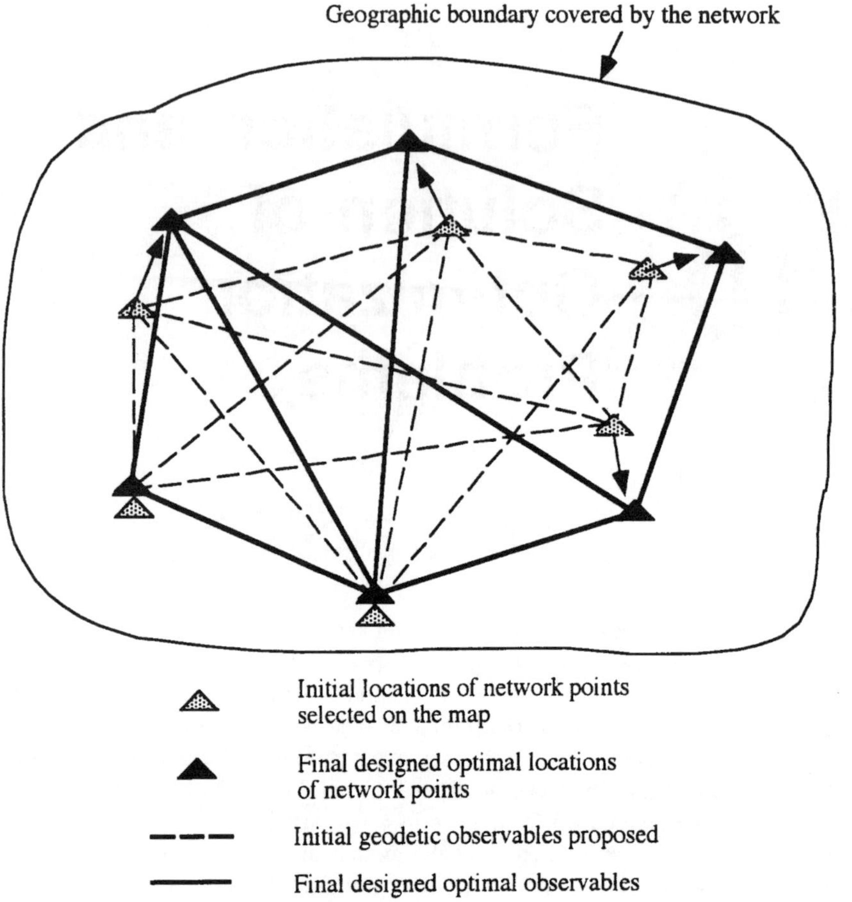

Figure 10.1. Designing a geodetic control network

Step 2. Perform a preliminary field reconnaissance and based on the available instrumentation determine the possible connections between network stations by geodetic observables. The dotted lines in Figure 10.1 represent the initially proposed geodetic observables.

Step 3. The proposed station locations and geodetic observables in Steps 1 and 2 constitute an initial design of the network. According to the initial design, hypothetical precisions or weights of observables are then used to simulate the quality of the network using the quality measures as given in Chapter 6. Then, ***improvements*** to the initial design are introduced, when necessary, by either increasing or decreasing the types and numbers of geodetic observables initially proposed at Step 2 until the simulated network quality meets or exceeds the project criteria set by the client(s) requesting the control.

Step 4. Perform a field reconnaissance to examine the physical possibilities of the simulated network. Control stations are temporarily marked on the ground. If conventional terrestrial geodetic observables are proposed, they require intervisibility between network stations. If the GPS technique is to be used, the station site should be wide open, i.e., no existing obstruction should be allowed to block the signals from the GPS satellites. In the case that a simulated network station is impossible to construct in nature, alternatives are sought and recorded on the map.

Step 5. Steps 3 and 4 are iterated until an appropriate network design is accomplished. The final designed locations of network points and geodetic observables are described in Figure 10.1 by the solid triangles and solid lines, respectively.

After Step 5, the network is then monumented and surveyed and finally analyzed using the theory of network analysis as described in Part I of this book and then presented to the network user.

The above conventional approach for network design has been used by surveyors/engineers for centuries and it is still widely used today. It is clear from the above outlined procedures that Step 3 is the most critical step in obtaining the best network design. That is, after an initial network design is proposed through field reconnaissance or other means, how to optimally introduce improvements to the initial design in order to achieve the pre-set network quality with the minimum effort remains the most critical task. As discussed in §8.2 and §8.3, the improvements could be introduced either by the "trial and error" method or solved for fully analytically through a mathematical procedure. When the "trial and error" approach is used, the initial station locations are rarely moved, and therefore the Second-Order Design (SOD) is usually performed, in which the designer has to either add to or delete some of the initially proposed observables manually, then compute and compare the resulting network quality with the pre-set criteria. This is done iteratively until the network quality meets the criteria. The success of the "trial and error" approach is to a large extent dependent on the experiences and knowledge of the designer. And as discussed in §8.2, the "trial and error" approach may never find the best network.

The word "**optimization**" has recently come into use in geodesy to indicate designing networks based on well-specified quantitative considerations and techniques; it suggests planning for the best solution. The concepts and procedures of the above conventional approach for network design are in fact well described by the following quotation from Alberda (1974):

> "... *in particular with respect to terrain difficulties and the choice of methods of measurement, the planning of networks means in practice that one starts with a solution that is feasible under the given circumstances*

and available material means, and then introduces ***improvements*** *until the plan is not too expensive and good enough.*"

In light of this quotation, it is the "***improvements***" that are the unknown quantities that have to be optimally solved for at the design stage. The purpose of this chapter is to develop a fully analytical mathematical procedure to solve for the optimal improvements to the initial design. The entire solution process can be fully automated by the computer without the need for human intervention. When the analytical approach is used, optimal improvements to the initial station locations can also be easily solved. This is especially of significance for geodetic networks established for engineering purposes, since the general configuration of a network is usually suggested by the specific project survey control needed and by the topography of the surface of the earth covered by the network area, or by the shape of the buildings or the constructions, or suggested by the position where the maximum deformation for object points or minimum deformation for reference points may be expected in the case of a deformation network. Certain changes of the position of netpoints, which may vary in magnitude from point to point are, however, always possible, and they should be used to optimize the configuration of the network (Koch, 1985).

Finally, in order to develop an analytical approach for network optimal design, the improvements to an initial design must be described by mathematical terms. Let us consider a network in three-dimensional space and assume that approximate locations of network points with approximate coordinates (x_i^0, y_i^0, z_i^0, i=1, ..., m) were selected from a reconnaissance according to the topography and/or project survey control needs. The configuration of the network can be optimized by introducing improvements (Δx_i, Δy_i, Δz_i, i=1, ..., m) in the coordinates of each point. As for the solution of weights of observations, we can also start with a set of approximate weights p_i^0 (i=1, ..., n) that can be realized with the least efforts, then introduce improvements (Δp_i, i=1, ..., n) in order to achieve the design criteria. The final optimal positions and weights are then obtained by adding the solved for optimal improvements to their corresponding initial values as follows:

$$\begin{aligned} x_i &= x_i^0 + \Delta x_i \\ y_i &= y_i^0 + \Delta y_i \quad i = 1, \ldots, m \\ z_i &= z_i^0 + \Delta z_i \end{aligned} \tag{10.1}$$

and

$$p_i = p_i^0 + \Delta p_i \quad i = 1, \ldots, n \tag{10.2}$$

10.2 Basic Requirements for an Optimal Network

As discussed above, the parameters to be optimized in the optimization of a geodetic network can be the improvements to the initial station positions and/or observation weights. We say that a solution for these parameters is optimal if it satisfies the optimality criteria adopted to define the quality of the network and it is physically realizable. In order to solve for the improvements analytically, the major problem is how to bring the quality criteria into a strong mathematical form, i.e., to establish the explicit relation between the preset design criteria of precision, reliability, and cost and the unknown parameters to be optimized, and that is accomplished in this section by using the technique of Taylor series expansion to linearize the nonlinear matrix equations related to network design, converting various network quality requirements into constraints on the unknown parameters to be optimally solved for.

10.2.1 Precision Requirement

Chapter 6 described a variety of network precision measures that may be used to set up precision optimality criteria for network optimization. Discussed in this section are some typical precision measures that include the global variance-covariance matrix, its eigenvalues, and the variance of a certain function of the estimated coordinates since any other precision measures can be exclusively derived from them.

In order to convert a precision criterion into constraints or requirements on the unknown parameters to be optimized, the precision measures are linearized using Taylor series expansion. At first, assuming the use of a minimum constrained network datum, the global variance-covariance matrix of coordinates is written from Equation (4.131) and (4.137) as follows

$$\mathbf{C}_X = \sigma_0^2\,\mathbf{Q}_X = \sigma_0^2\left[\left(\mathbf{A}^T\mathbf{P}\mathbf{A} + \mathbf{D}\mathbf{D}^T\right)^{-1} - \mathbf{H}\left(\mathbf{H}^T\mathbf{D}\mathbf{D}^T\mathbf{H}\right)^{-1}\mathbf{H}^T\right]$$

where σ_0^2 is the a priori variance factor, and it is usually assumed to be 1.0 at the design stage.

Note that in three-dimensional space elements of matrix $\mathbf{C}_x$ are nonlinear functions of both the observational weights (p_i, i = 1, ..., n) and station coordinates (x_i, y_i, z_i) (i=1, ..., m) that are embedded in the network configuration matrix **A**. Supplied with initial values of both the coordinates and observation weights, matrix $\mathbf{C}_x$ can be approximated by Taylor series restricted to linear term as follows

$$C_X = C_X^0 + \sum_1^m \frac{\partial C_X}{\partial x_i}\Delta x_i + \sum_1^m \frac{\partial C_X}{\partial y_i}\Delta y_i + \sum_1^m \frac{\partial C_X}{\partial z_i}\Delta z_i + \sum_1^n \frac{\partial C_X}{\partial p_i}\Delta p_i \qquad (10.3)$$

where

$$\mathbf{C}_X^0 = \sigma_0^2\left[\left(\mathbf{A}^T\mathbf{P}\mathbf{A} + \mathbf{D}\mathbf{D}^T\right)^{-1} - \mathbf{H}\left(\mathbf{H}^T\mathbf{D}\mathbf{D}^T\mathbf{H}\right)^{-1}\mathbf{H}^T\right]\Big|_{\mathbf{x}^0,\,\mathbf{y}^0,\,\mathbf{z}^0,\,\mathbf{p}^0} \qquad (10.4)$$

$$
\begin{aligned}
\frac{\partial \mathbf{C}_x}{\partial x_i} = \sigma_0^2 \Bigg\{ & -\left(\mathbf{A}^T\mathbf{P}\mathbf{A} + \mathbf{D}\mathbf{D}^T\right)^{-1} \left[\left(\frac{\partial \mathbf{A}}{\partial x_i}\right)^T \mathbf{P}\mathbf{A} + \mathbf{A}^T\mathbf{P}\left(\frac{\partial \mathbf{A}}{\partial x_i}\right) \right. \\
& \left. + \frac{\partial \mathbf{D}}{\partial x_i}\mathbf{D}^T + \mathbf{D}\left(\frac{\partial \mathbf{D}}{\partial x_i}\right)^T \right] \left(\mathbf{A}^T\mathbf{P}\mathbf{A} + \mathbf{D}\mathbf{D}^T\right)^{-1} - \left(\frac{\partial \mathbf{H}}{\partial x_i}\right) \\
& \left(\mathbf{H}^T\mathbf{D}\mathbf{D}^T\mathbf{H}\right)^{-1}\mathbf{H}^T - \mathbf{H}\left(\mathbf{H}^T\mathbf{D}\mathbf{D}^T\mathbf{H}\right)^{-1}\left(\frac{\partial \mathbf{H}}{\partial x_i}\right)^T \\
& + \mathbf{H}\left(\mathbf{H}^T\mathbf{D}\mathbf{D}^T\mathbf{H}\right)^{-1} \left[\left(\frac{\partial \mathbf{H}}{\partial x_i}\right)^T \mathbf{D}\mathbf{D}^T\mathbf{H} + \mathbf{H}^T\mathbf{D}\mathbf{D}^T \right. \\
& \left. \left(\frac{\partial \mathbf{H}}{\partial x_i}\right) + \mathbf{H}^T\left(\frac{\partial \mathbf{D}}{\partial x_i}\right)\mathbf{D}^T\mathbf{H} + \mathbf{H}^T\mathbf{D}\left(\frac{\partial \mathbf{D}}{\partial x_i}\right)^T \mathbf{H} \right] \\
& \left(\mathbf{H}^T\mathbf{D}\mathbf{D}^T\mathbf{H}\right)^{-1}\mathbf{H}^T \Bigg\} \Bigg|_{\mathbf{x}^0,\, \mathbf{y}^0,\, \mathbf{z}^0,\, \mathbf{P}^0}
\end{aligned}
\tag{10.5}
$$

$$
\begin{aligned}
\frac{\partial \mathbf{C}_x}{\partial y_i} = \sigma_0^2 \Bigg\{ & -\left(\mathbf{A}^T\mathbf{P}\mathbf{A} + \mathbf{D}\mathbf{D}^T\right)^{-1} \left[\left(\frac{\partial \mathbf{A}}{\partial y_i}\right)^T \mathbf{P}\mathbf{A} + \mathbf{A}^T\mathbf{P}\left(\frac{\partial \mathbf{A}}{\partial y_i}\right) \right. \\
& \left. + \frac{\partial \mathbf{D}}{\partial y_i}\mathbf{D}^T + \mathbf{D}\left(\frac{\partial \mathbf{D}}{\partial y_i}\right)^T \right] \left(\mathbf{A}^T\mathbf{P}\mathbf{A} + \mathbf{D}\mathbf{D}^T\right)^{-1} - \left(\frac{\partial \mathbf{H}}{\partial y_i}\right) \\
& \left(\mathbf{H}^T\mathbf{D}\mathbf{D}^T\mathbf{H}\right)^{-1}\mathbf{H}^T - \mathbf{H}\left(\mathbf{H}^T\mathbf{D}\mathbf{D}^T\mathbf{H}\right)^{-1}\left(\frac{\partial \mathbf{H}}{\partial y_i}\right)^T \\
& + \mathbf{H}\left(\mathbf{H}^T\mathbf{D}\mathbf{D}^T\mathbf{H}\right)^{-1} \left[\left(\frac{\partial \mathbf{H}}{\partial y_i}\right)^T \mathbf{D}\mathbf{D}^T\mathbf{H} + \mathbf{H}^T\mathbf{D}\mathbf{D}^T \right. \\
& \left. \left(\frac{\partial \mathbf{H}}{\partial y_i}\right) + \mathbf{H}^T\left(\frac{\partial \mathbf{D}}{\partial y_i}\right)\mathbf{D}^T\mathbf{H} + \mathbf{H}^T\mathbf{D}\left(\frac{\partial \mathbf{D}}{\partial y_i}\right)^T \mathbf{H} \right] \\
& \left(\mathbf{H}^T\mathbf{D}\mathbf{D}^T\mathbf{H}\right)^{-1}\mathbf{H}^T \Bigg\} \Bigg|_{\mathbf{x}^0,\, \mathbf{y}^0,\, \mathbf{z}^0,\, \mathbf{P}^0}
\end{aligned}
\tag{10.6}
$$

$$\frac{\partial \mathbf{C}_x}{\partial z_i} = \sigma_0^2 \Bigg\{ -\left(\mathbf{A}^T\mathbf{PA} + \mathbf{DD}^T\right)^{-1} \Bigg[\left(\frac{\partial \mathbf{A}}{\partial z_i}\right)^T \mathbf{PA} + \mathbf{A}^T\mathbf{P}\left(\frac{\partial \mathbf{A}}{\partial z_i}\right)$$

$$+ \frac{\partial \mathbf{D}}{\partial z_i}\mathbf{D}^T + \mathbf{D}\left(\frac{\partial \mathbf{D}}{\partial z_i}\right)^T \Bigg] \left(\mathbf{A}^T\mathbf{PA} + \mathbf{DD}^T\right)^{-1} - \left(\frac{\partial \mathbf{H}}{\partial z_i}\right)$$

$$\left(\mathbf{H}^T\mathbf{DD}^T\mathbf{H}\right)^{-1}\mathbf{H}^T - \mathbf{H}\left(\mathbf{H}^T\mathbf{DD}^T\mathbf{H}\right)^{-1}\left(\frac{\partial \mathbf{H}}{\partial z_i}\right)^T$$

$$+ \mathbf{H}\left(\mathbf{H}^T\mathbf{DD}^T\mathbf{H}\right)^{-1} \Bigg[\left(\frac{\partial \mathbf{H}}{\partial z_i}\right)^T \mathbf{DD}^T\mathbf{H} + \mathbf{H}^T\mathbf{DD}^T \tag{10.7}$$

$$\left(\frac{\partial \mathbf{H}}{\partial z_i}\right) + \mathbf{H}^T\left(\frac{\partial \mathbf{D}}{\partial z_i}\right)\mathbf{D}^T\mathbf{H} + \mathbf{H}^T\mathbf{D}\left(\frac{\partial \mathbf{D}}{\partial z_i}\right)^T \mathbf{H} \Bigg]$$

$$\left(\mathbf{H}^T\mathbf{DD}^T\mathbf{H}\right)^{-1}\mathbf{H}^T \Bigg\} \Bigg|_{\mathbf{x}^0,\, \mathbf{y}^0,\, \mathbf{z}^0,\, \mathbf{P}^0}$$

$$\frac{\partial \mathbf{C}_x}{\partial p_i} = \sigma_0^2 \Bigg\{ -\left(\mathbf{A}^T\mathbf{PA} + \mathbf{DD}^T\right)^{-1} \left[\mathbf{A}^T\left(\frac{\partial \mathbf{P}}{\partial p_i}\right)\mathbf{A} \right]$$

$$\left(\mathbf{A}^T\mathbf{PA} + \mathbf{DD}^T\right)^{-1} \Bigg\} \Bigg|_{\mathbf{x}^0,\, \mathbf{y}^0,\, \mathbf{z}^0,\, \mathbf{P}^0} \tag{10.8}$$

m is the total number of netpoints
n is the total number of observables
$\mathbf{x}^0$, $\mathbf{y}^0$, and $\mathbf{z}^0$ are vectors of initial coordinates of netpoints selected in reconnaissance
$\mathbf{P}^0$ consists of the approximate values of weights, and finally
(Δx_i, Δy_i, Δz_i) (i=1, ..., m) and Δp_i (i = 1, ..., n) are the improvements, which are to be optimally solved for, to the initial coordinates and initial weights, respectively.

The derivatives of matrices **A** and **P** with respect to (x_i, y_i, z_i) (i=1, ..., m) and p_i (i=1, ..., n) can be obtained fully analytically as discussed later in §10.3.

When the eigenvalues of $\mathbf{C}_x$ are used to establish precision risk functions, for instance, the N-, A-, E-, S-, and D- optimality precision criteria described in §9.1.1, they also have to be linearized. Note that the eigenvalue λ_i of $\mathbf{C}_x$ can be expressed by [cf. Equation (6.11)]:

$$\lambda_i = \mathbf{e}_i^T \mathbf{C}_x \mathbf{e}_i \quad (i = 1, \ldots, u) \tag{10.9}$$

with $\mathbf{e}_i$ being the eigenvector corresponding to λ_i of $\mathbf{C}_x$. Using Taylor series expansion, λ_i can be approximated by

$$\begin{aligned}\lambda_i = {} & \mathbf{e}_{i0}^T C_x^0 \mathbf{e}_{i0} + \sum_{j=1}^{m} \left(\mathbf{e}_{i0}^T \frac{\partial \mathbf{C}_x}{\partial x_i} \mathbf{e}_{i0} \right) \Delta x_j + \sum_{j=1}^{m} \left(\mathbf{e}_{i0}^T \frac{\partial \mathbf{C}_x}{\partial y_j} \mathbf{e}_{i0} \right) \Delta y_j \\ & + \sum_{j=1}^{m} \left(\mathbf{e}_{i0}^T \frac{\partial \mathbf{C}_x}{\partial z_j} \mathbf{e}_{i0} \right) \Delta z_j + \sum_{j=1}^{m} \left(\mathbf{e}_{i0}^T \frac{\partial \mathbf{C}_x}{\partial p_j} \mathbf{e}_{i0} \right) \Delta p_j \\ & (i = 1, \ldots, u)\end{aligned} \tag{10.10}$$

with $\mathbf{e}_{i0}$ being the approximate value of $\mathbf{e}_i$, calculated from matrix C_x^0. For examples of optimization models using the N-, A-, and E-optimality criteria, the interested reader is referred to Kuang (1992a).

Finally, if one demands that some certain function $f = \mathbf{c}^T \mathbf{x}$ of $\mathbf{x}$ to satisfy a certain precision requirement [cf. Equation (9.7)], the variance of f can be approximated by Taylor series of linear form as follows (cf. also Kuang, 1993f)

$$c_f = c_f^0 + \sum_{i=1}^{m} \frac{\partial c_f}{\partial x_i} \Delta x_i + \sum_{i=1}^{m} \frac{\partial c_f}{\partial y_i} \Delta y_i + \sum_{i=1}^{m} \frac{\partial c_f}{\partial z_i} \Delta z_i + \sum_{i=1}^{n} \frac{\partial c_f}{\partial p_i} \Delta p_i \tag{10.11}$$

where

$$c_f^0 = \left\{ \mathbf{c}^T \, \mathbf{C}_x \, \mathbf{c} \right\} \Big|_{\mathbf{x}^0, \mathbf{y}^0, \mathbf{z}^0, \mathbf{p}^0} \tag{10.12}$$

$$\frac{\partial c_f}{\partial x_i} = \left\{ \mathbf{c}^T \left(\frac{\partial \mathbf{C}_x}{\partial x_i} \right) \mathbf{c} \right\} \Big|_{\mathbf{x}^0, \mathbf{y}^0, \mathbf{z}^0, \mathbf{p}^0} \tag{10.13}$$

$$\frac{\partial c_f}{\partial y_i} = \left\{ \mathbf{c}^T \left(\frac{\partial \mathbf{C}_x}{\partial y_i} \right) \mathbf{c} \right\} \Big|_{\mathbf{x}^0, \mathbf{y}^0, \mathbf{z}^0, \mathbf{p}^0} \tag{10.14}$$

$$\frac{\partial c_f}{\partial z_i} = \left\{ \mathbf{c}^T \left(\frac{\partial \mathbf{C}_x}{\partial z_i} \right) \mathbf{c} \right\} \Big|_{\mathbf{x}^0, \mathbf{y}^0, \mathbf{z}^0, \mathbf{p}^0} \tag{10.15}$$

$$\frac{\partial c_f}{\partial p_i} = \left\{ \mathbf{c}^T \left(\frac{\partial \mathbf{C}_x}{\partial p_i} \right) \mathbf{c} \right\} \Big|_{\mathbf{x}^0, \mathbf{y}^0, \mathbf{z}^0, \mathbf{p}^0} \tag{10.16}$$

When a criterion matrix $\mathbf{C}_s$ is used as the precision criteria, the design problem seeks an optimal configuration matrix **A** and weight matrix **P** such that $\mathbf{C}_s$ can be best approximated by $\mathbf{C}_x$. The precision risk function in this case is [cf. (9.33)]

$$\| \mathbf{C}_x - \mathbf{C}_s \| = \text{minimum} \quad (\text{optimal precision}) \tag{10.17}$$

This function also intends to avoid that the precision of some of the coordinates becomes disproportionally "better" than the others. In addition, one may also desire that the solution of **A** and **P** yield a "better" variance-covariance matrix in some sense than the presupposed criterion matrix $\mathbf{C}_s$. This can be expressed by (cf. Schaffrin, 1981)

$$\text{vecdiag}\,(\mathbf{C}_x) \le \text{vecdiag}\,(\mathbf{C}_s) \quad (\text{precision control}) \tag{10.18}$$

where "vecdiag" is an operator that produces a vector consisting of the diagonal elements of a matrix.

Further, it should be noted that when a Taylor-Karman structured criterion matrix is used, it must be transferred into the same network datum as that used for calculating the variance-covariance matrix $\mathbf{C}_x$. In this case, $\mathbf{C}_s$ must also be linearized as follows:

$$\mathbf{C}_s = \mathbf{C}_s^0 + \sum_1^m \frac{\partial \mathbf{C}_s}{\partial x_i} \Delta x_i + \sum_1^m \frac{\partial \mathbf{C}_s}{\partial y_i} \Delta y_i + \sum_1^m \frac{\partial \mathbf{C}_s}{\partial z_i} \Delta z_i \tag{10.19}$$

where

$$C_s^0 = \left(\mathbf{S}\, \mathbf{C}_{TK}\, \mathbf{S}^T \right) \Big|_{\mathbf{x}^0, \mathbf{y}^0, \mathbf{z}^0} \tag{10.20}$$

$$\frac{\partial \mathbf{C}_s}{\partial x_i} = \left[\left(\frac{\partial \mathbf{S}}{\partial x_i} \right) \mathbf{C}_{TK}\, \mathbf{S}^T + \mathbf{S}\, \mathbf{C}_{TK} \left(\frac{\partial \mathbf{S}}{\partial x_i} \right)^T \right] \Big|_{\mathbf{x}^0, \mathbf{y}^0, \mathbf{z}^0, \mathbf{p}^0} \tag{10.21}$$

$$\frac{\partial \mathbf{C}_s}{\partial y_i} = \left[\left(\frac{\partial \mathbf{S}}{\partial y_i} \right) \mathbf{C}_{TK}\, \mathbf{S}^T + \mathbf{S}\, \mathbf{C}_{TK} \left(\frac{\partial \mathbf{S}}{\partial y_i} \right)^T \right] \Big|_{\mathbf{x}^0, \mathbf{y}^0, \mathbf{z}^0, \mathbf{p}^0} \tag{10.22}$$

$$\frac{\partial \mathbf{C}_s}{\partial z_i} = \left[\left(\frac{\partial \mathbf{S}}{\partial z_i} \right) \mathbf{C}_{TK}\, \mathbf{S}^T + \mathbf{S}\, \mathbf{C}_{TK} \left(\frac{\partial \mathbf{S}}{\partial z_i} \right)^T \right] \Big|_{\mathbf{x}^0, \mathbf{y}^0, \mathbf{z}^0, \mathbf{p}^0} \tag{10.23}$$

$$\frac{\partial \mathbf{S}}{\partial x_i} = -\left(\frac{\partial \mathbf{H}}{\partial x_i}\right)\left(\mathbf{D}^T\mathbf{H}\right)^{-1}\mathbf{D}^T - \mathbf{H}\left(\mathbf{D}^T\mathbf{H}\right)^{-1}\left(\frac{\partial \mathbf{D}}{\partial x_i}\right)^T$$
$$+\mathbf{H}\left(\mathbf{D}^T\mathbf{H}\right)^{-1}\left[\left(\frac{\partial \mathbf{D}}{\partial x_i}\right)^T\mathbf{H} + \mathbf{D}^T\left(\frac{\partial \mathbf{H}}{\partial x_i}\right)\right]\left(\mathbf{D}^T\mathbf{H}\right)^{-1}\mathbf{D}^T \tag{10.24}$$

$$\frac{\partial \mathbf{S}}{\partial y_i} = -\left(\frac{\partial \mathbf{H}}{\partial y_i}\right)\left(\mathbf{D}^T\mathbf{H}\right)^{-1}\mathbf{D}^T - \mathbf{H}\left(\mathbf{D}^T\mathbf{H}\right)^{-1}\left(\frac{\partial \mathbf{D}}{\partial y_i}\right)^T$$
$$+\mathbf{H}\left(\mathbf{D}^T\mathbf{H}\right)^{-1}\left[\left(\frac{\partial \mathbf{D}}{\partial y_i}\right)^T\mathbf{H} + \mathbf{D}^T\left(\frac{\partial \mathbf{H}}{\partial y_i}\right)\right]\left(\mathbf{D}^T\mathbf{H}\right)^{-1}\mathbf{D}^T \tag{10.25}$$

$$\frac{\partial \mathbf{S}}{\partial z_i} = -\left(\frac{\partial \mathbf{H}}{\partial z_i}\right)\left(\mathbf{D}^T\mathbf{H}\right)^{-1}\mathbf{D}^T - \mathbf{H}\left(\mathbf{D}^T\mathbf{H}\right)^{-1}\left(\frac{\partial \mathbf{D}}{\partial z_i}\right)^T$$
$$+\mathbf{H}\left(\mathbf{D}^T\mathbf{H}\right)^{-1}\left[\left(\frac{\partial \mathbf{D}}{\partial z_i}\right)^T\mathbf{H} + \mathbf{D}^T\left(\frac{\partial \mathbf{H}}{\partial z_i}\right)\right]\left(\mathbf{D}^T\mathbf{H}\right)^{-1}\mathbf{D}^T \tag{10.26}$$

Denote

$$\mathbf{u} = \text{vec}\left(\mathbf{C}_s^0\right) - \text{vec}\left(\mathbf{C}_x^0\right) \tag{10.27}$$

$$\begin{aligned} H = \Bigg[& \text{vec}\left(\frac{\partial \mathbf{C}_x}{\partial x_1} - \frac{\partial \mathbf{C}_s}{\partial x_1}\right) \text{vec}\left(\frac{\partial \mathbf{C}_x}{\partial y_1} - \frac{\partial \mathbf{C}_s}{\partial y_1}\right) \text{vec}\left(\frac{\partial \mathbf{C}_x}{\partial z_1} - \frac{\partial \mathbf{C}_s}{\partial z_1}\right) \cdots \\ & \text{vec}\left(\frac{\partial \mathbf{C}_x}{\partial x_m} - \frac{\partial \mathbf{C}_s}{\partial x_m}\right) \text{vec}\left(\frac{\partial \mathbf{C}_x}{\partial y_m} - \frac{\partial \mathbf{C}_s}{\partial y_m}\right) \text{vec}\left(\frac{\partial \mathbf{C}_x}{\partial z_m} - \frac{\partial \mathbf{C}_s}{\partial z_m}\right) \\ & \text{vec}\left(\frac{\partial \mathbf{C}_x}{\partial p_1}\right) \cdots \text{vec}\left(\frac{\partial \mathbf{C}_x}{\partial p_n}\right)\Bigg] \end{aligned} \tag{10.28}$$

$$\mathbf{w} = \left(\Delta x_1\ \Delta y_1\ \Delta z_1\ \ldots\ \Delta x_m\ \Delta y_m\ \Delta z_m\ \Delta p_1\ \ldots\ \Delta p_n\right)^T \tag{10.29}$$

where the operator "vec" produces a vector by stacking the columns of a quadratic matrix one under another in a single column. Considering Equations (10.27) to (10.29), one may reformulate precision criteria (10.17) and (10.18) in a compact matrix and vector form as follows

$$\| \mathbf{H}\,\mathbf{w} - \mathbf{u} \| = \min \quad \text{(optimal precision)} \tag{10.30}$$

$$\mathbf{H}_1\,\mathbf{w} - \mathbf{u}_1 \leq \mathbf{0} \quad \text{(precision control)} \tag{10.31}$$

where

$$\mathbf{H}_1 = \left(\mathbf{I}_u \,\Theta\, \mathbf{I}_u\right)^T \mathbf{H} \tag{10.32}$$

$$\mathbf{u}_1 = \left(\mathbf{I}_u \,\Theta\, \mathbf{I}_u\right)^T \mathbf{u} \tag{10.33}$$

$\mathbf{I}_u$ is the u by u unit matrix and Θ represents the Khatri-Rao product (see Appendix A).

With the linearized precision measures, precision requirements based on any other types of precision criteria as discussed in §9.1 can be treated similarly as above.

Finally, it should be noted that when the configuration matrix $\mathbf{A}$ is nonsingular, for instance, when the constraint equations are eliminated through the observation equations as discussed in §4.3.4, computation of the global variance-covariance matrix $\mathbf{C}_x$ is then simplified by

$$\mathbf{C}_x = \sigma_0^2 \left(\mathbf{A}^T \mathbf{P}\,\mathbf{A}\right)^{-1} \tag{10.34}$$

In this case, formulae for the partial derivatives of the variance-covariance matrix with respect to station coordinates and observational weights are also simplified as follows

$$\frac{\partial \mathbf{C}_x}{\partial x_i} = -\sigma_0^2 \left(\mathbf{A}^T \mathbf{P}\,\mathbf{A}\right)^{-1} \left(\frac{\partial \mathbf{A}^T}{\partial x_i} \mathbf{P}\,\mathbf{A} + \mathbf{A}^T \mathbf{P} \frac{\partial \mathbf{A}}{\partial x_i} \right) \left(\mathbf{A}^T \mathbf{P}\,\mathbf{A}\right)^{-1} \tag{10.35}$$

$$\frac{\partial \mathbf{C}_x}{\partial y_i} = -\sigma_0^2 \left(\mathbf{A}^T \mathbf{P}\,\mathbf{A}\right)^{-1} \left(\frac{\partial \mathbf{A}^T}{\partial y_i} \mathbf{P}\,\mathbf{A} + \mathbf{A}^T \mathbf{P} \frac{\partial \mathbf{A}}{\partial y_i} \right) \left(\mathbf{A}^T \mathbf{P}\,\mathbf{A}\right)^{-1} \tag{10.36}$$

$$\frac{\partial \mathbf{C}_x}{\partial z_i} = -\sigma_0^2 \left(\mathbf{A}^T \mathbf{P}\,\mathbf{A}\right)^{-1} \left(\frac{\partial \mathbf{A}^T}{\partial z_i} \mathbf{P}\,\mathbf{A} + \mathbf{A}^T \mathbf{P} \frac{\partial \mathbf{A}}{\partial z_i} \right) \left(\mathbf{A}^T \mathbf{P}\,\mathbf{A}\right)^{-1} \tag{10.37}$$

$$\frac{\partial \mathbf{C}_x}{\partial p_i} = -\sigma_0^2 \left(\mathbf{A}^T \mathbf{P}\,\mathbf{A}\right)^{-1} \left(\mathbf{A}^T \frac{\partial \mathbf{P}}{\partial p_i} \mathbf{A} \right) \left(\mathbf{A}^T \mathbf{P}\,\mathbf{A}\right)^{-1} \tag{10.38}$$

10.2.2 Reliability Requirement

From §9.2, a general requirement for both optimal internal and external reliability for geodetic networks can be written as

$$\| \mathbf{r} \| = \max \quad \text{(optimal reliability)} \tag{10.39}$$

or

$$\| \mathbf{r} \| \geq r_m \quad \text{(reliability control)} \tag{10.40}$$

where the vector **r** consists of the redundancy numbers as follows

$$\mathbf{r} = \begin{pmatrix} r_1 & r_2 & \cdots & r_n \end{pmatrix}^T \tag{10.41}$$

and r_m is a certain pre-set boundary value. The redundancy numbers are defined by the diagonal elements of matrix **R**, i.e.,

$$r_i = (\mathbf{R})_{ii} \quad (i = 1, \ldots, n) \tag{10.42}$$

with matrix **R** being defined by Equation (5.2) as follows

$$\mathbf{R} = \mathbf{I} - \mathbf{A}\left(\mathbf{A}^T\mathbf{P}\mathbf{A} + \mathbf{D}\mathbf{D}^T\right)^{-1}\mathbf{A}^T\mathbf{P}$$

Approximating matrix **R** by its Taylor series of linear term, we have

$$\mathbf{R} = \mathbf{R}^0 + \sum_1^m \left(\frac{\partial \mathbf{R}}{\partial x_i}\right)\Delta x_i + \sum_1^m \left(\frac{\partial \mathbf{R}}{\partial y_i}\right)\Delta y_i + \sum_1^m \left(\frac{\partial \mathbf{R}}{\partial z_i}\right)\Delta z_i + \sum_1^n \left(\frac{\partial \mathbf{R}}{\partial p_i}\right)\Delta p_i \tag{10.43}$$

where

$$\mathbf{R}^0 = \left[\mathbf{I} - \mathbf{A}\left(\mathbf{A}^T\mathbf{P}\mathbf{A} + \mathbf{D}\mathbf{D}^T\right)^{-1}\mathbf{A}^T\mathbf{P}\right]_{\mathbf{x}^0, \mathbf{y}^0, \mathbf{z}^0, \mathbf{p}^0} \tag{10.44}$$

$$\begin{aligned} \frac{\partial \mathbf{R}}{\partial x_i} = \Bigg[& -\frac{\partial \mathbf{A}}{\partial x_i}\left(\mathbf{A}^T\mathbf{P}\mathbf{A} + \mathbf{D}\mathbf{D}^T\right)^{-1}\mathbf{A}^T\mathbf{P} - \mathbf{A}\left(\mathbf{A}^T\mathbf{P}\mathbf{A} + \mathbf{D}\mathbf{D}^T\right)^{-1}\frac{\partial \mathbf{A}^T}{\partial x_i}\mathbf{P} \\ & + \mathbf{A}\left(\mathbf{A}^T\mathbf{P}\mathbf{A} + \mathbf{D}\mathbf{D}^T\right)^{-1}\left(\frac{\partial \mathbf{A}^T}{\partial x_i}\mathbf{P}\mathbf{A} + \mathbf{A}^T\mathbf{P}\frac{\partial \mathbf{A}}{\partial x_i} + \frac{\partial \mathbf{D}}{\partial x_i}\mathbf{D}^T \right. \\ & \left. + \mathbf{D}\frac{\partial \mathbf{D}^T}{\partial x_i}\right)\left(\mathbf{A}^T\mathbf{P}\mathbf{A} + \mathbf{D}\mathbf{D}^T\right)^{-1}\mathbf{A}^T\mathbf{P}\Bigg]_{\mathbf{x}^0, \mathbf{y}^0, \mathbf{z}^0, \mathbf{p}^0} \end{aligned} \tag{10.45}$$

$$\frac{\partial \mathbf{R}}{\partial y_i} = \Bigg[-\frac{\partial \mathbf{A}}{\partial y_i}\left(\mathbf{A}^T\mathbf{PA} + \mathbf{DD}^T\right)^{-1}\mathbf{A}^T\mathbf{P} - \mathbf{A}\left(\mathbf{A}^T\mathbf{PA} + \mathbf{DD}^T\right)^{-1}\frac{\partial \mathbf{A}^T}{\partial y_i}\mathbf{P}$$
$$+ \mathbf{A}\left(\mathbf{A}^T\mathbf{PA} + \mathbf{DD}^T\right)^{-1}\left(\frac{\partial \mathbf{A}^T}{\partial y_i}\mathbf{PA} + \mathbf{A}^T\mathbf{P}\frac{\partial \mathbf{A}}{\partial y_i} + \frac{\partial \mathbf{D}}{\partial y_i}\mathbf{D}^T\right. \tag{10.46}$$
$$\left. + \mathbf{D}\frac{\partial \mathbf{D}^T}{\partial y_i}\right)\left(\mathbf{A}^T\mathbf{PA} + \mathbf{DD}^T\right)^{-1}\mathbf{A}^T\mathbf{P}\Bigg]\Bigg|_{\mathbf{x}^0, \mathbf{y}^0, \mathbf{z}^0, \mathbf{P}^0}$$

$$\frac{\partial \mathbf{R}}{\partial z_i} = \Bigg[-\frac{\partial \mathbf{A}}{\partial z_i}\left(\mathbf{A}^T\mathbf{PA} + \mathbf{DD}^T\right)^{-1}\mathbf{A}^T\mathbf{P} - \mathbf{A}\left(\mathbf{A}^T\mathbf{PA} + \mathbf{DD}^T\right)^{-1}\frac{\partial \mathbf{A}^T}{\partial z_i}\mathbf{P}$$
$$+ \mathbf{A}\left(\mathbf{A}^T\mathbf{PA} + \mathbf{DD}^T\right)^{-1}\left(\frac{\partial \mathbf{A}^T}{\partial z_i}\mathbf{PA} + \mathbf{A}^T\mathbf{P}\frac{\partial \mathbf{A}}{\partial z_i} + \frac{\partial \mathbf{D}}{\partial z_i}\mathbf{D}^T\right. \tag{10.47}$$
$$\left. + \mathbf{D}\frac{\partial \mathbf{D}^T}{\partial z_i}\right)\left(\mathbf{A}^T\mathbf{PA} + \mathbf{DD}^T\right)^{-1}\mathbf{A}^T\mathbf{P}\Bigg]\Bigg|_{\mathbf{x}^0, \mathbf{y}^0, \mathbf{z}^0, \mathbf{P}^0}$$

$$\frac{\partial \mathbf{R}}{\partial p_i} = \Bigg\{ \mathbf{A}\left(\mathbf{A}^T\mathbf{PA} + \mathbf{DD}^T\right)^{-1}\left(\mathbf{A}^T\frac{\partial \mathbf{P}}{\partial p_i}\mathbf{A}\right)\left(\mathbf{A}^T\mathbf{PA} + \mathbf{DD}^T\right)^{-1}\mathbf{A}^T\mathbf{P}$$
$$- \mathbf{A}\left(\mathbf{A}^T\mathbf{PA} + \mathbf{DD}^T\right)^{-1}\mathbf{A}^T\frac{\partial \mathbf{P}}{\partial p_i}\Bigg\}\Bigg|_{\mathbf{x}^0, \mathbf{y}^0, \mathbf{z}^0, \mathbf{P}^0} \tag{10.48}$$

The vector $\mathbf{r}$ of redundancy numbers can then be expressed by

$$\mathbf{r} = \left(\mathbf{I}_n \ \Theta \ \mathbf{I}_n\right)^T \left(\mathbf{r}^0 + \mathbf{R}_1 \ \mathbf{w}\right) \tag{10.49}$$

where

$$\mathbf{r}^0 = \text{vec}\left(\mathbf{R}^0\right) \tag{10.50}$$

$$\mathbf{R}_1 = \left[\text{vec}\left(\frac{\partial \mathbf{R}}{\partial x_1}\right) \text{vec}\left(\frac{\partial \mathbf{R}}{\partial y_1}\right) \text{vec}\left(\frac{\partial \mathbf{R}}{\partial z_1}\right) \ldots \text{vec}\left(\frac{\partial \mathbf{R}}{\partial x_m}\right) \text{vec}\left(\frac{\partial \mathbf{R}}{\partial y_m}\right) \right.$$
$$\left. \text{vec}\left(\frac{\partial \mathbf{R}}{\partial z_m}\right) \text{vec}\left(\frac{\partial R}{\partial p_1}\right) \ldots \text{vec}\left(\frac{\partial \mathbf{R}}{\partial p_n}\right) \right] \tag{10.51}$$

$\mathbf{I}_n$ is the n by n unit matrix, Θ the Khatri-Rao product, and **w** is the same as before.

Denote

$$\mathbf{r}_{00} = \left(\mathbf{I}_n \,\Theta\, \mathbf{I}_n\right)^T \mathbf{r}^0 \tag{10.52}$$

$$\mathbf{R}_{11} = \left(\mathbf{I}_n \,\Theta\, \mathbf{I}_n\right)^T \mathbf{R}_1 \tag{10.53}$$

A general criteria for optimum network reliability can then be expressed by

$$\left\| \left(\mathbf{r}_{00} + \mathbf{R}_{11}\,\mathbf{w}\right) \right\| = \max \quad \text{(optimal reliability)} \tag{10.54}$$

$$\left\| \left(\mathbf{r}_{00} + \mathbf{R}_{11}\,\mathbf{w}\right) \right\| \geq r_m \quad \text{(reliability control)} \tag{10.55}$$

Finally, similar to the case of the global variance-covariance matrix, when the configuration matrix **A** is nonsingular, matrix **R** is calculated by

$$\mathbf{R} = \mathbf{I} - \mathbf{A}\left(\mathbf{A}^T\mathbf{P}\mathbf{A}\right)^{-1}\mathbf{A}^T\mathbf{P} \tag{10.56}$$

The partial derivatives of matrix **R** with respect to station coordinates and observational weights are then

$$\begin{aligned}\frac{\partial \mathbf{R}}{\partial x_i} = \Bigg[& -\frac{\partial \mathbf{A}}{\partial x_i}\left(\mathbf{A}^T\mathbf{P}\mathbf{A}\right)^{-1}\mathbf{A}^T\mathbf{P} - \mathbf{A}\left(\mathbf{A}^T\mathbf{P}\mathbf{A}\right)^{-1}\frac{\partial \mathbf{A}^T}{\partial x_i}\mathbf{P} \\ & + \mathbf{A}\left(\mathbf{A}^T\mathbf{P}\mathbf{A}\right)^{-1}\left(\frac{\partial \mathbf{A}^T}{\partial x_i}\mathbf{P}\mathbf{A} + \mathbf{A}^T\mathbf{P}\frac{\partial \mathbf{A}}{\partial x_i}\right)\left(\mathbf{A}^T\mathbf{P}\mathbf{A}\right)^{-1}\mathbf{A}^T\mathbf{P}\Bigg]\end{aligned} \tag{10.57}$$

$$\begin{aligned}\frac{\partial \mathbf{R}}{\partial y_i} = \Bigg[& -\frac{\partial \mathbf{A}}{\partial y_i}\left(\mathbf{A}^T\mathbf{P}\mathbf{A}\right)^{-1}\mathbf{A}^T\mathbf{P} - \mathbf{A}\left(\mathbf{A}^T\mathbf{P}\mathbf{A}\right)^{-1}\frac{\partial \mathbf{A}^T}{\partial y_i}\mathbf{P} \\ & + \mathbf{A}\left(\mathbf{A}^T\mathbf{P}\mathbf{A}\right)^{-1}\left(\frac{\partial \mathbf{A}^T}{\partial y_i}\mathbf{P}\mathbf{A} + \mathbf{A}^T\mathbf{P}\frac{\partial \mathbf{A}}{\partial y_i}\right)\left(\mathbf{A}^T\mathbf{P}\mathbf{A}\right)^{-1}\mathbf{A}^T\mathbf{P}\Bigg]\end{aligned} \tag{10.58}$$

$$\begin{aligned}\frac{\partial \mathbf{R}}{\partial z_i} = \Bigg[& -\frac{\partial \mathbf{A}}{\partial z_i}\left(\mathbf{A}^T\mathbf{P}\mathbf{A}\right)^{-1}\mathbf{A}^T\mathbf{P} - \mathbf{A}\left(\mathbf{A}^T\mathbf{P}\mathbf{A}\right)^{-1}\frac{\partial \mathbf{A}^T}{\partial z_i}\mathbf{P} \\ & + \mathbf{A}\left(\mathbf{A}^T\mathbf{P}\mathbf{A}\right)^{-1}\left(\frac{\partial \mathbf{A}^T}{\partial z_i}\mathbf{P}\mathbf{A} + \mathbf{A}^T\mathbf{P}\frac{\partial \mathbf{A}}{\partial z_i}\right)\left(\mathbf{A}^T\mathbf{P}\mathbf{A}\right)^{-1}\mathbf{A}^T\mathbf{P}\Bigg]\end{aligned} \tag{10.59}$$

$$\frac{\partial \mathbf{R}}{\partial p_i} = \mathbf{A}\left(\mathbf{A}^T\mathbf{P}\mathbf{A}\right)^{-1}\left(\mathbf{A}^T\frac{\partial \mathbf{P}}{\partial p_i}\mathbf{A}\right)\left(\mathbf{A}^T\mathbf{P}\mathbf{A}\right)^{-1}\mathbf{A}^T\mathbf{P}$$
$$-\mathbf{A}\left(\mathbf{A}^T\mathbf{P}\mathbf{A}\right)^{-1}\mathbf{A}^T\frac{\partial \mathbf{P}}{\partial p_i} \tag{10.60}$$

10.2.3 Cost Requirement

According to §9.3, a simplified cost criterion for optimization can be

$$\| \mathbf{P} \| = \min \quad \text{(minimum cost)} \tag{10.61}$$

or

$$\| \mathbf{P} \| \le c_m \quad \text{(cost control)} \tag{10.62}$$

where **P** is the weight matrix of observations, c_m is a pre-set upper boundary value, and $\|\cdot\|$ represents the norm of a matrix. When using Taylor series linearization, matrix **P** can also be expressed by

$$\mathbf{P} = \mathbf{P}^0 + \sum_1^n \frac{\partial \mathbf{P}}{\partial p_i}\Delta p_i \tag{10.63}$$

where $\mathbf{P}^0$ is the approximate weight matrix. When considering **P** as a diagonal matrix, we have

$$\| \mathbf{P} \| = \| \operatorname{vecdiag}(\mathbf{P}) \|$$
$$= \left\| \left(\mathbf{I}_n \Theta \mathbf{I}_n\right)^T \left[\operatorname{vec}\left(\mathbf{P}^0\right) + \sum_1^n \operatorname{vec}\left(\frac{\partial \mathbf{P}}{\partial p_i}\right)\Delta p_i\right] \right\| \tag{10.64}$$

where "vecdiag" is an operator that produces a vector from the diagonal elements of a matrix.

Denote

$$\mathbf{c}_{00} = \left(\mathbf{I}_n \Theta \mathbf{I}_n\right)^T \operatorname{vec}\left(\mathbf{P}^0\right) \tag{10.65}$$

$$\mathbf{C}_{11} = \left[\mathbf{0}\ \left(\mathbf{I}_n \Theta \mathbf{I}_n\right)^T \operatorname{vec}\left(\frac{\partial \mathbf{P}}{\partial p_1}\right) \ldots \left(\mathbf{I}_n \Theta \mathbf{I}_n\right)^T \operatorname{vec}\left(\frac{\partial \mathbf{P}}{\partial p_n}\right)\right] \tag{10.66}$$

where $\mathbf{I}_n$ is the n by n unit matrix, Θ the Khatri-Rao product, and $\mathbf{0}$ a sub-matrix of zeros corresponding to position improvements. Equations (10.61) and (10.62) can then be reformulated as follows

$$\| \mathbf{c}_{00} + \mathbf{C}_{11}\mathbf{w} \| = \min \quad (\text{minimum cost}) \tag{10.67}$$

or

$$\| \mathbf{c}_{00} + \mathbf{C}_{11}\mathbf{w} \| \leq c_m \quad (\text{cost control}) \tag{10.68}$$

with **w** meaning the same as before.

10.2.4 Physical Constraints

In addition to satisfying precision, reliability, and cost criteria, solution of the improvements to an initial design should also satisfy some physical constraints. These include network datum considerations and realizability.

Datum Consideration

Since a translation, rotation and scaling of a network does not change its shape, the position improvements to be introduced for the netpoints should be constrained such that no translation, scaling and differential rotation with respect to the network to be optimized are introduced (cf. also Koch, 1982), i.e.

$$\left(\mathbf{D}^T \ \mathbf{0}\right)\mathbf{w} = \mathbf{0} \tag{10.69}$$

where **D** is the datum matrix used to calculate the coordinates of netpoints; $\mathbf{0}$ is a sub-matrix of zeros corresponding to weight improvements; and **w** is the same as before.

Realizability

At first, the position improvements to be introduced should be bounded by the topography and/or other considerations, i.e.,

$$\begin{aligned} a_{1i} &\leq \Delta x_i \leq a_{2i} \\ b_{1i} &\leq \Delta y_i \leq b_{2i} \quad (i = 1, \ldots, m) \\ c_{1i} &\leq \Delta z_i \leq c_{2i} \end{aligned} \tag{10.70}$$

where the boundary values $[a_{1i}, a_{2i}]$, $[b_{1i}, b_{2i}]$, and $[c_{1i}, c_{2i}]$ may be established in the field by reconnaissance.

As for the weight improvements Δp_i (i=1, ..., n) to be introduced, the weights of observations should be non-negative and be bounded by the maximum achievable accuracy of the available instruments, i.e.,

$$0 \le p_i \le \frac{\sigma_0^2}{(\sigma_i)_{min}^2} \tag{10.71}$$

or

$$-p_i^0 \le \Delta p_i \le \frac{\sigma_0^2}{(\sigma_i)_{min}^2} - p_i^0 = (\Delta p_i)_{max} \quad (i = 1, \ldots, n) \tag{10.72}$$

where p_i^0 (i=1, ..., n) are the approximate values for weights p_i, σ_0^2 is the a priori variance factor that is usually set to be 1.0 at the design stage, and $(\sigma_i)_{min}^2$ (i=1, ..., n) are the minimum variance that can be achieved for each observable l_i (i=1, ..., n).

Combining Equations (10.70) and (10.72), one obtains

$$\mathbf{A}_{00}\,\mathbf{w} \le \mathbf{b}_{00} \tag{10.73}$$

where

$$\mathbf{A}_{00} = \begin{pmatrix} \mathbf{I} \\ -\mathbf{I} \end{pmatrix} \tag{10.74}$$

$$\begin{aligned} b_{00} = \big[&a_{21}\ b_{21}\ c_{21} \ldots a_{2m}\ b_{2m}\ c_{2m}\ (\Delta p_1)_{max} \cdots (\Delta p_n)_{max} \\ &-a_{11}\ -b_{11}\ -c_{11} \ldots -a_{1m}\ -b_{1m}\ -c_{1m}\ p_1^0 \ldots p_n^0\big]^T \end{aligned} \tag{10.75}$$

I is a (3m+n) by (3m+n) unit matrix.

Finally, it should be noted that although formulae developed in this section refer to the three-dimensional (3D) space, these general formulations also apply to the two-dimensional (2D) case simply by dropping the terms associated with coordinate z. Specific formulations for the calculation of differentials involved are, however, totally different for the 3D and 2D case, and they will be discussed below in §10.3.

10.3 Evaluation of Differentials

Derived in this section are the partial derivatives of the configuration matrix **A**, and the datum matrices **D** and **H** with respect to the coordinates of network points, and those of the weight matrix **P** with respect to individual weights, which are needed in linearizing network quality measures discussed in the above section.

The elements of the configuration matrix **A** are formed as linear or nonlinear functions of the approximate coordinates. The form of functions that constitute the elements of the configuration matrix **A** can be determined geometrically depending on the specific types of observables proposed. Matrix **P** contains the individual observation weights of the observables. It characterizes the precisions of the proposed geodetic observables, and theoretically, it can be either a fully populated or a diagonal matrix depending on whether the observables are correlated or not. For uncorrelated observations, matrix **P** will be a diagonal matrix, and this is the situation we will consider here. A fully populated weight matrix **P** for raw observations will be extremely difficult to realize in practice, if not impossible.

10.3.1 Differentiation of the Configuration Matrix in the Three-Dimensional RLG System

Partial derivatives of the configuration matrix **A** with respect to the coordinates of network points depend on the types of observables proposed. In the three-dimensional RLG system, differentiation of the configuration matrix related to some basic geodetic observables can be derived according to the observational modeling given in §4.1.1 as follows:

Spatial Distances:

$$\frac{\partial\left(\mathbf{a}_{s_{ij}}^{T}\mathbf{U}\right)}{\partial x_i^0}=\left[-\frac{s_{ij}^0+\left(\Delta x_{ij}^0\right)\dfrac{\partial s_{ij}^0}{\partial x_i^0}}{\left(s_{ij}^0\right)^2}\quad-\frac{\left(\Delta y_{ij}^0\right)}{\left(s_{ij}^0\right)^2}\left(\frac{\partial s_{ij}^0}{\partial x_i^0}\right)\right.$$
$$\left.-\frac{\left(\Delta z_{ij}^0\right)}{\left(s_{ij}^0\right)^2}\left(\frac{\partial s_{ij}^0}{\partial x_i^0}\right)\right]\mathbf{U} \tag{10.76}$$

$$\frac{\partial\left(\mathbf{a}_{s_{ij}}^{T}\mathbf{U}\right)}{\partial y_i^0}=\left[-\frac{\left(\Delta x_{ij}^0\right)}{\left(s_{ij}^0\right)^2}\left(\frac{\partial s_{ij}^0}{\partial y_i^0}\right)\quad-\frac{s_{ij}^0+\left(\Delta y_{ij}^0\right)\dfrac{\partial s_{ij}^0}{\partial y_i^0}}{\left(s_{ij}^0\right)^2}\right.$$
$$\left.-\frac{\left(\Delta z_{ij}^0\right)}{\left(s_{ij}^0\right)^2}\left(\frac{\partial s_{ij}^0}{\partial y_i^0}\right)\right]\mathbf{U} \tag{10.77}$$

$$\frac{\partial\left(\mathbf{a}_{s_{ij}}^{T}\mathbf{U}\right)}{\partial z_i^0}=\left[-\frac{\left(\Delta x_{ij}^0\right)}{\left(s_{ij}^0\right)^2}\left(\frac{\partial s_{ij}^0}{\partial z_i^0}\right) \quad -\frac{\left(\Delta y_{ij}^0\right)}{\left(s_{ij}^0\right)^2}\left(\frac{\partial s_{ij}^0}{\partial z_i^0}\right) \quad -\frac{s_{ij}^0+\left(\Delta z_{ij}^0\right)\dfrac{\partial s_{ij}^0}{\partial z_i^0}}{\left(s_{ij}^0\right)^2}\right]\mathbf{U} \tag{10.78}$$

$$\frac{\partial\left(\mathbf{a}_{s_{ij}}^{T}\mathbf{U}\right)}{\partial x_j^0}=-\frac{\partial\left(\mathbf{a}_{s_{ij}}^{T}\mathbf{U}\right)}{\partial x_i^0} \tag{10.79}$$

$$\frac{\partial\left(\mathbf{a}_{s_{ij}}^{T}\mathbf{U}\right)}{\partial y_j^0}=-\frac{\partial\left(\mathbf{a}_{s_{ij}}^{T}\mathbf{U}\right)}{\partial y_i^0} \tag{10.80}$$

$$\frac{\partial\left(\mathbf{a}_{s_{ij}}^{T}\mathbf{U}\right)}{\partial z_j^0}=-\frac{\partial\left(\mathbf{a}_{s_{ij}}^{T}\mathbf{U}\right)}{\partial z_i^0} \tag{10.81}$$

Geodetic Azimuth and Horizontal Directions

$$\frac{\partial\left(\mathbf{a}_{\alpha_{ij}}^{T}\mathbf{U}\right)}{\partial x_i^0}=\left[-2\frac{\left(\Delta y_{ij}^0\right)}{\left(L_{ij}^0\right)^3}\left(\frac{\partial L_{ij}^0}{\partial x_i^0}\right) \quad \frac{\left(L_{ij}^0\right)^2+2L_{ij}^0\Delta x_{ij}^0\left(\dfrac{\partial L_{ij}^0}{\partial x_i^0}\right)}{\left(L_{ij}^0\right)^4} \quad 0\right]\mathbf{U} \tag{10.82}$$

$$\frac{\partial\left(\mathbf{a}_{\alpha_{ij}}^{T}\mathbf{U}\right)}{\partial y_i^0}=\left[-\frac{\left(L_{ij}^0\right)^2+2L_{ij}^0\Delta y_{ij}^0\left(\dfrac{\partial L_{ij}^0}{\partial y_i^0}\right)}{\left(L_{ij}^0\right)^4} \quad 2\frac{\left(\Delta x_{ij}^0\right)}{\left(L_{ij}^0\right)^3}\left(\frac{\partial L_{ij}^0}{\partial y_i^0}\right) \quad 0\right]\mathbf{U} \tag{10.83}$$

$$\frac{\partial\left(\mathbf{a}_{\alpha_{ij}}^{T}\mathbf{U}\right)}{\partial z_i^0} = 0 \tag{10.84}$$

$$\frac{\partial\left(\mathbf{a}_{\alpha_{ij}}^{T}\mathbf{U}\right)}{\partial x_j^0} = -\frac{\partial\left(\mathbf{a}_{\alpha_{ij}}^{T}\mathbf{U}\right)}{\partial x_i^0} \tag{10.85}$$

$$\frac{\partial\left(\mathbf{a}_{\alpha_{ij}}^{T}\mathbf{U}\right)}{\partial y_j^0} = -\frac{\partial\left(\mathbf{a}_{\alpha_{ij}}^{T}\mathbf{U}\right)}{\partial y_i^0} \tag{10.86}$$

$$\frac{\partial\left(\mathbf{a}_{\alpha_{ij}}^{T}\mathbf{U}\right)}{\partial z_j^0} = 0 \tag{10.87}$$

Geodetic Zenith Distances

$$\begin{aligned}\frac{\partial\left(\mathbf{a}_{Z_{ij}}^{T}\mathbf{U}\right)}{\partial x_i^0} = \Bigg\{ & -\frac{\left(\Delta z_{ij}^0\right)L_{ij}^0\left(s_{ij}^0\right)^2 + \left[\left(\frac{\partial L_{ij}^0}{\partial x_i^0}\right)\left(s_{ij}^0\right)^2 + 2s_{ij}^0 L_{ij}^0\left(\frac{\partial s_{ij}^0}{\partial x_i^0}\right)\right]\Delta x_{ij}^0 \Delta z_{ij}^0}{\left(L_{ij}^0\right)^2\left(s_{ij}^0\right)^4} \\ & -\frac{\left[\left(\frac{\partial L_{ij}^0}{\partial x_i^0}\right)\left(s_{ij}^0\right)^2 + 2s_{ij}^0 L_{ij}^0\left(\frac{\partial s_{ij}^0}{\partial x_i^0}\right)\right]\Delta y_{ij}^0 \Delta z_{ij}^0}{\left(L_{ij}^0\right)^2\left(s_{ij}^0\right)^4} \\ & -\frac{\left[\left(\frac{\partial L_{ij}^0}{\partial x_i^0}\right)\left(s_{ij}^0\right)^2 - 2s_{ij}^0 L_{ij}^0\left(\frac{\partial s_{ij}^0}{\partial x_i^0}\right)\right]}{\left(s_{ij}^0\right)^4} \Bigg\}\mathbf{U}\end{aligned} \tag{10.88}$$

$$\frac{\partial\left(\mathbf{a}_{Z_{ij}}^{T}\mathbf{U}\right)}{\partial y_i^0}=\left\{-\frac{\left[\left(\frac{\partial L_{ij}^0}{\partial y_i^0}\right)\left(s_{ij}^0\right)^2+2s_{ij}^0L_{ij}^0\left(\frac{\partial s_{ij}^0}{\partial y_i^0}\right)\right]\Delta x_{ij}^0\Delta z_{ij}^0}{\left(L_{ij}^0\right)^2\left(s_{ij}^0\right)^4}\right.$$

$$-\frac{\left(\Delta z_{ij}^0\right)L_{ij}^0\left(s_{ij}^0\right)^2+\left[\left(\frac{\partial L_{ij}^0}{\partial y_i^0}\right)\left(s_{ij}^0\right)^2+2s_{ij}^0L_{ij}^0\left(\frac{\partial s_{ij}^0}{\partial y_i^0}\right)\right]\Delta y_{ij}^0\Delta z_{ij}^0}{\left(L_{ij}^0\right)^2\left(s_{ij}^0\right)^4}$$

$$\left.-\frac{\left[\left(\frac{\partial L_{ij}^0}{\partial y_i^0}\right)\left(s_{ij}^0\right)^2-2s_{ij}^0L_{ij}^0\left(\frac{\partial s_{ij}^0}{\partial y_i^0}\right)\right]}{\left(s_{ij}^0\right)^4}\right\}\mathbf{U} \tag{10.89}$$

$$\frac{\partial\left(\mathbf{a}_{Z_{ij}}^{T}\mathbf{U}\right)}{\partial z_i^0}=\left\{-\frac{\left(\Delta x_{ij}^0\right)L_{ij}^0\left(s_{ij}^0\right)^2+2s_{ij}^0L_{ij}^0\left(\frac{\partial s_{ij}^0}{\partial z_i^0}\right)\Delta x_{ij}^0\Delta z_{ij}^0}{\left(L_{ij}^0\right)^2\left(s_{ij}^0\right)^4}\right.$$

$$\left.-\frac{\left(\Delta y_{ij}^0\right)L_{ij}^0\left(s_{ij}^0\right)^2+2s_{ij}^0L_{ij}^0\left(\frac{\partial s_{ij}^0}{\partial z_i^0}\right)\Delta y_{ij}^0\Delta z_{ij}^0}{\left(L_{ij}^0\right)^2\left(s_{ij}^0\right)^4}\quad\frac{2L_{ij}^0\left(\frac{\partial s_{ij}^0}{\partial z_i^0}\right)}{\left(s_{ij}^0\right)^3}\right\}\mathbf{U} \tag{10.90}$$

$$\frac{\partial\left(\mathbf{a}_{Z_{ij}}^{T}\mathbf{U}\right)}{\partial x_j^0}=-\frac{\partial\left(\mathbf{a}_{Z_{ij}}^{T}\mathbf{U}\right)}{\partial x_i^0} \tag{10.91}$$

$$\frac{\partial\left(\mathbf{a}_{Z_{ij}}^{T}\mathbf{U}\right)}{\partial y_j^0} = -\frac{\partial\left(\mathbf{a}_{Z_{ij}}^{T}\mathbf{U}\right)}{\partial y_i^0} \tag{10.92}$$

$$\frac{\partial\left(\mathbf{a}_{Z_{ij}}^{T}\mathbf{U}\right)}{\partial z_j^0} = -\frac{\partial\left(\mathbf{a}_{Z_{ij}}^{T}\mathbf{U}\right)}{\partial z_i^0} \tag{10.93}$$

In Equations (10.76) to (10.93), some relevant quantities are defined below:

$$\frac{\partial s_{ij}^0}{\partial x_i^0} = -\frac{\Delta x_{ij}^0}{s_{ij}^0} \tag{10.94}$$

$$\frac{\partial s_{ij}^0}{\partial y_i^0} = -\frac{\Delta y_{ij}^0}{s_{ij}^0} \tag{10.95}$$

$$\frac{\partial s_{ij}^0}{\partial z_i^0} = -\frac{\Delta z_{ij}^0}{s_{ij}^0} \tag{10.96}$$

$$\frac{\partial L_{ij}^0}{\partial x_i^0} = -\frac{\Delta x_{ij}^0}{L_{ij}^0} \tag{10.97}$$

$$\frac{\partial L_{ij}^0}{\partial y_i^0} = -\frac{\Delta y_{ij}^0}{L_{ij}^0} \tag{10.98}$$

$$\mathbf{U} = (-\mathbf{I}\ \mathbf{I}) \text{ with } \mathbf{I} \text{ being a 3 by 3 identity matrix} \tag{10.99}$$

All the other relevant quantities have been defined in §4.1.1.2.

10.3.2 Differentiation of the Configuration Matrix in a Mapping Plane Coordinate System

Differentiation of the network configuration matrix related to some basic geodetic observables in a mapping plane coordinate system can be derived according to the observation modeling given in §4.1.2 as follows (cf. also Kuang, 1992d):

Grid Distances:

$$\frac{\partial\left(\mathbf{a}_{l_{ij}}^{T}\right)}{\partial x_i^0}=\left[\frac{l_{ij}^0+\left(\Delta x_{ij}^0\right)\frac{\partial l_{ij}^0}{\partial x_i^0}}{\left(l_{ij}^0\right)^2} \quad \frac{\left(\Delta y_{ij}^0\right)}{\left(l_{ij}^0\right)^2}\left(\frac{\partial l_{ij}^0}{\partial x_i^0}\right) \right.$$
$$\left. -\frac{l_{ij}^0+\left(\Delta x_{ij}^0\right)\frac{\partial l_{ij}^0}{\partial x_i^0}}{\left(l_{ij}^0\right)^2} \quad -\frac{\left(\Delta y_{ij}^0\right)}{\left(l_{ij}^0\right)^2}\left(\frac{\partial l_{ij}^0}{\partial x_i^0}\right)\right] \tag{10.100}$$

$$\frac{\partial\left(\mathbf{a}_{l_{ij}}^{T}\right)}{\partial y_i^0}=\left[\frac{\left(\Delta x_{ij}^0\right)}{\left(l_{ij}^0\right)^2}\left(\frac{\partial l_{ij}^0}{\partial y_i^0}\right) \quad \frac{l_{ij}^0+\left(\Delta y_{ij}^0\right)\frac{\partial l_{ij}^0}{\partial y_i^0}}{\left(l_{ij}^0\right)^2}\right.$$
$$\left. -\frac{\left(\Delta x_{ij}^0\right)}{\left(l_{ij}^0\right)^2}\left(\frac{\partial l_{ij}^0}{\partial y_i^0}\right) \quad -\frac{l_{ij}^0+\left(\Delta y_{ij}^0\right)\frac{\partial l_{ij}^0}{\partial y_i^0}}{\left(l_{ij}^0\right)^2}\right] \tag{10.101}$$

$$\frac{\partial\left(\mathbf{a}_{l_{ij}}^{T}\right)}{\partial x_j^0}=-\frac{\partial\left(\mathbf{a}_{l_{ij}}^{T}\right)}{\partial x_i^0} \tag{10.102}$$

$$\frac{\partial\left(\mathbf{a}_{l_{ij}}^{T}\right)}{\partial y_j^0}=-\frac{\partial\left(\mathbf{a}_{l_{ij}}^{T}\right)}{\partial y_i^0} \tag{10.103}$$

Grid Azimuth and Horizontal Directions

$$\frac{\partial\left(\mathbf{a}_{t_{ij}}^{T}\right)}{\partial x_i^0}=\left[\begin{array}{llll} 2\frac{\left(\Delta y_{ij}^0\right)}{\left(l_{ij}^0\right)^3}\left(\frac{\partial l_{ij}^0}{\partial x_i^0}\right) & -\frac{\left(l_{ij}^0\right)^2+2l_{ij}^0\Delta x_{ij}^0\left(\frac{\partial l_{ij}^0}{\partial x_i^0}\right)}{\left(l_{ij}^0\right)^4} & -2\frac{\left(\Delta y_{ij}^0\right)}{\left(l_{ij}^0\right)^3}\left(\frac{\partial l_{ij}^0}{\partial x_i^0}\right) & \frac{\left(l_{ij}^0\right)^2+2l_{ij}^0\Delta x_{ij}^0\left(\frac{\partial l_{ij}^0}{\partial x_i^0}\right)}{\left(l_{ij}^0\right)^4} \end{array}\right] \tag{10.104}$$

$$\frac{\partial\left(\mathbf{a}_{t_{ij}}^{T}\right)}{\partial y_i^0}=\left[\begin{array}{llll} \frac{\left(l_{ij}^0\right)^2+2l_{ij}^0\Delta y_{ij}^0\left(\frac{\partial l_{ij}^0}{\partial y_i^0}\right)}{\left(l_{ij}^0\right)^4} & -2\frac{\left(\Delta x_{ij}^0\right)}{\left(l_{ij}^0\right)^3}\left(\frac{\partial l_{ij}^0}{\partial y_i^0}\right) & -\frac{\left(l_{ij}^0\right)^2+2l_{ij}^0\Delta y_{ij}^0\left(\frac{\partial l_{ij}^0}{\partial y_i^0}\right)}{\left(l_{ij}^0\right)^4} & 2\frac{\left(\Delta x_{ij}^0\right)}{\left(l_{ij}^0\right)^3}\left(\frac{\partial l_{ij}^0}{\partial y_i^0}\right) \end{array}\right] \tag{10.105}$$

$$\frac{\partial\left(\mathbf{a}_{t_{ij}}^{T}\right)}{\partial x_j^0}=-\frac{\partial\left(\mathbf{a}_{t_{ij}}^{T}\right)}{\partial x_i^0} \tag{10.106}$$

$$\frac{\partial\left(\mathbf{a}_{t_{ij}}^{T}\right)}{\partial y_j^0}=-\frac{\partial\left(\mathbf{a}_{t_{ij}}^{T}\right)}{\partial y_i^0} \tag{10.107}$$

where

$$\frac{\partial\left(l_{ij}^{0}\right)}{\partial x_i^0} = -\frac{\Delta x_{ij}^0}{l_{ij}^0} \tag{10.108}$$

$$\frac{\partial\left(l_{ij}^{0}\right)}{\partial y_i^0} = -\frac{\Delta y_{ij}^0}{l_{ij}^0} \tag{10.109}$$

All the other relevant quantities have been defined in §4.1.2.

10.3.3 Differentiation of Matrices D and H

First, the partial derivatives of a minimum constrained datum matrix **D** with respect to station coordinates are dependent on how the datum constraints are formed. If the network datum is provided by conventional minimum constraints, the derivatives can be obtained from the differentiation of the configuration matrix of the specific observable(s) held fixed; for instance, an azimuth, a distance, and/or zenith distances, etc. Otherwise, the differentiation must be made according to the specific form of the functions that constitute the elements of matrix **D**.

Second, the partial derivatives of the inner constrained datum matrix **H** with respect to station coordinates can be in general expressed as follows:

$$\frac{\partial \mathbf{H}}{\partial x_i} = \begin{bmatrix} \mathbf{0} & \cdots & \mathbf{0} & \left(\frac{\partial \mathbf{E}_i}{\partial x_i}\right)^T & \mathbf{0} & \cdots & \mathbf{0} \end{bmatrix}^T \tag{10.110}$$

$$\frac{\partial \mathbf{H}}{\partial y_i} = \begin{bmatrix} \mathbf{0} & \cdots & \mathbf{0} & \left(\frac{\partial \mathbf{E}_i}{\partial y_i}\right)^T & \mathbf{0} & \cdots & \mathbf{0} \end{bmatrix}^T \tag{10.111}$$

$$\frac{\partial \mathbf{H}}{\partial z_i} = \begin{bmatrix} \mathbf{0} & \cdots & \mathbf{0} & \left(\frac{\partial \mathbf{E}_i}{\partial z_i}\right)^T & \mathbf{0} & \cdots & \mathbf{0} \end{bmatrix}^T \tag{10.112}$$

$$(i=1, \ldots, m)$$

where for three-dimensional networks with seven datum parameters **0** is a 7 by 3 submatrix of zeros, and $\partial \mathbf{E}/\partial x_i$, $\partial \mathbf{E}/\partial y_i$, $\partial \mathbf{E}/\partial z_i$ are evaluated as follows

$$\frac{\partial \mathbf{E}_i}{\partial x_i} = \begin{pmatrix} 0 & 0 & 0 & 0 & 0 & 0 & 1 \\ 0 & 0 & 0 & 0 & 0 & 1 & 0 \\ 0 & 0 & 0 & 0 & -1 & 0 & 0 \end{pmatrix} \tag{10.113}$$

$$\frac{\partial \mathbf{E}_i}{\partial y_i} = \begin{pmatrix} 0 & 0 & 0 & 0 & 0 & -1 & 0 \\ 0 & 0 & 0 & 0 & 0 & 0 & 1 \\ 0 & 0 & 0 & 1 & 0 & 0 & 0 \end{pmatrix} \tag{10.114}$$

$$\frac{\partial \mathbf{E}_i}{\partial z_i} = \begin{pmatrix} 0 & 0 & 0 & 0 & 1 & 0 & 0 \\ 0 & 0 & 0 & -1 & 0 & 0 & 0 \\ 0 & 0 & 0 & 0 & 0 & 0 & 1 \end{pmatrix} \tag{10.115}$$

while for two-dimensional networks with four datum parameters **0** is a 4 by 2 matrix of zeros and $\partial \mathbf{E}/\partial x_i$, $\partial \mathbf{E}/\partial y_i$ calculated by

$$\frac{\partial \mathbf{E}_i}{\partial x_i} = \begin{pmatrix} 0 & 0 & 0 & 1 \\ 0 & 0 & 1 & 0 \end{pmatrix} \tag{10.116}$$

$$\frac{\partial \mathbf{E}_i}{\partial y_i} = \begin{pmatrix} 0 & 0 & -1 & 0 \\ 0 & 0 & 0 & 1 \end{pmatrix} \tag{10.117}$$

Some columns in Equations (10.113) to (10.117) may be removed depending on the specific datum defects of the network.

10.3.4 Differentiation of the Weight Matrix

Considering the case of uncorrelated observations, the partial derivatives of the weight matrix **P** with respect to the individual weight p_i is a matrix with zeros except for identity at the (i,i)-th position, i.e.,

$$\frac{\partial \mathbf{P}_i}{\partial p_i} = \left(a_{kj}\right) \text{ with } a_{kj} = \begin{cases} 1.0 \text{ if } k = j = i \\ 0.0 \text{ elsewhere} \end{cases} \tag{10.118}$$

10.3.5 Differentials Related to Correlated Angles Derived from Independent Directions

As discussed in §4.1.3 of Chapter 4, in order to eliminate the orientation parameters from the observation equations related to direction observations, the set of independent directions, denoted by $\boldsymbol{l}_{di}$, observed at station i with observational matrix $\mathbf{P}_{l di}$ can be replaced rigorously by a set of correlated angles, denoted by $\boldsymbol{l}_{ai}$, corresponding to the same station with observational weight matrix $\mathbf{P}_{l ai}$.

At first, since an individual angle $\beta_{i\text{-}j\text{-}k}$ can be expressed as the difference of two azimuths or directions, the partial derivatives of its configuration matrix with respect to station coordinates can be obtained from those of azimuths (or directions). For instance, in the three-dimensional space, since a horizontal angle $\beta_{i\text{-}j\text{-}k}$ involves three points i, j, and k, let us denote the vector of coordinate corrections for these three points as follows

$$\delta \mathbf{p}_{i\text{-}j\text{-}k} = \left(\delta x_i \;\delta y_i \;\delta z_i \;\delta x_j \;\delta y_j \;\delta z_j \;\delta x_k \;\delta y_k \;\delta z_k\right)^T \tag{10.119}$$

Considering Equation (4.40), the configuration matrix of a horizontal angle can be expressed by

$$\mathbf{a}^T_{\beta_{i\text{-}j\text{-}k}} = \tilde{\mathbf{a}}^T_{\alpha_{i\text{-}k}} - \tilde{\mathbf{a}}^T_{\alpha_{i\text{-}j}} \tag{10.120}$$

where

$$\tilde{\mathbf{a}}^T_{\alpha_{i\text{-}j}} = \left(\mathbf{a}^T_{\alpha_{i\text{-}j}}\mathbf{U}\mathbf{E}_1 \quad \mathbf{a}^T_{\alpha_{i\text{-}j}}\mathbf{U}\mathbf{E}_2 \quad 0\,0\,0\right) \tag{10.121}$$

$$\tilde{\mathbf{a}}^T_{\alpha_{i\text{-}k}} = \left(\mathbf{a}^T_{\alpha_{i\text{-}k}}\mathbf{U}\mathbf{E}_1 \quad 0\,0\,0 \quad \mathbf{a}^T_{\alpha_{i\text{-}k}}\mathbf{U}\mathbf{E}_2\right) \tag{10.122}$$

$$\mathbf{E}_1 = \begin{pmatrix} \mathbf{I}_3 \\ \mathbf{0}_3 \end{pmatrix} \tag{10.123}$$

$$\mathbf{E}_2 = \begin{pmatrix} \mathbf{0}_3 \\ \mathbf{I}_3 \end{pmatrix} \tag{10.124}$$

with $\mathbf{I}_3$ and $\mathbf{0}_3$ being a 3 by 3 identity matrix and a 3 by 3 matrix of zeros, respectively.

The partial derivatives of $a^T_{\beta_{i\text{-}j\text{-}k}}$ with respect to station coordinates (x_i, y_i, z_i), (x_j, y_j, z_j), and (x_k, y_k, z_k) are therefore

$$\frac{\partial \mathbf{a}^T_{\beta_{i\text{-}j\text{-}k}}}{\partial x_{i;\,j;\,k}} = \frac{\partial \tilde{\mathbf{a}}^T_{\alpha_{i\text{-}k}}}{\partial x_{i;\,j;\,k}} - \frac{\partial \tilde{\mathbf{a}}^T_{\alpha_{i\text{-}j}}}{\partial x_{i;\,j;\,k}} \tag{10.125}$$

$$\frac{\partial \mathbf{a}^{T}_{\beta_{i\text{-}j\text{-}k}}}{\partial y_{i;\,j;\,k}} = \frac{\partial \tilde{\mathbf{a}}^{T}_{\alpha_{i\text{-}k}}}{\partial y_{i;\,j;\,k}} - \frac{\partial \tilde{\mathbf{a}}^{T}_{\alpha_{i\text{-}j}}}{\partial y_{i;\,j;\,k}} \tag{10.126}$$

$$\frac{\partial \mathbf{a}^{T}_{\beta_{i\text{-}j\text{-}k}}}{\partial z_{i;\,j;\,k}} = \frac{\partial \tilde{\mathbf{a}}^{T}_{\alpha_{i\text{-}k}}}{\partial z_{i;\,j;\,k}} - \frac{\partial \tilde{\mathbf{a}}^{T}_{\alpha_{i\text{-}j}}}{\partial z_{i;\,j;\,k}} \tag{10.127}$$

where

$$\frac{\partial \tilde{\mathbf{a}}^{T}_{\alpha_{i\text{-}j}}}{\partial x_{i;\,j;\,k}} = \left[\frac{\partial \left(\mathbf{a}^{T}_{\alpha_{i\text{-}j}} \mathbf{U} \right)}{\partial x_{i;\,j;\,k}} \mathbf{E}_1 \quad \frac{\partial \left(\mathbf{a}^{T}_{\alpha_{i\text{-}j}} \mathbf{U} \right)}{\partial x_{i;\,j;\,k}} \mathbf{E}_2 \quad 0\,0\,0 \right] \tag{10.128}$$

$$\frac{\partial \tilde{\mathbf{a}}^{T}_{\alpha_{i\text{-}j}}}{\partial y_{i;\,j;\,k}} = \left[\frac{\partial \left(\mathbf{a}^{T}_{\alpha_{i\text{-}j}} \mathbf{U} \right)}{\partial y_{i;\,j;\,k}} \mathbf{E}_1 \quad \frac{\partial \left(\mathbf{a}^{T}_{\alpha_{i\text{-}j}} \mathbf{U} \right)}{\partial y_{i;\,j;\,k}} \mathbf{E}_2 \quad 0\,0\,0 \right] \tag{10.129}$$

$$\frac{\partial \tilde{\mathbf{a}}^{T}_{\alpha_{i\text{-}j}}}{\partial z_{i;\,j;\,k}} = \left[\frac{\partial \left(\mathbf{a}^{T}_{\alpha_{i\text{-}j}} \mathbf{U} \right)}{\partial z_{i;\,j;\,k}} \mathbf{E}_1 \quad \frac{\partial \left(\mathbf{a}^{T}_{\alpha_{i\text{-}j}} \mathbf{U} \right)}{\partial z_{i;\,j;\,k}} \mathbf{E}_2 \quad 0\,0\,0 \right] \tag{10.130}$$

$$\frac{\partial \tilde{\mathbf{a}}^{T}_{\alpha_{i\text{-}k}}}{\partial x_{i;\,j;\,k}} = \left[\frac{\partial \left(\mathbf{a}^{T}_{\alpha_{i\text{-}k}} \mathbf{U} \right)}{\partial x_{i;\,j;\,k}} \mathbf{E}_1 \quad 0\,0\,0 \quad \frac{\partial \left(\mathbf{a}^{T}_{\alpha_{i\text{-}k}} \mathbf{U} \right)}{\partial x_{i;\,j;\,k}} \mathbf{E}_2 \right] \tag{10.131}$$

$$\frac{\partial \tilde{\mathbf{a}}^{T}_{\alpha_{i\text{-}k}}}{\partial y_{i;\,j;\,k}} = \left[\frac{\partial \left(\mathbf{a}^{T}_{\alpha_{i\text{-}k}} \mathbf{U} \right)}{\partial y_{i;\,j;\,k}} \mathbf{E}_1 \quad 0\,0\,0 \quad \frac{\partial \left(\mathbf{a}^{T}_{\alpha_{i\text{-}k}} \mathbf{U} \right)}{\partial y_{i;\,j;\,k}} \mathbf{E}_2 \right] \tag{10.132}$$

$$\frac{\partial \tilde{\mathbf{a}}^{T}_{\alpha_{i\text{-}k}}}{\partial z_{i;\,j;\,k}} = \left[\frac{\partial \left(\mathbf{a}^{T}_{\alpha_{i\text{-}k}} \mathbf{U} \right)}{\partial z_{i;\,j;\,k}} \mathbf{E}_1 \quad 0\,0\,0 \quad \frac{\partial \left(\mathbf{a}^{T}_{\alpha_{i\text{-}k}} \mathbf{U} \right)}{\partial z_{i;\,j;\,k}} \mathbf{E}_2 \right] \tag{10.133}$$

Computation of derivatives in Equations (10.128) to (10.133) can be directly obtained from Equations (10.82) to (10.87). Similar expressions can be obtained for the two-dimensional case.

Finally, from Equation (4.73) the derivative of $\mathbf{P}_{lai}$ with respect p_i is

$$\frac{\partial P_{lai}}{\partial p_i} = \left(\mathbf{BB}^{\mathrm{T}}\right)^{-1} \tag{10.134}$$

10.4 Formulation of Mathematical Models for Optimization

Mathematically, **optimization** means determining the maximum or minimum of a target function under a number of constraints (equalities or inequalities or both). For our purpose—the optimization of a geodetic network, the target function will be one that represents the network quality. As discussed before, the three general measures used to evaluate this quality are precision, reliability, and economy. A network should be designed in such a way that it has good precision (i.e., it can realize the required precision of the station coordinates), high reliability (i.e., it is robust to residual gross errors in the observations), and low cost. Therefore, our target function will be of the type (cf. Schaffrin, 1985):

$$\alpha_p(\text{precision}) + \alpha_r(\text{reliability}) + \alpha_c(\text{cost})^{-1} = \max \tag{10.135}$$

for suitably chosen weight coefficients α_p, α_r, and α_c. This is a multi-objective optimization problem. If we let one of the coefficients go to infinity, we obtain some extreme cases of the target function leading to single-objective optimization models as follows:

- Case (1): $\alpha_p \longrightarrow \infty \Longrightarrow$ precision = max
  reliability $\geq$ constant
  cost $\leq$ constant
  Here the precision is optimized while we have to control reliability and cost.
- Case (2): $\alpha_r \longrightarrow \infty \Longrightarrow$ reliability = max
  precision $\geq$ constant
  cost $\leq$ constant
  Here the reliability is optimized while cost and precision are controlled.
- Case (3): $\alpha_c \longrightarrow \infty \Longrightarrow$ cost = min
  precision $\geq$ constant
  reliability $\geq$ constant
  Here the cost is optimized while we have to control the precision and reliability.

Thus, depending on which alternative we choose, the mathematical models for the optimization of a geodetic network can be established by combining a number of the network quality requirements as discussed in §10.2 above as follows.

Model I: ***Best approximation of the criterion matrix***

$$\| \mathbf{H}\,\mathbf{w} - \mathbf{u} \| = \min$$

Subject to:

$$\mathbf{H}_1\mathbf{w} - \mathbf{u}_1 \le \mathbf{0} \tag{I-1}$$

$$\| \mathbf{r}_{00} + \mathbf{R}_{11}\mathbf{w} \| \ge r_m \tag{I-2}$$

$$\| \mathbf{c}_{00} + \mathbf{C}_{11}\mathbf{w} \| \le c_m \tag{I-3}$$

$$\left(\mathbf{D}^T \;\; 0\right)\mathbf{w} = \mathbf{0} \tag{I-4}$$

$$\mathbf{A}_{00}\mathbf{w} \le \mathbf{b}_{00} \tag{I-5}$$

Model II: ***Maximizing the reliability***

$$-\| \mathbf{r}_{00} + \mathbf{R}_{11}\mathbf{w} \| = \min$$

Subject to:

$$\mathbf{H}_1\mathbf{w} - \mathbf{u}_1 \le \mathbf{0} \tag{II-1}$$

$$\| \mathbf{c}_{00} + \mathbf{C}_{11}\mathbf{w} \| \le c_m \tag{II-2}$$

$$\left(\mathbf{D}^T \;\; \mathbf{0}\right)\mathbf{w} = \mathbf{0} \tag{II-3}$$

$$\mathbf{A}_{00}\mathbf{w} \le \mathbf{b}_{00} \tag{II-4}$$

Model III: ***Minimizing the cost of observations***

$$\| \mathbf{c}_{00} + \mathbf{C}_{11}\mathbf{w} \| = \min$$

Subject to:

$$\mathbf{H}_1\mathbf{w} - \mathbf{u}_1 \le \mathbf{0} \tag{III-1}$$

$$\| \mathbf{r}_{00} + \mathbf{R}_{11}\mathbf{w} \| \ge r_m \tag{III-2}$$

$$\left(\mathbf{D}^T \;\; \mathbf{0}\right)\mathbf{w} = \mathbf{0} \tag{III-3}$$

$$\mathbf{A}_{00}\mathbf{w} \leq \mathbf{b}_{00} \tag{III-4}$$

Optimization models based on any other optimality criteria may be established similarly.

10.5 Solution of the Mathematical Models

Up to now, the single-objective mathematical models for the optimization of a geodetic network have been established except that all the norms of matrices and vectors in the models were not defined. Depending on the choice of norm, the whole problem gets a different formulation and, as a consequence, we have to apply a different algorithm in order to find its solution. If, for instance, we allow for general norms within the models, we would obtain general mathematical programming problems with linear and nonlinear constraints (cf. Schaffrin, 1981).

10.5.1 Definition of Matrix and Vector Norms

A mapping of a linear vector space **X** over **R** into **R** is called "norm," and its value for $\mathbf{x} \in \mathbf{X}$ is denoted by $\| \mathbf{x} \|$, if it holds:

(1) $\| \mathbf{x} \| \geq \mathbf{0}$ for $\mathbf{x} \in \mathbf{X}$
(2) $\| \mathbf{x} + \mathbf{y} \| \leq \| \mathbf{x} \| + \| \mathbf{y} \|$ for $\mathbf{x}, \mathbf{y} \in \mathbf{X}$
(3) $\| \alpha \mathbf{x} \| = |\alpha| \, \| \mathbf{x} \|$ for $\mathbf{x} \in \mathbf{X}$, $\alpha \in \mathbf{R}$
(4) $\| \mathbf{x} \| = 0 \Longrightarrow \mathbf{x} = \mathbf{0}$

A "matrix norm" can be defined in terms of a vector norm. Given a vector norm $\| \cdot \|$ and a matrix **A**, consider $\| \mathbf{A} \mathbf{x} \|$ for all vectors such that $\| \mathbf{x} \| = 1$. The matrix norm induced by the vector norm is given by

$$\| \mathbf{A} \| = \max_{\| \mathbf{x} \| = 1} \| \mathbf{A} \mathbf{x} \| \tag{10.136}$$

Typical examples are the L_ρ-norms for finite dimensional vector spaces as defined with respect to some algebraic basis by

$$\left\| (x_1, x_2, \ldots, x_n) \right\|_\rho = \left(\sum_1^n | x_i |^\rho \right)^{1/\rho} \tag{10.137}$$

for any arbitrary chosen number ρ with $1 \leq \rho \leq \infty$. For $\rho = 1, 2$, and ∞, the L_1, L_2, and L_∞-norms are defined respectively as

$$\left\| (x_1, x_2, \ldots, x_n) \right\|_1 = \sum_1^n | x_i | \tag{10.138}$$

$$\left\| (x_1, x_2, \ldots, x_n) \right\|_2 = \left(\sum_1^n x_i^2 \right)^{1/2} \quad (10.139)$$

$$\left\| (x_1, x_2, \ldots, x_n) \right\|_\infty = \max\left\{ |x_i|, \ i = 1, \ldots, n \right\} \quad (10.140)$$

The L_∞-norm is also called "uniform norm" or "Tschebycheff norm."

On the other hand, for any space of matrices with arbitrary, but specified size we may obtain a matrix norm simply by identifying

$$\| \mathbf{A} \| = \| \mathrm{vec}\,(\mathbf{A}) \| \quad (10.141)$$

for a certain vector norm, where the "vec" operator stacks one column of the matrix under the other.

10.5.2 Choice of Norm

As for the "choice of norm" for network optimization, it has been argued by Schaffrin (1981) that a choice of ρ with $1 \leq \rho < 2$ would be less suitable for fitting a criterion matrix because of the loss of smoothing power while for cost requirement ρ should not exceed two. Mathematical formulations corresponding to the choice of the L_ρ-norms with r=1, 2, and ∞ for best fitting a criterion matrix have been given in Kuang (1992e). A comparative study has shown that the choice of L_2-norm has advantages over that of the other norms from the point of view of computational cost. Van Mierlo (1981) suggested L_∞-norm for reliability. Therefore, as a special case, we will use L_2, L_1, and L_∞-norms for the best fitting of the criterion matrix, for the minimum cost, and for the maximum reliability requirements respectively. Under these considerations, the above proposed mathematical models (I) to (III) become

Model I: ***Best approximation of the criterion matrix***

Minimize

$$\left(\mathbf{w}^T \mathbf{H}^T \mathbf{H}\, \mathbf{w} - 2\, \mathbf{u}^T \mathbf{H}\, \mathbf{w} + \mathbf{u}^T \mathbf{u} \right)$$

Subject to:

$$\mathbf{H}_1 \mathbf{w} - \mathbf{u}_1 \leq \mathbf{0} \quad \text{(I-1)}$$

$$\left(\mathbf{r}_{00} + \mathbf{R}_{11} \mathbf{w} \right) \geq \mathbf{r}_m \quad \text{(I-2)}$$

$$\gamma^T \mathbf{c}_{00} + \gamma^T \mathbf{C}_{11}\, \mathbf{w} \leq c_m \quad \text{(I-3)}$$

$$\left(\mathbf{D}^{\mathrm{T}} \;\; \mathbf{0}\right)\mathbf{w} = \mathbf{0} \tag{I-4}$$

$$\mathbf{A}_{00}\mathbf{w} \le \mathbf{b}_{00} \tag{I-5}$$

where $\boldsymbol{\gamma}$ is a (3m+n) by one vector of constants, and $\mathbf{r}_{\mathrm{m}}$ the n by one vector with all elements being r_m.

Model II: ***Maximizing the internal reliability***

Minimize

$$\max\left\{-\left(\mathbf{r}_{00} + \mathbf{R}_{11}\,\mathbf{w}\right)_i,\ i = 1,\ \ldots,\ n\right\}$$

Subject to

$$\mathbf{H}_1\mathbf{w} - \mathbf{u}_1 \le \mathbf{0} \tag{II-1}$$

$$\gamma^{\mathrm{T}}\,\mathbf{c}_{00} + \gamma^{\mathrm{T}}\,\mathbf{C}_{11}\,\mathbf{w} \le c_{\mathrm{m}} \tag{II-2}$$

$$\left(\mathbf{D}^{\mathrm{T}} \;\; \mathbf{0}\right)\mathbf{w} = \mathbf{0} \tag{II-3}$$

$$\mathbf{A}_{00}\mathbf{w} \le \mathbf{b}_{00} \tag{II-4}$$

Model III: ***Minimizing the cost of observations***

Minimize

$$\gamma^{\mathrm{T}}\,\mathbf{c}_{00} + \gamma^{\mathrm{T}}\,\mathbf{C}_{11}\,\mathbf{w}$$

Subject to

$$\mathbf{H}_1\mathbf{w} - \mathbf{u}_1 \le \mathbf{0} \tag{III-1}$$

$$\left(\mathbf{r}_{00} + \mathbf{R}_{11}\mathbf{w}\right) \ge \mathbf{r}_{\mathrm{m}} \tag{III-2}$$

$$\left(\mathbf{D}^{\mathrm{T}} \;\; \mathbf{0}\right)\mathbf{w} = \mathbf{0} \tag{III-3}$$

$$\mathbf{A}_{00}\mathbf{w} \le \mathbf{b}_{00} \tag{III-4}$$

10.5.3 The Solution Methods

Model I is the L_2-approximation solvable by Quadratic Programming (QP); Model II is the L_∞-approximation solvable by Linear Programming (LP); and finally Model III is the L_1-approximation solvable by Linear Programming (LP). Basic problems of QP and LP will be discussed in Appendix B. Standard software for Linear or Quadratic Programming, if available, can be used to obtain the solutions for the mathematical models. A software package using FORTRAN 77 has been developed by the author (Kuang, 1991) to solve the above mathematical models, and they are available to the reader(s) when requested.

10.6 The Multi-Objective Optimization Model (MOOM)

In the above sections, the single-objective mathematical models for the optimization of a geodetic network have been established. For practical applications, depending on the aspects of the problem we prefer, we may choose one of the Models I to III to obtain the optimal configuration and observation plan of a geodetic network. Since the objective function in Model I is strictly convex for a positive-definite matrix $\mathbf{H}^T\mathbf{H}$, it has a unique minimizing solution if and only if the constraints in (I-1) to (I-5) are consistent. In Models II and III, the objective functions are convex, they have unique minimizing solutions if and only if their respective constraints in (i-1) to (i-4) (i = II, III) are consistent and no rank deficiency exists for the coefficient matrix of the whole set of constraints of each model. Therefore, the consistency of constraints is decisive for the solution of each of the mathematical models. The possibility of inconsistency of constraints in (I-1) was discussed by Schaffrin et al. (1980). He proposed to approach the inconsistency by "parameterization," i.e., by adding a certain negative value at the right-hand side in order to generate consistency and thus the existence of the desired solution. The problem with that approach is that through the "parameterization," the original constraints in (I-1) become in fact redundant.

An analysis of the constraints in (I-1) to (I-5) indicates that the physical constraints in (I-4) to (I-5) are compulsory. They represent the physical environment in which we can optimize our network. In most cases, the precision, reliability, and cost constraints ((I-1) to (I-3)) may be contradictory, i.e., the best network that would simultaneously satisfy the objective function and all the constraints may not exist. Physically, that can be understood in this way: once the maximum precision of instrumentation, the total expenditure, and the topographic conditions are given, the maximum precision and reliability attainable are limited. Thus if the criterion matrix $\mathbf{C}_s$ of station coordinates, the criterion vector for reliability $\mathbf{r}_m$ are not defined properly, i.e., if they *exceed* the reality, the constraints in (I-1) to (I-3) will be inconsistent, and no solution exists. This analysis is applicable to all the other two Models II and III. Therefore, to ensure a universally applicable optimization procedure, a unified mathematical model, called the Multi-Objective Optimization Model (MOOM), is proposed, in which instead of approximating the constraints in (I-1) to (I-3) "from one side," an approximation "from both sides" is permitted. In the following subsections, the concept of multi-objective optimization is reviewed first, and the multi-objective optimization model for geodetic networks is then established.

10.6.1 The Concept of Multi-Objective Optimization

The study of the multi-objective optimization is a well-established branch in operations research. The mathematical theories involved are quite sophisticated. In the following, a brief explanation of the problem statement and the solution concepts is given in order to apply this theory to solving our problems.

A **M**ulti-**O**bjective **O**ptimization **P**roblem (**MOOP**) is defined as:

Minimize

$$\mathbf{f}(\mathbf{x}) = \left[f_1(\mathbf{x}), \; \ldots, \; f_p(\mathbf{x})\right]^T \tag{10.142}$$

Subject to

$$\mathbf{x} \quad \varepsilon\, \mathbf{X} = \left\{\mathbf{x}\text{: } \mathbf{g}(\mathbf{x}) = \left[g_1(\mathbf{x}), \; \ldots, \; g_m(\mathbf{x})\right]^T \le \mathbf{0}\right\} \tag{10.143}$$

where f_i (i=1, ..., p) are the individual objective functions and g_j (j=1, ..., m) the constraint functions. Here all the functions are assumed to be continuously differentiable. This kind of problem is also called a ***vector optimization***.

Unlike traditional mathematical programming with a single objective function, an optimal solution that minimizes all the objective functions simultaneously does not necessarily exist. The final decision should be made by taking the total balance of objectives into account. Thus the decision maker's value is usually represented by saying whether or not an alternative **x** is preferred to another alternative **x′**, or equivalently whether or not **f**(**x**) is preferred to **f**(**x′**). For this purpose, a scalar-valued function $u(f_1, \ldots, f_p)$ representing the decision maker's preference called ***preference function*** has to be defined. Once we obtain such a preference function, our problem reduces to the traditional mathematical programming:

Maximize

$$u\left(f_1, \; \ldots, \; f_p\right) \tag{10.144}$$

Subject to

$$\mathbf{x} \quad \varepsilon\, \mathbf{X} = \left\{\mathbf{x}\text{: } \mathbf{g}(\mathbf{x}) = \left[g_1(\mathbf{x}), \; \ldots, \; g_m(\mathbf{x})\right]^T \le \mathbf{0}\right\}$$

Instead of strict optimality, the notion of efficiency is introduced in multi-objective optimization. A vector $\mathbf{f}(\hat{\mathbf{x}})$ is said to be efficient if there is no **f**(**x**) (**x** ε **X**) preferred to $\mathbf{f}(\hat{\mathbf{x}})$. Mathematically, the most fundamental kind of efficient solution is usually called a Pareto optimal solution or noninferior solution. A point $\hat{\mathbf{x}} \varepsilon \mathbf{X}$ is said to be a Pareto optimal solution (or noninferior solution) to the problem MOOP if there is no $\mathbf{x} \varepsilon \mathbf{X}$ such that $\mathbf{f}(\mathbf{x}) \leq \mathbf{f}(\hat{\mathbf{x}})$. The final decision is usually made among the set of efficient solutions. There are a number of ways to characterize the efficiency of a solution. One characterization is taken as the best approximation to the ideal point. Consider the multi-objective optimization problem MOOP, an ideal point $\tilde{\mathbf{f}} = (\tilde{\mathbf{f}}_1, \ldots, \tilde{\mathbf{f}}_p)^T$ is defined as:

$$\tilde{f}_i = \inf\{f_i(\mathbf{x}) \colon \mathbf{x} \,\varepsilon\, \mathbf{X}\} > -\infty, \ i = 1, \ \ldots, \ p \tag{10.145}$$

However, we might take another ideal point $\bar{\mathbf{f}}$ if it satisfies

$$\bar{\mathbf{f}} \leq \tilde{\mathbf{f}} \tag{10.146}$$

A solution $\hat{\mathbf{x}}$ is efficient if

$$d = \left\| \mathbf{f}(\hat{\mathbf{x}}) - \bar{\mathbf{f}} \right\| = \inf\left\{ \left\| \mathbf{f}(\mathbf{x}) - \bar{\mathbf{f}} \right\| \right\} \mathbf{x} \,\varepsilon\, \mathbf{X} \tag{10.147}$$

where $\| \bullet \|$ represents the norm of a vector. It can be the L_ρ-norms as defined by

$$\| \mathbf{f} \|_\rho = \left[\sum_{i=1}^{p} | \mu_i \, f_i |^\rho \right]^{1/\rho}, \ \rho \,\varepsilon\, [1, \infty) \tag{10.148}$$

or the one introduced by Dinkelbach and Isermann (1973)

$$\| \mathbf{f} \|_\rho = \| \mathbf{f} \|_\infty + \frac{1}{\rho} \left(\sum_{i=1}^{p} | \mu_i \, f_i | \right), \ \rho \,\varepsilon\, [1, \infty) \tag{10.149}$$

where $\boldsymbol{\mu} = (\mu_1, \mu_2, \ldots, \mu_p)^T$ is the weighting vector. Intuitively, if the ideal point is unattainable, $d = \| \mathbf{f}(\mathbf{x}) - \bar{\mathbf{f}} \|$ represents the distance between $\mathbf{f}(\mathbf{x})$ and $\bar{\mathbf{f}}$ and can be considered as a measure of regret resulted from unattainability of $\mathbf{f}(\mathbf{x})$ to $\bar{\mathbf{f}}$. An efficient solution is to minimize this regret. In order to apply this idea in practice, the following problems have to be solved:

(i) How to choose the distance function
(ii) How to decide the weighting vector and
(iii) How to make a common scale for the different objective functions.

At first, Sawaragi et al. (1985) suggest that it suffices to use the L_ρ-norms with $\rho \geq 1$ to evaluate the distance. The solution to problem (ii) is quite subjective, since that is closely related to the decision maker's *preference attitudes* to some specific objective functions. Thus, one used to adopt the weight based on his experience. If the same weight is selected for each objective function, then the resulting solution is the one that improves *equally*, in some sense, each criterion as much as possible. Dyer (1972) suggested a numerical method for deciding the weights under interaction with decision makers. The method is complicated and difficult to use in practice. Finally, if the L_p-norms distance function is used, it is important to make a common scale for the objective functions. For example, if the positive and negative values of the objective functions are mixed, or if their physical units and numerical orders are different from each other, then some of the objective functions are sensitive and others not. One approach to make a common scale for objective functions is to use the relative degree of the non-attainability of $f_i(\mathbf{x})$ to its ideal value $\bar{f}_i$, i.e., $(f_i(\mathbf{x}) - \bar{f}_i)/\bar{f}_i$ as new objective functions. This enables us to ignore the need to pay extra attention to the difference among the dimension and the numerical order of the different objective functions.

10.6.2 The Multi-Objective Optimization Model (MOOM) for Geodetic Networks

According to the concepts of the multi-objective optimization discussed above, a mathematical model corresponding to the solution of the multi-objective optimization problem (10.135) is established under the concept of "ideal point" as follows:

Minimize

$$\left[\frac{\| \mathbf{H}\,\mathbf{w} - \mathbf{u} \|}{\| \mathrm{vec}\,(\mathbf{C}_s) \|} + \frac{\| \mathbf{R}_{11}\,\mathbf{w} - (\mathbf{r}_m - \mathbf{r}_{00}) \|}{\| \mathbf{r}_m \|} + \frac{\| \gamma^T\,\mathbf{C}_{11}\,\mathbf{w} - (c_m - \gamma^T\,\mathbf{c}_{00}) \|}{\| c_m \|} \right] \tag{10.150}$$

Subject to:

$$\begin{pmatrix} \mathbf{D}^T & \mathbf{0} \end{pmatrix} \mathbf{w} = \mathbf{0}$$

$$\mathbf{A}_{00}\mathbf{w} \le \mathbf{b}_{00}$$

where the "ideal point" representing precision, reliability, and cost are $\mathbf{C}_s$, $\mathbf{r}_m$, and c_m, respectively. This model tries to minimize the differences among the precision, reliability, and cost and their respective ideal counterparts simultaneously, subject to the physical constraints, and therefore it includes all the intentions of Models I to III. Interpreting this intuitively, this model tries to best approximate *equally* the precision, reliability, and cost criteria under the given geography and instrumentation condition in the sense of minimum norms. If appropriate weighting coefficients α_p, α_r, and α_c for precision, reliability, and cost respectively are selected, the decision maker desires to improve some of the criteria with larger weights more strongly than the others. This practical meaning encourages us to accept easily the obtained solution. The selection of the weighting coefficients is, however, to a large extent subjective, and it should be based on the specific problems to be solved.

By applying the L_2-norm to the above model from the point of view of computational benefit, the objective function (10.150) becomes

Minimize

$$\left(\mathbf{w}^T \mathbf{H}_0^T \mathbf{H}_0 \mathbf{w} - 2\,\mathbf{u}_0^T \mathbf{H}_0 \mathbf{w} + \mathbf{u}_0^T \mathbf{u}_0\right) \tag{10.151}$$

where

$$\mathbf{H}_0 = \begin{pmatrix} \mathbf{H} / \sqrt{\mathbf{u}_s^T \mathbf{u}_s} \\ \mathbf{R}_{11} / \sqrt{\mathbf{r}_m^T \mathbf{r}_m} \\ \gamma^T \mathbf{C}_{11} / c_m \end{pmatrix} \tag{10.152}$$

$$\mathbf{u}_0 = \begin{pmatrix} \mathbf{u} / \sqrt{[\mathrm{vec}(\mathbf{C}_s)]^T\ \mathrm{vec}(\mathbf{C}_s)} \\ (\mathbf{r}_m - \mathbf{r}_{00}) / \sqrt{\mathbf{r}_m^T \mathbf{r}_m} \\ \left(c_m - \gamma^T \mathbf{c}_{00}\right) / c_m \end{pmatrix} \tag{10.153}$$

This model has a unique minimizing solution if and only if the matrix $\mathbf{H}_0^T\mathbf{H}_0$ is positive definite. In practical applications, some of the quality criteria may be omitted in the objective function, depending on one's specific needs. In this case the model tries to best approximate the quality criteria that are included in the model.

Finally, it should be noted that the above optimization model approximates the network quality criteria "from both sides;" that is, it cannot always guarantee that the resulting network quality will be equal to or better than the required values, although this situation may certainly happen in certain cases. If there is a case where all the design criteria of precision and reliability must be satisfied whatever the total expenditure of the project is, there are two ways to accomplish this goal. The first is to add some additional constraints to the above model so that the design criteria are approximated "from one side," i.e.,

$$\mathbf{A}_{11}\mathbf{w} \le \mathbf{b}_{11} \tag{10.154}$$

where

$$\mathbf{A}_{11} = \begin{pmatrix} \mathbf{H}_1 \\ -\mathbf{R}_{11} \end{pmatrix} \tag{10.155}$$

$$\mathbf{b}_{11} = \begin{pmatrix} \mathbf{u}_1 \\ \mathbf{r}_{00} - \mathbf{r}_{\mathrm{m}} \end{pmatrix} \tag{10.156}$$

and all the other relevant quantities in (10.155) and (10.156) mean the same as before.

However, as discussed before, since the system of constraints (10.154) may be contradictory, leading to no solution, another way is to establish new ideal points $\overline{\mathbf{C}}_{\mathrm{s}}$ and $\overline{\mathbf{r}}_{\mathrm{m}}$ for precision and reliability, respectively, which according to Equation (10.146) must satisfy

$$\mathrm{vecdiag}\left(\overline{\mathbf{C}}_{\mathrm{s}}\right) < \mathrm{vecdiag}\left(\mathbf{C}_{\mathrm{s}}\right) \tag{10.157}$$

$$\overline{\mathbf{r}_{\mathrm{m}}} > \mathbf{r}_{\mathrm{m}} \tag{10.158}$$

where "vecdiag" means the same as before. Equations (10.157) and (10.158) mean that the precision criteria and ideal redundancy numbers used for optimization are better than what is actually required. By adjusting the values of the ideal points, the optimization algorithm may produce a network in which the achievable precision and reliability will be better or at least equal to the required values. However, here the upper limits regarding the weights in vector $\mathbf{b}_{00}$ have to be put large enough so that it is physically possible to achieve the required network quality.

It should be noted that all the solutions to the above mathematical models should be iterative since Taylor series of linear form is being used. A computer software package has been developed by the author using FORTRAN 77 to solve both the proposed single-objective and multi-objective mathematical models. Applications of the optimization models to solving different network design problems in one-, two-, and three-dimensional space have been elaborated on by the author in a series of publications (cf. Kuang et al., 1991; Kuang, 1992a-f, 1993a-h, 1994; Kuang and Chrzanowski, 1992a,b, 1994). In particular, the application of the multi-objective optimization model has been discussed in Kuang (1993c,h), and Kuang and Chrzanowski (1992a). Figure 10.2 shows the computation flowchart. The network quality criteria have to be defined first according to the purpose of the network. They are usually given by the network user, i.e., the client. The optimization computation starts with an initial design characterized by initially determined station coordinates $\mathbf{x}^0$, and geodetic observables with initial observation weights $\mathbf{P}^0$ through field reconnaissance and/or other means. The easiest way is to start with a "*maximum*" network; that is, to include all the possible geodetic observables in the initial design, and the selection of the initial weights may be arbitrary. The optimal coordinate changes to be introduced to the initially selected survey points and optimal weight changes to be introduced to the initially adopted approximate weights are then solved iteratively through a properly chosen optimization model discussed in this chapter. After the solution converges, the optimal values of both coordinates and weights are obtained by adding the solved for optimal changes to their corresponding approximate values [cf. (10.1) and (10.2)]. Finally, the survey marks can

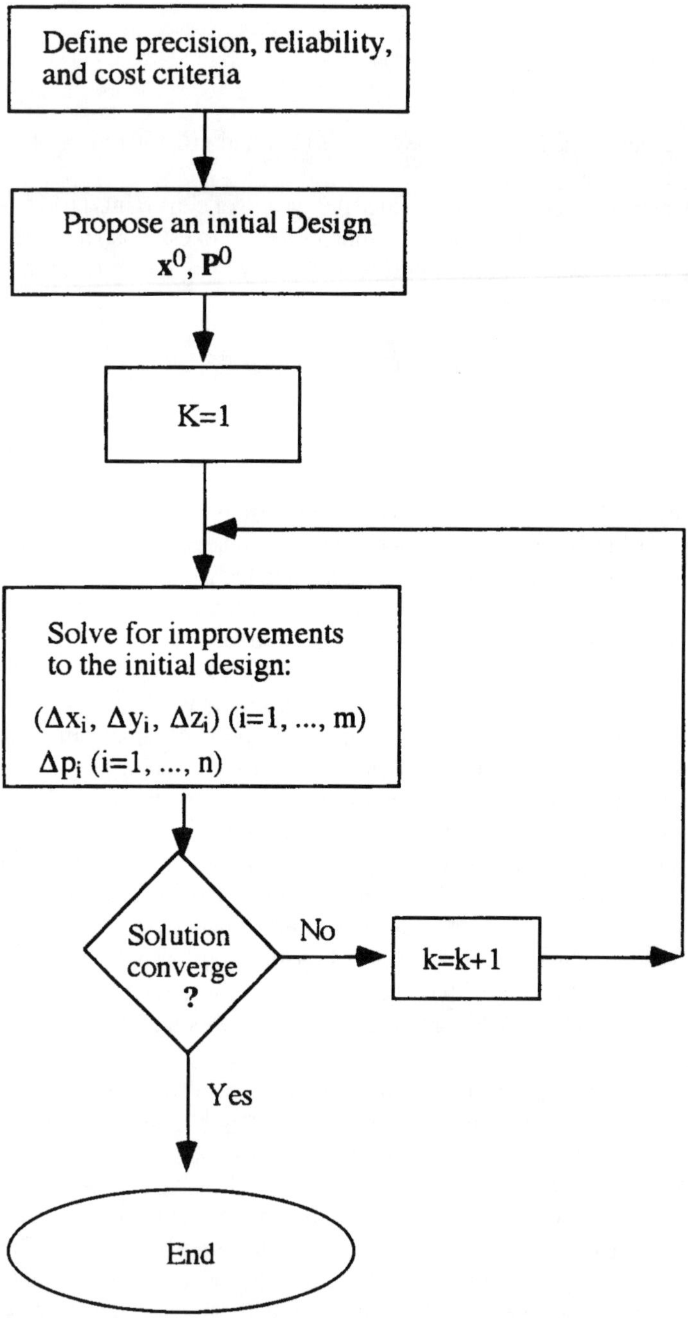

Figure 10.2. The computation flowchart

then be monumented according to the solved optimal coordinates of all the initially selected points. As for the final observing scheme, all the observables that obtained zero or insignificant weights can be eliminated from the initial design. For observables that obtained significant optimal weights, instrumentation needed to take the measurements can be chosen according to the required precision calculated from the optimal weights. Examples given in Chapter 12 demonstrate in detail how the optimization methodology is used in practice.

11 Deformation Monitoring Network Design

This chapter discusses the optimal design of deformation monitoring networks. As with geodetic positioning networks, network design must precede the field campaign in order to prevent the project from failing. As discussed in §7.3, there are now hundreds of available models of various geodetic and non-geodetic instruments for deformation measurements. Selection of a monitoring technique depends heavily on the type, the magnitude, and the rate of the deformation. The decision about which instruments should be used and where they should be located leads to the need for a proper design and optimization of a proposed measuring scheme that should be based on the best possible combination of all the available measuring instrumentation. In discussing geotechnical instrumentation for performance monitoring, Peck (in Dunnicliff, 1988) states that

> "*...every instrument on a project should be selected and placed to assist answering a specific question, the wrong instruments in the wrong places provide information that may at best be confusing and at worst divert attention from telltale signs of trouble. Too much instrumentation is wasteful and may disillusion those who pay the bills, while too little, arising from a desire to save money, can be more than false economy: it can even be dangerous....*"

Therefore, the design of a monitoring scheme should satisfy not only the best geometrical strength of the network of the observation stations, as is the case in geodetic positioning surveys, but should primarily satisfy the needs of the subsequent physical interpretation of the monitoring results, i.e., should give optimal results when solving for the deformation parameters of the selected deformation model (Chrzanowski et al., 1986).

A successful design of monitoring networks requires the a priori knowledge of the expected deformation, i.e., the deformation model **Be** [cf.(7.1)] must be known. That is to say, the specific purpose of the network must be known. Such information comes from a study of the relevant physical, geomechanic, and/or structural properties of the object. In this book we concentrate on deformations of local or regional scales. Usually, such a monitoring network consists of a geodetic network, plus some isolated non-geodetic observables that may not be geometrically connected with the geodetic network. Theory and applications of optimization of integrated deformation monitoring schemes with geodetic and non-geodetic observables have been studied in Kuang (1991, 1992f). This chapter gives a comprehensive review of the theory for the optimal design of a conventional geodetic network established for monitoring deformations. For details on some other aspects of the developed methodology, the interested reader is referred to the given references.

11.1 Design Orders of a Monitoring Network

Similar to the optimal design of geodetic positioning networks, the main task of the optimization design of monitoring networks can be stated as follows

- Selection of a monitoring methodology
- Determination of the optimal locations of both reference and object points, and
- Computation of the optimal distribution of required observational accuracies among heterogeneous observables.

Following the convention of design orders for geodetic networks discussed in §8.1, one may consider the same classification of the design orders for deformation monitoring networks. There are, however, significant differences in the contents of design problems in positioning networks versus monitoring networks.

First of all, there is no Zero-Order Design (ZOD) problem in monitoring networks. For instance, in the case of a reference network, correct displacements can be obtained only by comparing coordinates at different epochs with respect to the same reference datum. Hence not only the reference datum at the initial epoch has to be defined, but also its motion at the subsequent epochs has to be identified. Therefore, the main problem here is not to define an optimal datum for the initial epoch but to confirm the stability of the reference frame at the initial epoch, and only the one that maintains the same position and orientation at the subsequent epochs can be considered as the "optimal" datum for it. Various techniques were developed for this purpose (cf. Chen and Chrzanowski, 1986). In the case of a relative network, the deformation parameters are in general datum independent, and therefore the choice of the reference frame for the deformation analysis is of no practical importance.

In regard to the First-Order Design (FOD), the shape of the network largely depends on the topography covered by the network or the shape of the structures to be monitored. Certain changes, however, of the positions of the points in the monitoring network are always possible, and they should be used to optimize the configuration of the network with respect to an optimality criterion. Generally, the location of the reference points should be outside of the deformable body and outside of the zone of the acting forces that produce the deformation. In contrast, the object points should be distributed over the area that is considered important from the safety point of view and in which the most obvious deformations (maximum values of the deformation parameters) are expected to occur. The existing prediction theories and methods such as the finite element method can help in identifying the stable versus the most unstable areas. Since the parameters of interest in monitoring networks are not the coordinates (positions) themselves but their variation (displacements) with time and other deformation parameters, all other isolated observations such as tiltmeter, strainmeter, pendulum, and repeated accurate measurements of baselines and angles not belonging to a geodetic network, must be integrated in order to obtain an optimally designed monitoring network. The geodetic network can be seen in this case not as a means of obtaining the absolute positions of the displaced points with the higher accuracy, but rather as a means of providing good spatial correlation between various observables and providing information on the global deformation trend. The location of geotechnical and other non-geodetic instrumentation should be selected where it represents the general trend of the deformation or where the maximum deformations are expected. Therefore in the FOD for monitoring networks, the optimal positions for both geodetic and non-geodetic instrumentation should be solved for simultaneously (cf. Kuang, 1991).

The Second-Order Design (SOD) problem for monitoring networks is to find accuracies of both the geodetic and non-geodetic observables or weight matrix **P** that leads to accuracies of estimates of all unknown parameters as close as possible to some given idealized counterparts, e.g., the criterion matrix. The matrix **P** derived from the solution of the SOD problem gives directions for the choice of instrumentation and observational procedures.

The Third-Order Design (THOD) problem of improvement of existing networks might be very useful for monitoring networks. When at some epoch it is realized that accuracies of displacements or deformation parameters are unsatisfactory or the observed deformations do not follow the expected trend, that is, the a priori postulated deformation model, the monitoring network must be improved by additional observables or additional points in future epochs. Finally, the Combined First-Order and Second-Order Design problem appears if neither the monitoring configuration nor the weights of observations are known and they have to be optimally solved for simultaneously.

Theory and applications of optimization of integrated deformation monitoring schemes with geodetic and non-geodetic observables have been investigated in Kuang (1991, 1992f). This chapter gives a comprehensive review of the theory for the optimal design of a conventional geodetic network for monitoring deformations. For details on some other aspects of the developed methodology, the interested reader is referred to the given references.

11.2 Quality Control Measures and Optimality Criteria for Deformation Monitoring Networks

This section deals with the quality measures and optimization criteria for deformation networks. In addition to such measures as precision, reliability, and economy, which characterize the quality of a geodetic positioning network, a fourth measure, i.e., sensitivity, should be added to evaluate the quality of a deformation monitoring network. Sensitivity describes the network's ability to detect postulated displacements or deformation parameters of certain magnitude. As with geodetic positioning networks, in order to perform a fully analytical optimal network design, quantitative expressions for these quality measures and criteria are derived first, and they are subsequently converted into requirements on the unknown parameters to be optimized, i.e., the ***improvements*** to an initial design.

11.2.1 Measures and Criteria for Precision

In contrast to geodetic positioning networks where the parameters of concern are the station coordinates, in deformation monitoring networks the parameters of interest are deformation parameters (**e**) or displacements (**d**) derived from observations. Precision measures of deformation networks are, therefore, based on the variance-covariance matrix of the deformation parameters. Referring to Equation (7.28), in the design phase, it is justified to assume that the observation networks are the same for all the epochs (i.e., $\mathbf{A}_i = \mathbf{A}$, $\mathbf{P}_i = \mathbf{P}$, for all i). Considering only two epochs, the expressions for the solution of deformation parameters $\hat{\mathbf{e}}$ and their associated variance-covariance matrix $\mathbf{C}_e$ are as follows

$$\hat{\mathbf{e}} = \left(\mathbf{B}^T\,\mathbf{A}^T\,\mathbf{P}\,\mathbf{A}\,\mathbf{B}\right)^{-1}\mathbf{B}^T\,\mathbf{A}^T\,\mathbf{P}(l_2 - l_1) \tag{11.1}$$

$$\mathbf{C}_e = \sigma_0^2\mathbf{Q}_e = 2\sigma_0^2\left(\mathbf{B}^T\,\mathbf{A}^T\,\mathbf{P}\,\mathbf{A}\,\mathbf{B}\right)^{-1} \tag{11.2}$$

where σ_0^2 is the a priori variance factor. When **B=I** (identity matrix), the deformation monitoring reduces to determine the displacements of object points. In this case, the estimated displacements $\hat{\mathbf{d}}$ and their associated variance-covariance matrix $\mathbf{C}_d$ can be expressed by

$$\begin{aligned}\hat{\mathbf{d}} &= \left(\mathbf{A}^T\,\mathbf{P}\,\mathbf{A} + \mathbf{D}\mathbf{D}^T\right)^{-1}\mathbf{A}^T\,\mathbf{P}(l_2 - l_1)\\ &= \hat{\mathbf{x}}_2 - \hat{\mathbf{x}}_1\end{aligned} \tag{11.3}$$

$$\mathbf{C}_d = \sigma_0^2\mathbf{Q}_d = 2\sigma_0^2\left[\left(\mathbf{A}^T\mathbf{P}\mathbf{A} + \mathbf{D}\mathbf{D}^T\right)^{-1} - \mathbf{H}\left(\mathbf{H}^T\mathbf{D}\mathbf{D}^T\mathbf{H}\right)^{-1}\mathbf{H}^T\right] \tag{11.4}$$

where **D** and **H** represent the minimum constrained and inner constrained datum matrices as discussed before in §4.2.3 and §4.2.4, and $\hat{\mathbf{x}}_1$ and $\hat{\mathbf{x}}_2$ are the coordinate vectors determined for the first and second measurement epoch, respectively.

It should be noted that although deformation strictly means change of shape and dimension, detection of displacements, scale changes, and rotations, etc., are usually prompted in practice. In many cases, the deformation parameters are derived from changes in coordinates of the object points as a result of survey observations made at two or more epochs. The changes in coordinates (i.e., displacements) of points are the quantities that geodesists and surveyors like to work with because they allow for an easy identification of the deformation model for further geometrical analysis (Chrzanowski et al., 1983). However, the use of "coordinates of points" introduces problems associated with the way in which those coordinates are defined, i.e., the coordinates are datum-dependent. Careful selection of a datum will provide a picture of the displacement field, which makes it easier to identify the deformation model in the space domain or to identify the suspected unstable points of a reference network. At the design phase, the a priori knowledge of the expected deformation is of much help for the definition of an appropriate datum. Such information comes from a study of the relevant physical, geomechanic, and/or structural properties of the object. In the absence of any a priori reliable data about the likely behavior of points on and around the object it is necessary in the first instance to assume that all points are subject to deformation, and some sophisticated techniques can be used to define an appropriate datum at the stage of post-processing of the data (cf. Chrzanowski et al., 1983; Chen, 1983; Chrzanowski et al., 1986).

11.2.1.1 The Precision Criteria for Deformation Networks

As with geodetic positioning networks, the precision criteria for deformation networks can be either a scalar risk function as discussed in §9.1.1 or a criterion matrix. However, the formation of a criterion matrix for displacements and deformation parameters may require some special attention, since a good precision of coordinates may in general not ensure a good precision of the deformation parameters. A suitable criterion matrix for deformation parameters can be a diagonal matrix with its diagonal elements being defined according to the required precision for each individual parameter. In the case that the deformation parameters to be determined are simply the displacements of the object points, the approaches discussed in §9.1.2 may be used to construct a criterion matrix for displacements. Another approach for defining a criterion matrix for displacements monitoring was discussed in Sprinsky (1978), Wimmer (1981), and Crosilla (1982; 1983b). They proposed to construct a criterion matrix for displacements to be monitored by contracting the eigenvalue spectrum of the covariance matrix of the adjusted network coordinates and/or by rotating the h-dimensional error ellipsoid to redistribute the allocated variance in order to detect displacements with certain predicted magnitudes and orientations. A review of this technique is given below:

According to (11.3) and (11.4), the confidence hyper-ellipsoid, centered at **d**, for the estimated displacements can be expressed by

$$(\mathbf{d}-\hat{\mathbf{d}})^{T}(\mathbf{C}_{\hat{d}})^{-}(\mathbf{d}-\hat{\mathbf{d}}) \leq \xi_{\chi^2_{h,\,1-\alpha}} \tag{11.5}$$

where an unspecified generalized inverse for matrix $\mathbf{C}_{\hat{d}}$ is used since it is assumed to have a rank defect of r_d, which is equal to the number of datum defects of the network, i.e.,

$$\mathrm{rank}(\mathbf{C}_{\hat{d}}) = h < u - r_d \tag{11.6}$$

where u is the dimension of the coordinate vector.

In this case, (11.5) corresponds to the equation of a h-dimensional error ellipsoid, in which the length of the semi-axes and their directions are given respectively by the square root of the eigenvalues greater than zero, and their corresponding eigenvectors of $\mathbf{C}_{\hat{d}}$, which is decomposed as follows [cf. (6.11)]:

$$\mathbf{C}_{\hat{d}} = (V_1 \quad V_2)\begin{pmatrix} \Lambda_1 & 0 \\ 0 & 0 \end{pmatrix}\begin{pmatrix} V_1^T \\ V_2^T \end{pmatrix} = V_1\,\Lambda_1\,V_1^T \tag{11.7}$$

where

$$\begin{aligned} &\Lambda_1 = \mathrm{diag}\,(\lambda_1, \lambda_2, \ldots, \lambda_h) \\ &\text{with eigenvalues } \lambda_1 \geq \lambda_2 \geq \ldots \geq \lambda_h > 0 \end{aligned} \tag{11.8}$$

V_1 and V_2 are the submatrices of the normalized eigenvectors relating to the non-zero eigenvalues and zero eigenvalues of $\mathbf{C}_{\hat{d}}$, respectively.

First, the contraction of the eigenvalue spectrum may be accomplished using a "contraction parameter t," with $0 \leq t \leq 1$, arbitrary but fixed in two different formulations

$$\begin{aligned} &\text{i) } \tilde{\lambda}_1 = \tilde{\lambda}_1 - t(\lambda_1 - \lambda_h) \text{ and} \\ &\tilde{\lambda}_i = \min(\tilde{\lambda}_i, \lambda_1) \text{ for } i = 2, \ldots, h \text{ (Sprinsky, 1978; Wimmer, 1982a)} \end{aligned} \tag{11.9}$$

$$\text{ii) } \tilde{\lambda}_i = \lambda_i - t(\lambda_i - \lambda_h) \text{ for } i = 1, 2, \ldots, h \text{ (Crosilla, 1983a)} \tag{11.10}$$

Second, if a deformation has been defined by means of a prediction method, the worst directions to detect point movements are those that correspond to the direction of the greater semi-axes of the error ellipsoid; that is, the eigenvectors relating to the greater eigenvalues of $\mathbf{C}_{\hat{d}}$. According to the conceived deformation model, it might be useful to proceed with some rotations of the component pairs of the essential eigenvector in such a way that they would be oriented as close as possible along a direction orthogonal to that of the movement predicted by the deformation model. For a sound statistical study of the displacement vector $\hat{\mathbf{d}}$, based on the analysis of the principal components (Niemeier, 1982) it is necessary to investigate a total variance percentage equal to 40–60% of tr($\mathbf{C}_{\hat{d}}$). Whenever λ_1 alone represents such a value it is sufficient to rotate only the component pairs of the first eigenvector. Otherwise, account must be taken of all the eigenvectors relating to the greatest eigenvalues for which $\Sigma\lambda_i$ is equal to 40–60% of tr($\mathbf{C}_{\hat{d}}$).

One rotation technique involves the rotation, equal to an angle $\varphi_{x,y}$, of all the pairs of eigenvector components relating to the coordinates x_i, y_i of m netpoints:

$$\begin{bmatrix} v^0_{x_i1} \\ v^0_{y_i1} \end{bmatrix} = \begin{bmatrix} \cos(\varphi_{x_iy_i}) & \sin(\varphi_{x_iy_i}) \\ -\sin(\varphi_{x_iy_i}) & \cos(\varphi_{x_iy_i}) \end{bmatrix} \begin{bmatrix} v_{x_i1} \\ v_{y_i1} \end{bmatrix} \quad (i = 1, \ldots, h) \tag{11.11}$$

This follows from the fact that in order to ensure the orthogonality of all the h vectors in $\boldsymbol{V}_1$ of $\boldsymbol{V}$ it is necessary to perform the same rotation for all the pairs of elements of the h eigenvectors relating to the coordinates x_i, y_i of the netpoints.

Another rotation technique is the so-called procrustean transformation of the $\boldsymbol{V}_1$ eigenvector matrix. With the procrustean transformation each pair of h eigenvector components can be rotated independently. This is of particular importance in case the first eigenvalue λ_1 does not represent 40–60% of the total variance and it is, therefore, necessary to take more than one eigenvector into consideration in order to satisfy, from the statistical point of view, the condition of orthogonality between the pairs of essential eigenvector components and the predicted direction of the movement. From the independent rotations of each couple of eigenvector components, the resulting matrix, that is the target matrix $\boldsymbol{V}_0$, is no longer constituted by orthogonal eigenvectors. The procrustean transformation (Schönemann, 1966) is to find an orthogonal matrix $\mathbf{T}$ that satisfies the approximation

$$\boldsymbol{V}_1\,\mathbf{T} \cong \boldsymbol{V}_0 \tag{11.12}$$

from the least squares point of view.

Finally, the criterion matrix for the displacement vector $\hat{\mathbf{d}}$ resulting from the procrustean transformation technique and from the contraction of the eigenvalue spectrum can be written as

$$\mathbf{C}^c_{\hat{d}} = \boldsymbol{V}_1\,\mathrm{T}\,\tilde{\Lambda}_1\,\mathbf{T}^{\mathrm{T}}\boldsymbol{V}_1^{\mathrm{T}} \tag{11.13}$$

where $\tilde{\Lambda}_1$ is the matrix of contracted eigenvalues, and the matrix $\mathbf{T}$ is calculated by

$$\mathbf{T} = \mathbf{X}\,\mathbf{Y}^{\mathrm{T}} \tag{11.14}$$

with $\mathbf{X}$, $\mathbf{Y}$ being from the results of the singular value decomposition of matrix $\boldsymbol{V}_1^{\mathrm{T}}\,\boldsymbol{V}_0$, i.e.,

$$\boldsymbol{V}_1^{\mathrm{T}}\,\boldsymbol{V}_0 = \mathbf{X}\,\mathbf{Z}\,\mathbf{Y}^{\mathrm{T}} \tag{11.15}$$

For more details of the method, the interested reader is referred to Crosilla (1983b).

11.2.1.2 Precision Requirement on the Network Configuration and Observation Weights

As with geodetic positioning networks, in order to attain the required accuracies of the deformation parameters with an optimal design of the network configuration (**A**) and the observation weights (**P**), a properly chosen precision criterion has to be converted into requirements on the unknown parameters to be optimally solved for. Let us consider the case in which a criterion matrix $\mathbf{C}_e^c$ for deformation parameters has been chosen as the precision criterion, the design problem then seeks an optimal configuration (**A**) and weights (**P**) such that it can be best approximated by $\mathbf{C}_e$, i.e.,

$$\left\| \mathbf{Ce} - \mathbf{C}_e^c \right\| = \min \quad (\text{optimal precision}) \tag{11.16}$$

and

$$\text{vecdiag}(\mathbf{C}_e) \leq \text{vecdiag}(\mathbf{C}_e^c) \quad (\text{precision control}) \tag{11.17}$$

Note that elements of matrix $\mathbf{C}_e$ are nonlinear functions of both the observation weights (p_i, i=1, ..., n) and station coordinates (x_i, y_i, z_i, i=1, ..., m) embedded in the configuration matrix (**A**) and the deformation model matrix **B**. When an initial design is given, matrix $\mathbf{C}_e$ may be approximated by Taylor series restricted to linear term as follows:

$$\begin{aligned} \mathbf{C}_e = \mathbf{C}_e^0 + \sum_{i=1}^{m}\left(\frac{\partial \mathbf{C}_e}{\partial x_i}\right)\Delta x_i + \sum_{i=1}^{m}\left(\frac{\partial \mathbf{C}_e}{\partial y_i}\right)\Delta y_i \\ + \sum_{i=1}^{m}\left(\frac{\partial \mathbf{C}_e}{\partial z_i}\right)\Delta z_i + \sum_{i=1}^{n}\left(\frac{\partial \mathbf{C}_e}{\partial p_i}\right)\Delta p_i \end{aligned} \tag{11.18}$$

where

$$\mathbf{C}_e^0 = 2\sigma_0^2\left(\mathbf{B}^T\mathbf{A}^T\,\mathbf{P}\,\mathbf{A}\,\mathbf{B}\right)^{-1}\Big|_{\mathbf{x}^0,\,\mathbf{y}^0,\,\mathbf{z}^0,\,\mathbf{p}^0} \tag{11.19}$$

$$\begin{aligned} \frac{\partial \mathbf{C}_e}{\partial x_i} = -\frac{1}{2\sigma_0^2}\Bigg\{\mathbf{C}_e^0\Bigg[\mathbf{B}^T\mathbf{A}^T\,\mathbf{P}\left(\frac{\partial \mathbf{A}}{\partial x_i}\mathbf{B} + \mathbf{A}\frac{\partial \mathbf{B}}{\partial x_i}\right) \\ + \left(\mathbf{B}^T\frac{\partial \mathbf{A}^T}{\partial x_i} + \frac{\partial \mathbf{B}^T}{\partial x_i}\mathbf{A}^T\right)\mathbf{P}\,\mathbf{A}\,\mathbf{B}\Bigg]\mathbf{C}_e^0\Bigg\}\Bigg|_{\mathbf{x}^0,\,\mathbf{y}^0,\,\mathbf{z}^0,\,\mathbf{p}^0} \end{aligned} \tag{11.20}$$

$$\frac{\partial \mathbf{C}_e}{\partial y_i} = -\frac{1}{2\sigma_0^2}\left\{ \mathbf{C}_e^0 \left[\mathbf{B}^T \mathbf{A}^T \mathbf{P} \left(\frac{\partial \mathbf{A}}{\partial y_i} \mathbf{B} + \mathbf{A} \frac{\partial \mathbf{B}}{\partial y_i} \right) + \left(\mathbf{B}^T \frac{\partial \mathbf{A}^T}{\partial y_i} + \frac{\partial \mathbf{B}^T}{\partial y_i} \mathbf{A}^T \right) \mathbf{P}\,\mathbf{A}\,\mathbf{B} \right] \mathbf{C}_e^0 \right\} \Bigg|_{\mathbf{x}^0, \mathbf{y}^0, \mathbf{z}^0, \mathbf{P}^0} \tag{11.21}$$

$$\frac{\partial \mathbf{C}_e}{\partial z_i} = -\frac{1}{2\sigma_0^2}\left\{ \mathbf{C}_e^0 \left[\mathbf{B}^T \mathbf{A}^T \mathbf{P} \left(\frac{\partial \mathbf{A}}{\partial z_i} \mathbf{B} + \mathbf{A} \frac{\partial \mathbf{B}}{\partial z_i} \right) + \left(\mathbf{B}^T \frac{\partial \mathbf{A}^T}{\partial z_i} + \frac{\partial \mathbf{B}^T}{\partial z_i} \mathbf{A}^T \right) \mathbf{P}\,\mathbf{A}\,\mathbf{B} \right] \mathbf{C}_e^0 \right\} \Bigg|_{\mathbf{x}^0, \mathbf{y}^0, \mathbf{z}^0, \mathbf{P}^0} \tag{11.22}$$

$$\frac{\partial \mathbf{C}_e}{\partial p_i} = -\frac{1}{2\sigma_0^2}\left[\mathbf{C}_e^0 \left(\mathbf{B}^T \mathbf{A}^T \frac{\partial \mathbf{P}}{\partial p_i} \mathbf{A}\,\mathbf{B} \right) \mathbf{C}_e^0 \right] \Bigg|_{\mathbf{x}^0, \mathbf{y}^0, \mathbf{z}^0, \mathbf{P}^0} \tag{11.23}$$

m is the total number of netpoints
n is the total number of observables
$\mathbf{x}^0$, $\mathbf{y}^0$, $\mathbf{z}^0$ are vectors of initial coordinates of network stations as dictated by topography and accessibility, and $\mathbf{P}^0$ consists of approximate values of weights.

For basic geodetic observables, the partial derivatives of matrices **A** and **P** with respect to station coordinates and observational weights in both two- and three-dimensional space can be obtained directly from §10.3 of Chapter 10. The differentiation of matrix **B** depends on the form of functions that constitute the elements of the deformation model. For instance, when the deformation model has been assumed to be a homogeneous strain in the whole body with a differential rotation [cf. (7.11)], the differentiation of matrix **B** in the two-dimensional space will be as follows:

$$\frac{\partial \mathbf{B}}{\partial x_i} = \begin{pmatrix} \mathbf{0} & \cdots & \mathbf{0} & \frac{\partial \mathbf{B}_i^T}{\partial x_i} & \mathbf{0} & \cdots & \mathbf{0} \end{pmatrix}^T \tag{11.24}$$

$$\frac{\partial \mathbf{B}}{\partial y_i} = \begin{pmatrix} \mathbf{0} & \cdots & \mathbf{0} & \frac{\partial \mathbf{B}_i^T}{\partial y_i} & \mathbf{0} & \cdots & \mathbf{0} \end{pmatrix}^T \tag{11.25}$$

where **0** is a 4 by 2 matrix of zeros, and

$$\frac{\partial \mathbf{B}_i}{\partial x_i} = \begin{pmatrix} 1 & 0 & 0 & 0 \\ 0 & 1 & 0 & 1 \end{pmatrix} \tag{11.26}$$

$$\frac{\partial \mathbf{B}_i}{\partial y_i} = \begin{pmatrix} 0 & 1 & 0 & -1 \\ 0 & 0 & 1 & 0 \end{pmatrix} \tag{11.27}$$

In matrix and vector form, the precision criteria (11.16) and (11.17) can then be expressed by

$$\| \mathbf{H}_e \mathbf{w} - \mathbf{u}_e \| = \min \quad \text{(optimal precision)} \tag{11.28}$$

and

$$\mathbf{H}_{1e}\, \mathbf{w} - \mathbf{u}_{1e} \leq 0 \quad \text{(precision control)} \tag{11.29}$$

where

$$\mathbf{u}_e = \mathrm{vec}\left(\mathbf{C}_e^c\right) - \mathrm{vec}\left(\mathbf{C}_e^0\right) \tag{11.30}$$

$$\begin{aligned} \mathrm{He} = \Bigg[& \mathrm{vec}\left(\frac{\partial \mathbf{C}_e}{\partial x_1}\right) \mathrm{vec}\left(\frac{\partial \mathbf{C}_e}{\partial y_1}\right) \mathrm{vec}\left(\frac{\partial \mathbf{C}_e}{\partial z_1}\right) \cdots \mathrm{vec}\left(\frac{\partial \mathbf{C}_e}{\partial x_m}\right) \\ & \mathrm{vec}\left(\frac{\partial \mathbf{C}_e}{\partial y_m}\right) \mathrm{vec}\left(\frac{\partial \mathbf{C}_e}{\partial z_m}\right) \mathrm{vec}\left(\frac{\partial \mathbf{C}_e}{\partial p_1}\right) \cdots \mathrm{vec}\left(\frac{\partial \mathbf{C}_e}{\partial p_n}\right) \Bigg] \end{aligned} \tag{11.31}$$

$$\mathbf{H}_{1e} = \left(\mathbf{I}_u \,\Theta\, \mathbf{I}_u\right)^T \mathbf{H}_e \tag{11.32}$$

$$\mathbf{u}_{1e} = \left(\mathbf{I}_u \,\Theta\, \mathbf{I}_u\right)^T \mathbf{u}_e \tag{11.33}$$

$$\mathbf{w} = \left(\Delta x_1\ \Delta y_1\ \Delta z_1\ \ldots\ \Delta x_m\ \Delta y_m\ \Delta z_m\ \Delta p_1\ \ldots\ \Delta p_n\right)^T \tag{11.34}$$

$\mathbf{I}_u$ is the u by u unit matrix and Θ represents the Khatri-Rao product.

Finally, it should be noted that if a criterion matrix for datum-independent deformation parameters, such as relative block movements, strain parameters and their derivatives, is defined in the optimization procedure, it can be directly used to fit the variance-covariance matrix of the deformation parameters. Otherwise, if a criterion matrix for datum-dependent displacements is used as the network precision criterion, it has to be transformed, when necessary, to refer to the same datum as that used for calculating the displacements by the well-known S-transformation as described by (9.31) (cf. also §4.3.5).

11.2.2 Measures and Criteria for Reliability

In order to avoid a misinterpretation of gross errors and/or systematic effects in the observations as deformation phenomena, screening of the observations for gross errors and/or systematic effects should be done prior to the estimation of the deformation parameters. This is usually accomplished by performing a least squares network adjustment with station coordinates being the unknown parameters, when possible. In this regard, both the pre-adjustment and post-adjustment data screening techniques as discussed in Chapters 3 and 5 are applied. As with the geodetic positioning networks, internal reliability and external reliability are distinguished in a deformation network. At the design stage, a monitoring network should be designed with both high internal and external reliability. The former means that gross errors can be detected and eliminated as completely as possible, and an undetected gross error is small in comparison with the standard deviation of the observation, and the latter that the effect of an undetected residual gross error on the solution of deformation parameters is kept to a minimum.

Internal Reliability

First, a general internal reliability criteria for a monitoring network can be the same as that for a conventional geodetic positioning network as follows [cf. (10.39) and (10.40)]:

$$\| \mathbf{r} \| = \text{minimum} \quad (\text{optimal internal reliability})$$

or

$$\| \mathbf{r} \| \geq r_m \quad (\text{internal reliability control})$$

where $\mathbf{r}$ is the vector of redundancy numbers, r_m a given lower bound, and $\| \cdot \|$ represents the norm of a vector. Using Taylor series expansion, the above requirements can be converted into requirements on the ***improvements*** to an initial design as follows [cf. (10.54) and (10.55)]:

$$\left\| \left(\mathbf{r}_{00} + \mathbf{R}_{11}\, \mathbf{w} \right) \right\| = \max \quad (\text{optimal internal reliability})$$

$$\left\| \left(\mathbf{r}_{00} + \mathbf{R}_{11}\, \mathbf{w} \right) \right\| \geq r_m \quad (\text{internal reliability control})$$

where $\mathbf{r}_{00}$ and $\mathbf{R}_{11}$ are defined in (10.54) and (10.55), respectively.

External Reliability

Second, the maximum external reliability criteria for deformation networks intend to minimize the influence of undetectable gross errors on the solution of deformation parameters, instead of station coordinates as in the case of geodetic positioning networks. When considering one gross error at the time, influence of an undetectable gross error in the observations on the solution of deformation parameters can be obtained in the worst case from (11.1) as follows (Kuang, 1991)

$$\nabla_{0,\,i}\,\hat{\mathbf{e}} = 2\left(\mathbf{B}^{T}\mathbf{A}^{T}\,\mathbf{P}\,\mathbf{A}\,\mathbf{B}\right)^{-1}\mathbf{B}^{T}\mathbf{A}^{T}\,\mathbf{P}\mathbf{c}_{i}\nabla_{0,\,i}l \quad (i = 1, \ldots, n) \tag{11.35}$$

where $\mathbf{c}_i$ is a vector of zeros except for identity at the i-th position, and $\nabla_{0,i}l$ represents the maximum undetectable gross error in observation l_i (i=1, ..., n) given by (6.129).

A standardized measure of $\nabla_{0,i}\,\hat{\mathbf{e}}$ can be written by

$$\begin{aligned}\lambda_{0,\,i} &= \frac{1}{\sigma_0^2}\left(\nabla_{0,\,i}\,\hat{\mathbf{e}}\right)^{T}\mathbf{Q}_{e}^{-1}\left(\nabla_{0,\,i}\,\hat{\mathbf{e}}\right)\\ &= 2\,\delta_0^2\,\frac{\left[\mathbf{A}\,\mathbf{B}\left(\mathbf{B}^{T}\mathbf{A}^{T}\mathbf{P}\,\mathbf{A}\,\mathbf{B}\right)^{-1}\mathbf{B}^{T}\mathbf{A}^{T}\mathbf{P}\right]_{ii}}{r_i} \quad (i = 1, \ldots, n)\end{aligned} \tag{11.36}$$

The maximum value of $\bar{\lambda}_{0,i}$, i.e.,

$$\bar{\lambda}_0 = \max\left(\bar{\lambda}_{0,\,i},\ i = 1, \ldots, n\right) \tag{11.37}$$

can be used as a measure of the external reliability of the estimated deformation parameters.

From Kuang (1991), a general criterion for the external reliability can be stated as

$$\left\|\bar{\lambda}\right\| = \text{minimum} \quad (\text{optimal external reliability}) \tag{11.38}$$

or

$$\left\|\bar{\lambda}\right\| \le \bar{\lambda}_m \quad (\text{external reliability control}) \tag{11.39}$$

where $\bar{\lambda}_m$ is a given upper bound, and

$$\bar{\lambda} = \left(\bar{\lambda}_{0,\,1}, \bar{\lambda}_{0,\,2}, \ldots, \bar{\lambda}_{0,\,n}\right)^{T} \tag{11.40}$$

Denote

$$\mathbf{F} = \mathbf{A}\,\mathbf{B}\left(\mathbf{B}^T\mathbf{A}^T\,\mathbf{P}\,\mathbf{A}\,\mathbf{B}\right)^{-1}\mathbf{B}^T\,\mathbf{A}^T\,\mathbf{P} \tag{11.41}$$

$\bar{\lambda}_{0,\,i}$ can be expressed by (Kuang, 1992f)

$$\begin{aligned} \bar{\lambda}_{0,\,i} &= 2\delta_0^2\,\frac{[\mathbf{F}]_{ii}}{r_i} \\ &= 2\delta_0^2\,\frac{\mathbf{c}_i^T\,\mathbf{F}\,\mathbf{c}_i}{\mathbf{c}_i^T\,\mathbf{R}\,\mathbf{c}_i} \quad (i = 1,\ \ldots,\ n) \end{aligned} \tag{11.42}$$

where $\mathbf{c}_i$ means the same as that in (10.35), and matrix $\mathbf{R}$ is defined by (5.2). $\bar{\lambda}_{0,\,i}$ can then be approximated by the Taylor series of linear form as

$$\begin{aligned} \bar{\lambda}_{0,\,i} = \frac{2\delta_0^2}{\mathbf{c}_i^T\,\mathbf{R}^0\mathbf{c}_i}\Bigg\{ & \mathbf{c}_i^T\,\mathbf{F}^0\mathbf{c}_i \\ & + \sum_{j=1}^{m}\left[\mathbf{c}_i^T\left(\frac{\partial\mathbf{F}}{\partial x_j}\right)\mathbf{c}_i - \frac{\mathbf{c}_i^T\,\mathbf{F}^0\mathbf{c}_i}{\mathbf{c}_i^T\,\mathbf{R}^0\mathbf{c}_i}\,\mathbf{c}_i^T\left(\frac{\partial\mathbf{R}}{\partial x_j}\right)\mathbf{c}_i\right]\Delta x_j \\ & + \sum_{j=1}^{m}\left[\mathbf{c}_i^T\left(\frac{\partial\mathbf{F}}{\partial y_j}\right)\mathbf{c}_i - \frac{\mathbf{c}_i^T\,\mathbf{F}^0\mathbf{c}_i}{\mathbf{c}_i^T\,\mathbf{R}^0\mathbf{c}_i}\,\mathbf{c}_i^T\left(\frac{\partial\mathbf{R}}{\partial y_j}\right)\mathbf{c}_i\right]\Delta y_j \\ & + \sum_{j=1}^{m}\left[\mathbf{c}_i^T\left(\frac{\partial\mathbf{F}}{\partial z_j}\right)\mathbf{c}_i - \frac{\mathbf{c}_i^T\,\mathbf{F}^0\mathbf{c}_i}{\mathbf{c}_i^T\,\mathbf{R}^0\mathbf{c}_i}\,\mathbf{c}_i^T\left(\frac{\partial\mathbf{R}}{\partial z_j}\right)\mathbf{c}_i\right]\Delta z_j \\ & + \sum_{j=1}^{n}\left[\mathbf{c}_i^T\left(\frac{\partial\mathbf{F}}{\partial p_j}\right)\mathbf{c}_i - \frac{\mathbf{c}_i^T\,\mathbf{F}^0\mathbf{c}_i}{\mathbf{c}_i^T\,\mathbf{R}^0\mathbf{c}_i}\,\mathbf{c}_i^T\left(\frac{\partial\mathbf{R}}{\partial p_j}\right)\mathbf{c}_i\right]\Delta p_j\Bigg\} \end{aligned} \tag{11.43}$$

where

$$\mathbf{F}^0 = \left[\mathbf{A}\,\mathbf{B}\left(\mathbf{B}^T\mathbf{A}^T\mathbf{P}\mathbf{A}\mathbf{B}\right)^{-1}\mathbf{B}^T\,\mathbf{A}^T\,\mathbf{P}\right]\Big|_{\mathbf{x}^0,\,\mathbf{y}^0,\,\mathbf{z}^0,\,\mathbf{p}^0} \tag{11.44}$$

$$
\begin{aligned}
\frac{\partial \mathbf{F}}{\partial x_j} = \Bigg\{ & \left(\frac{\partial \mathbf{A}}{\partial x_j} \mathbf{B} + \mathbf{A} \frac{\partial \mathbf{B}}{\partial x_j} \right) \mathbf{M}_1 \mathbf{B}^T \mathbf{A}^T \mathbf{P} \\
& + \mathbf{A}\,\mathbf{B}\,\mathbf{M}_1 \left(\mathbf{B}^T \frac{\partial \mathbf{A}^T}{\partial x_j} + \frac{\partial \mathbf{B}^T}{\partial x_j} \mathbf{A}^T \right) \mathbf{P} \\
& - \mathbf{A}\,\mathbf{B}\,\mathbf{M}_1 \Bigg[\left(\mathbf{B}^T \frac{\partial \mathbf{A}^T}{\partial x_j} + \frac{\partial \mathbf{B}^T}{\partial x_j} \mathbf{A}^T \right) \mathbf{P}\,\mathbf{A}\,\mathbf{B} \\
& + \mathbf{B}^T \mathbf{A}^T \mathbf{P} \left(\frac{\partial \mathbf{A}}{\partial x_j} \mathbf{B} + \mathbf{A} \frac{\partial \mathbf{B}}{\partial x_j} \right) \Bigg] \mathbf{M}_1 \mathbf{B}^T \mathbf{A}^T \mathbf{P} \Bigg\} \Bigg|_{\mathbf{x}^0,\, \mathbf{y}^0,\, \mathbf{z}^0,\, \mathbf{P}^0}
\end{aligned}
\tag{11.45}
$$

$$
\begin{aligned}
\frac{\partial \mathbf{F}}{\partial y_j} = \Bigg\{ & \left(\frac{\partial \mathbf{A}}{\partial y_j} \mathbf{B} + \mathbf{A} \frac{\partial \mathbf{B}}{\partial y_j} \right) \mathbf{M}_1 \mathbf{B}^T \mathbf{A}^T \mathbf{P} \\
& + \mathbf{A}\,\mathbf{B}\,\mathbf{M}_1 \left(\mathbf{B}^T \frac{\partial \mathbf{A}^T}{\partial y_j} + \frac{\partial \mathbf{B}^T}{\partial y_j} \mathbf{A}^T \right) \mathbf{P} \\
& - \mathbf{A}\,\mathbf{B}\,\mathbf{M}_1 \Bigg[\left(\mathbf{B}^T \frac{\partial \mathbf{A}^T}{\partial y_j} + \frac{\partial \mathbf{B}^T}{\partial y_j} \mathbf{A}^T \right) \mathbf{P}\,\mathbf{A}\,\mathbf{B} \\
& + \mathbf{B}^T \mathbf{A}^T \mathbf{P} \left(\frac{\partial \mathbf{A}}{\partial y_j} \mathbf{B} + \mathbf{A} \frac{\partial \mathbf{B}}{\partial y_j} \right) \Bigg] \mathbf{M}_1 \mathbf{B}^T \mathbf{A}^T \mathbf{P} \Bigg\} \Bigg|_{\mathbf{x}^0,\, \mathbf{y}^0,\, \mathbf{z}^0,\, \mathbf{P}^0}
\end{aligned}
\tag{11.46}
$$

$$
\begin{aligned}
\frac{\partial \mathbf{F}}{\partial z_j} = \Bigg\{ & \left(\frac{\partial \mathbf{A}}{\partial z_j} \mathbf{B} + \mathbf{A} \frac{\partial \mathbf{B}}{\partial z_j} \right) \mathbf{M}_1 \mathbf{B}^T \mathbf{A}^T \mathbf{P} \\
& + \mathbf{A}\,\mathbf{B}\,\mathbf{M}_1 \left(\mathbf{B}^T \frac{\partial \mathbf{A}^T}{\partial z_j} + \frac{\partial \mathbf{B}^T}{\partial z_j} \mathbf{A}^T \right) \mathbf{P} \\
& - \mathbf{A}\,\mathbf{B}\,\mathbf{M}_1 \Bigg[\left(\mathbf{B}^T \frac{\partial \mathbf{A}^T}{\partial z_j} + \frac{\partial \mathbf{B}^T}{\partial z_j} \mathbf{A}^T \right) \mathbf{P}\,\mathbf{A}\,\mathbf{B} \\
& + \mathbf{B}^T \mathbf{A}^T \mathbf{P} \left(\frac{\partial \mathbf{A}}{\partial z_j} \mathbf{B} + \mathbf{A} \frac{\partial \mathbf{B}}{\partial z_j} \right) \Bigg] \mathbf{M}_1 \mathbf{B}^T \mathbf{A}^T \mathbf{P} \Bigg\} \Bigg|_{\mathbf{x}^0,\, \mathbf{y}^0,\, \mathbf{z}^0,\, \mathbf{P}^0}
\end{aligned}
\tag{11.47}
$$

$$\frac{\partial \mathbf{F}}{\partial p_j} = \left[\mathbf{A B M}_1 \, \mathbf{B}^T \mathbf{A}^T \frac{\partial \mathbf{P}}{\partial p_j} - \mathbf{A\, B\, M}_1 \left(\mathbf{B}^T \mathbf{A}^T \frac{\partial \mathbf{P}}{\partial p_j} \mathbf{A\, B} \right) \mathbf{M}_1 \mathbf{B}^T \mathbf{A}^T \mathbf{P} \right] \Bigg|_{\mathbf{x}^0, \mathbf{y}^0, \mathbf{z}^0, \mathbf{P}^0} \tag{11.48}$$

$$\mathbf{M}_1 = \left(\mathbf{B}^T \mathbf{A}^T \mathbf{P\, A\, B} \right)^{-1} \tag{11.49}$$

Again, if written in matrix and vector form, the vector $\bar{\lambda}$ in (11.40) can be expressed by

$$\bar{\lambda} = \mathbf{g}_{00} + \mathbf{G}_{11}\, \mathbf{w} \tag{11.50}$$

where **w** is the same as before, and

$$g_{00} = \left(g_{01}\; g_{02} \cdots g_{0n} \right)^T \tag{11.51}$$

$$\mathbf{G}_{11} = \begin{pmatrix} g_{x11} & g_{y_{11}} & g_{z11} & \cdots & g_{x1m} & g_{y_{1m}} & g_{z1m} & g_{p_{11}} & \cdots & g_{p_{1n}} \\ g_{x21} & g_{y_{21}} & g_{z21} & \cdots & g_{x2m} & g_{y_{2m}} & g_{z2m} & g_{p_{21}} & \cdots & g_{p_{2n}} \\ \vdots & & & & \cdots & & & & & \vdots \\ g_{xn1} & g_{y_{n1}} & g_{zn1} & \cdots & g_{xnm} & g_{y_{nm}} & g_{znm} & g_{p_{n1}} & \cdots & g_{p_{nn}} \end{pmatrix} \tag{11.52}$$

with

$$g_{0i} = \frac{2\delta_0^2}{\mathbf{c}_i^T \mathbf{R}^0 \mathbf{c}_i} \mathbf{c}_i^T \mathbf{F}^0 \mathbf{c}_i \quad (i = 1, \ldots, n) \tag{11.53}$$

$$g_{x_{ij}} = \frac{2\delta_0^2}{\mathbf{c}_i^T \mathbf{R}^0 \mathbf{c}_i} \left[\mathbf{c}_i^T \left(\frac{\partial \mathbf{F}}{\partial x_j} \right) \mathbf{c}_i - \frac{\mathbf{c}_i^T \mathbf{F}^0 \mathbf{c}_i}{\mathbf{c}_i^T \mathbf{R}^0 \mathbf{c}_i} \mathbf{c}_i^T \left(\frac{\partial \mathbf{R}}{\partial x_j} \right) \mathbf{c}_i \right] \quad \left(i = 1, \ldots n;\; j = 1, \ldots, m \right) \tag{11.54}$$

$$g_{y_{ij}} = \frac{2\delta_0^2}{\mathbf{c}_i^T \mathbf{R}^0 \mathbf{c}_i} \left[\mathbf{c}_i^T \left(\frac{\partial \mathbf{F}}{\partial y_j} \right) \mathbf{c}_i - \frac{\mathbf{c}_i^T \mathbf{F}^0 \mathbf{c}_i}{\mathbf{c}_i^T \mathbf{R}^0 \mathbf{c}_i} \mathbf{c}_i^T \left(\frac{\partial \mathbf{R}}{\partial y_j} \right) \mathbf{c}_i \right] \quad \left(i = 1, \ldots n;\; j = 1, \ldots, m \right) \tag{11.55}$$

$$g_{z_{ij}} = \frac{2\delta_0^2}{\mathbf{c}_i^T \mathbf{R}^0 \mathbf{c}_i} \left[\mathbf{c}_i^T \left(\frac{\partial \mathbf{F}}{\partial z_j} \right) \mathbf{c}_i - \frac{\mathbf{c}_i^T \mathbf{F}^0 \mathbf{c}_i}{\mathbf{c}_i^T \mathbf{R}^0 \mathbf{c}_i} \mathbf{c}_i^T \left(\frac{\partial \mathbf{R}}{\partial z_j} \right) \mathbf{c}_i \right] \quad (11.56)$$

$$(i = 1, \ldots n;\ j = 1, \ldots, m)$$

$$g_{p_{ij}} = \frac{2\delta_0^2}{\mathbf{c}_i^T \mathbf{R}^0 \mathbf{c}_i} \left[\mathbf{c}_i^T \left(\frac{\partial \mathbf{F}}{\partial p_j} \right) \mathbf{c}_i - \frac{\mathbf{c}_i^T \mathbf{F}^0 \mathbf{c}_i}{\mathbf{c}_i^T \mathbf{R}^0 \mathbf{c}_i} \mathbf{c}_i^T \left(\frac{\partial \mathbf{R}}{\partial p_j} \right) \mathbf{c}_i \right] \quad (11.57)$$

$$(i,\ j = 1, \ldots n)$$

The differentiation of matrix **R** is done using (10.44) to (10.48) or (10.56) to (10.60) depending on whether the network configuration matrix **A** is singular or not.

Finally, a general external reliability criterion for deformation networks can be expressed in the form of (cf. also Kuang, 1991, 1992f):

$$\left\| \left(\mathbf{g}_{00} + \mathbf{G}_{11}\, \mathbf{w} \right) \right\| = \min \quad \text{(optimal external reliability)} \quad (11.58)$$

or

$$\left\| \left(\mathbf{g}_{00} + \mathbf{G}_{11}\, \mathbf{w} \right) \right\| \le \overline{\lambda}_m \quad \text{(external reliability control)} \quad (11.59)$$

11.2.3 Measures and Criteria for Sensitivity

At the design stage of a monitoring network, postulated displacements or deformation parameters, based on the particular mechanism causing the deformation, can be made. The network should then be designed so that the postulated displacements or deformation parameters, if they occur, can be detected with specified confidence level (1-α) and power (1-β). This is called the sensitivity of the design.

One way to approach the sensitivity problem for displacement monitoring was described by Niemeier (1982). It is based on the global congruency test, i.e., testing of the significance of displacements, using the following statistic (Pelzer, 1971)

$$\omega = \frac{\left[\hat{\mathbf{d}} - E(\hat{\mathbf{d}}) \right]^T \mathbf{Q}_{\hat{d}}^- \left[\hat{\mathbf{d}} - E(\hat{\mathbf{d}}) \right]}{h\, \sigma_0^2} \quad (11.60)$$

where $E\{\cdot\}$ represents the statistical expectation operator, $\hat{\mathbf{d}}$ and $\mathbf{Q}_{\hat{d}}$ are the estimated displacements and their associated cofactor matrix, respectively. h is the rank of $\mathbf{Q}_{\hat{d}}$, and σ_0^2 the a priori variance factor.

The global congruency test deals with testing the following null hypothesis H_0

$$H_0: E(\hat{\mathbf{d}}) = \mathbf{0} \tag{11.61}$$

against the alternative hypothesis

$$H_a: E(\hat{\mathbf{d}}) = \mathbf{d}_i \neq \mathbf{0} \tag{11.62}$$

Under H_0 and H_a the statistic ω follows the central Fisher distribution $F_{h,\infty}$ and the non-central Fisher distribution $F_{h,\omega_i,\infty}$, respectively, i.e.,

$$\omega \,|\, H_0 \in F_{h,\infty} \tag{11.63}$$

$$\omega \,|\, H_a \in F_{h,\omega_i,\infty} \tag{11.64}$$

where the non-centrality parameter ω_i is calculated by

$$\omega_i = \frac{\mathbf{d}_i^T \mathbf{Q}_{\hat{d}}^- \mathbf{d}_i}{\sigma_0^2} \tag{11.65}$$

As discussed in §3.2, the power of a test is defined as the probability (1-β) that $\mathbf{d}_i$ will result in a rejection of the null hypothesis H_0 with a significance level α. Having fixed values of α and β, the critical value ω_0 for the non-centrality parameter is given by a complicated function

$$\omega_0 = f(h, \infty, \alpha_0, \beta_0) \tag{11.66}$$

The values of ω_0 can be found in tables given in Pearson and Hartley (1951) or Pelzer (1971). Thus all the alternative hypotheses, which have non-centrality parameters

$$\omega_i = \frac{\mathbf{d}_i^T \mathbf{Q}_{\hat{d}}^- \mathbf{d}_i}{\sigma_0^2} \geq \omega_0 \tag{11.67}$$

will lead to a rejection of the null hypothesis with the preset probabilities α_0 and β_0. We say, then the corresponding displacements $\mathbf{d}_i$ are detectable.

Note from (11.4) that

$$\mathbf{Q}_{\hat{d}}^- = \frac{1}{2}\mathbf{Q}_{\hat{x}}^- = \frac{1}{2}\left(\mathbf{A}^T \mathbf{P} \mathbf{A}\right) \tag{11.68}$$

Substituting (11.68) into (11.67), one obtains

$$\mathbf{d}_i^T \mathbf{A}^T \mathbf{P}\, \mathbf{A}\, \mathbf{d}_i \geq 2\, \sigma_0^2 \omega_0 \tag{11.69}$$

This inequality gives a requirement for the geometry of the network (i.e., the configuration matrix **A**) and the weight matrix **P** in order that the displacement vector $\mathbf{d}_i$ be detectable.

Similarly, the sensitivity criteria for the detection of deformation parameters can be established based on the significance test of the estimated deformation parameters discussed in §7.4.3. In order that an estimated scale of deformation parameters **e** be detectable, the following inequality should be satisfied

$$\frac{1}{\sigma_0^2} \mathbf{e}^T \mathbf{Q}_e^{-1}\, \mathbf{e} \geq \tilde{\omega}_0 \tag{11.70}$$

where $\tilde{\omega}_0$ is a critical value for the non-centrality parameter calculated by

$$\tilde{\omega}_0 = f(u,\ \infty,\ \alpha_0,\ \beta_0) \tag{11.71}$$

with u being the dimension of the vector of the deformation parameters.

Replacing $\mathbf{Q}_e^{-1}$ by ½($\mathbf{B}^T\mathbf{A}^T\mathbf{P}\ \mathbf{A}\ \mathbf{B}$), we have

$$\mathbf{e}^T \left(\mathbf{B}^T \mathbf{A}^T\ \mathbf{P}\ \mathbf{A}\ \mathbf{B}\right) \mathbf{e} \geq 2\, \sigma_0^2\, \tilde{\omega}_0 \tag{11.72}$$

An optimal sensitivity criterion for the determination of deformation parameters can be expressed in the form of

$$\mathbf{e}^T \left(\mathbf{B}^T \mathbf{A}^T\ \mathbf{P}\ \mathbf{A}\ \mathbf{B}\right) \mathbf{e} = \max \tag{11.73}$$

Using Taylor series expansion Equations (11.72) and (11.73) become

$$\mathbf{e}^T \left[\mathbf{M}^0 + \sum_{i=1}^{m} \left(\frac{\partial \mathbf{M}}{\partial x_i} \right) \Delta x_i + \sum_{i=1}^{m} \left(\frac{\partial \mathbf{M}}{\partial y_i} \right) \Delta y_i + \sum_{i=1}^{m} \left(\frac{\partial \mathbf{M}}{\partial z_i} \right) \Delta z_i + \sum_{i=1}^{n} \left(\frac{\partial \mathbf{M}}{\partial p_i} \right) \Delta p_i \right] \mathbf{e} \geq 2\sigma_0^2\, \tilde{\omega}_0 \tag{11.74}$$

and

$$\mathbf{e}^T \left[\mathbf{M}^0 + \sum_{i=1}^{m} \left(\frac{\partial \mathbf{M}}{\partial x_i} \right) \Delta x_i + \sum_{i=1}^{m} \left(\frac{\partial \mathbf{M}}{\partial y_i} \right) \Delta y_i + \sum_{i=1}^{m} \left(\frac{\partial \mathbf{M}}{\partial z_i} \right) \Delta z_i + \sum_{i=1}^{n} \left(\frac{\partial \mathbf{M}}{\partial p_i} \right) \Delta p_i \right] \mathbf{e} = \max \tag{11.75}$$

where

$$\mathbf{M}^0 = \left(\mathbf{B}^T\mathbf{A}^T\ \mathbf{P}\ \mathbf{A}\ \mathbf{B}\right)\Big|_{\mathbf{x}^0,\,\mathbf{y}^0,\,\mathbf{z}^0,\,\mathbf{P}^0} \tag{11.76}$$

$$\frac{\partial \mathbf{M}}{\partial x_i} = \left[\mathbf{B}^T\mathbf{A}^T\mathbf{P}\left(\frac{\partial \mathbf{A}}{\partial x_i}\mathbf{B} + \mathbf{A}\frac{\partial \mathbf{B}}{\partial x_i}\right) + \left(\mathbf{B}^T\frac{\partial \mathbf{A}^T}{\partial x_i} + \frac{\partial \mathbf{B}^T}{\partial x_i}\mathbf{A}^T\right)\mathbf{P}\ \mathbf{A}\ \mathbf{B}\right]\Bigg|_{\mathbf{x}^0,\,\mathbf{y}^0,\,\mathbf{z}^0,\,\mathbf{P}^0} \tag{11.77}$$

$$\frac{\partial \mathbf{M}}{\partial y_i} = \left[\mathbf{B}^T\mathbf{A}^T\mathbf{P}\left(\frac{\partial \mathbf{A}}{\partial y_i}\mathbf{B} + \mathbf{A}\frac{\partial \mathbf{B}}{\partial y_i}\right) + \left(\mathbf{B}^T\frac{\partial \mathbf{A}^T}{\partial y_i} + \frac{\partial \mathbf{B}^T}{\partial y_i}\mathbf{A}^T\right)\mathbf{P}\ \mathbf{A}\ \mathbf{B}\right]\Bigg|_{\mathbf{x}^0,\,\mathbf{y}^0,\,\mathbf{z}^0,\,\mathbf{P}^0} \tag{11.78}$$

$$\frac{\partial \mathbf{M}}{\partial z_i} = \left[\mathbf{B}^T\mathbf{A}^T\mathbf{P}\left(\frac{\partial \mathbf{A}}{\partial z_i}\mathbf{B} + \mathbf{A}\frac{\partial \mathbf{B}}{\partial z_i}\right) + \left(\mathbf{B}^T\frac{\partial \mathbf{A}^T}{\partial z_i} + \frac{\partial \mathbf{B}^T}{\partial z_i}\mathbf{A}^T\right)\mathbf{P}\ \mathbf{A}\ \mathbf{B}\right]\Bigg|_{\mathbf{x}^0,\,\mathbf{y}^0,\,\mathbf{z}^0,\,\mathbf{P}^0} \tag{11.79}$$

$$\frac{\partial \mathbf{M}}{\partial p_i} = \left(\mathbf{B}^T\mathbf{A}^T\frac{\partial \mathbf{P}}{\partial pi}\mathbf{A}\mathbf{B}\right)\Bigg|_{\mathbf{x}^0,\,\mathbf{y}^0,\,\mathbf{z}^0,\,\mathbf{P}^0} \tag{11.80}$$

Denote

$$s_0 = \mathbf{e}^T\ \mathbf{M}^0\ \mathbf{e} \tag{11.81}$$

$$\mathbf{s}^T = \left(\mathbf{e}^T\frac{\partial \mathbf{M}}{\partial x_1}\mathbf{e} \quad \mathbf{e}^T\frac{\partial \mathbf{M}}{\partial y_1}\mathbf{e} \quad \mathbf{e}^T\frac{\partial \mathbf{M}}{\partial z_1}\mathbf{e} \quad \cdots \quad \mathbf{e}^T\frac{\partial \mathbf{M}}{\partial x_m}\mathbf{e} \quad \mathbf{e}^T\frac{\partial \mathbf{M}}{\partial y_m}\mathbf{e} \quad \mathbf{e}^T\frac{\partial \mathbf{M}}{\partial z_m}\mathbf{e} \quad \mathbf{e}^T\frac{\partial \mathbf{M}}{\partial p_1}\mathbf{e} \quad \cdots \quad \mathbf{e}^T\frac{\partial \mathbf{M}}{\partial p_n}\mathbf{e}\right) \tag{11.82}$$

Equations (11.74) and (11.75) can then be expressed more compactly as follows:

$$s_0 + \mathbf{s}^T \mathbf{w} \geq s_m \quad \text{(sensitivity control)} \tag{11.83}$$

and

$$s_0 + \mathbf{s}^T \mathbf{w} = \max \quad \text{(optimal sensitivity)} \tag{11.84}$$

where $\mathbf{w}$ is the same as before, and

$$s_m = 2\,\sigma_0^2\,\tilde{\omega}_0 \tag{11.85}$$

Similarly, if the deformation parameters to be determined are displacements of the object points, the requirements of the sensitivity criteria on the network configuration and observation weights can be established by linearizing (11.69), or simply by replacing matrix $\mathbf{B}$ in (11.83) and (11.84) with an identity matrix $\mathbf{I}$.

11.2.4 Measures and Criteria for Economy

As with geodetic positioning networks, the cost of deformation monitoring surveys varies from one project to another. It is very difficult, if not impossible, to relate the total cost of a survey campaign solely to the network configuration and observational weights. A simplified cost criterion is then to minimize the norm of the observation weight matrix, i.e.,

$$\| \mathbf{P} \| = \text{minimum}$$

where $\| \cdot \|$ represents the norm of the weight matrix $\mathbf{P}$.

In terms of position and weight improvements, the cost criteria for deformation networks can be expressed from (10.67) and (10.68) as follows

$$\| \mathbf{c}_{00} + \mathbf{C}_{11}\mathbf{w} \| = \min \quad \text{(minimum cost)} \tag{11.86}$$

or

$$\| \mathbf{c}_{00} + \mathbf{C}_{11}\mathbf{w} \| \leq c_m \quad \text{(cost control)} \tag{11.87}$$

where $\mathbf{c}_{00}$ and $\mathbf{C}_{11}$ are described by (10.65) and (10.66), respectively.

11.3 Formulation and Solution of Optimization Problems for Monitoring Networks

The optimization of a monitoring network may be said to design a precise-, reliable-, and sensitive-enough network that can also be realized in an economical way (Kuang, 1991). A monitoring network should be designed in such a way that it can realize the required accuracy of the deformation parameters; that it will help in identifying and eliminating gross errors existing in the observation data prior to the estimation of deformation parameters and minimizing their effects on the deformation parameters in order to avoid misinterpreting measuring errors as deformation phenomena; and that it will ensure the detection of predicted deformations according to a selected tolerance criterion, in a most economic way. Similar to the case of geodetic positioning networks, the multi-objective target function for deformation networks can in general be expressed as follows:

$$\begin{aligned} &\alpha_p\,(\text{precision}) + \alpha_r\,(\text{reliability}) + \alpha_s\,(\text{sensitivity}) \\ &\qquad + \alpha_c\,(\text{cost})^{-1} = \max \end{aligned} \tag{11.88}$$

for suitably chosen weight coefficients α_p, α_r, α_s, and α_c for precision, reliability, sensitivity, and cost, respectively. If we let one of the coefficients go to infinity, we then obtain a single objective optimization model that tries to maximize one quality aspect of the network while the rest of the quality criteria enter into the model in the form of equality or inequality constraints. Single- and multi-objective optimization models for monitoring schemes have been discussed in detail in Kuang (1991, 1992f). As with geodetic positioning networks, single objective optimization models are not solvable in the case of inconsistency of the constraints. A universally applicable optimization model can be established using the theory of the multi-objective optimization as discussed in §10.6.1. This model tries to best approximate all the network quality criteria, i.e., to minimize the differences between the precision, reliability, sensitivity, cost, and their respective ideal counterparts simultaneously, subject to the physical constraints, i.e.,

Minimize

$$\begin{aligned} &\alpha_p \frac{\left\| \mathbf{H}_e\,\mathbf{w} - \mathbf{u}_e \right\|}{\left\| \mathrm{vec}\left(\mathbf{C}_e^c\right) \right\|} + \alpha_{ir} \frac{\left\| \mathbf{R}_{11}\,\mathbf{w} - \left(\mathbf{r}_m - \mathbf{r}_{00}\right) \right\|}{\left\| \mathbf{r}_m \right\|} \\ &+ \alpha_{er} \frac{\left\| \mathbf{G}_{11}\,\mathbf{w} - \left(\bar{\lambda}_m - \mathbf{g}_{00}\right) \right\|}{\left\| \bar{\lambda}_m \right\|} + \alpha_c \frac{\left\| \gamma^T\,\mathbf{C}_{11}\,\mathbf{w} - \left(\mathbf{c}_m - \gamma^T\,\mathbf{c}_{00}\right) \right\|}{\left\| c_m \right\|} \\ &+ \alpha_s \frac{\left\| \mathbf{s}^T\,\mathbf{w} - \left(s_m - s_0\right) \right\|}{\left\| s_m \right\|} \end{aligned} \tag{11.89}$$

Subject to:

$$\left(\mathbf{D}^T\ \mathbf{0}\right)\mathbf{w} = \mathbf{0} \tag{11.90}$$

$$\mathbf{A}_{00}\mathbf{w} \le \mathbf{b}_{00} \tag{11.91}$$

where the physical constraints (11.90) and (11.91) are the same as (10.69) and (10.73). $\mathbf{r}_m$ and $\bar{\lambda}_m$ are vectors with elements being r_m and $\bar{\lambda}_m$, respectively. The weighting coefficients α_p, α_{ir}, α_{er}, α_s, and α_c are for precision, internal reliability, external reliability, sensitivity, and cost, respectively. The selection of the weighting coefficients is based on the specific problems to be solved and is to a large extent subjective. The decision maker assigns larger weights to those quality criteria that he desires to improve more strongly than the others. When all the weighting coefficients are set to be equal to unity, this model tries to best approximate *equally* the precision, reliability, sensitivity, and cost criteria under the given geography and instrumentation condition. This mathematical model corresponds to the solution of the multi-objective optimization problem (11.88) under the concept of "ideal point" with specified "ideal point" representing precision, internal and external reliability, sensitivity, and cost being $\mathbf{C}_e^c$, $\mathbf{r}_m$, $\bar{\lambda}_m$, s_m, and c_m, respectively.

When applying the L_2 norm to (11.89), it becomes

Minimize

$$\left(\mathbf{w}^T\, \mathbf{H}_{0e}^T\, \mathbf{H}_{0e}\, \mathbf{w} - 2\, \mathbf{u}_{0e}^T\, \mathbf{H}_{0e}\, \mathbf{w} + \mathbf{u}_{0e}^T\, \mathbf{u}_{0e}\right) \tag{11.92}$$

where

$$\mathbf{H}_{0e} = \begin{pmatrix} \sqrt{\alpha_p}\mathbf{H}_e / \sqrt{\left[\mathrm{vec}\left(\mathbf{C}_e^c\right)\right]^T \mathrm{vec}\left(\mathbf{C}_e^c\right)} \\ \sqrt{\alpha_{ir}}\mathbf{R}_{11} / \sqrt{\mathbf{r}_m^T\mathbf{r}_m} \\ \sqrt{\alpha_{er}}\mathbf{G}_{11} / \sqrt{\bar{\lambda}_m^T\bar{\lambda}_m} \\ \sqrt{\alpha_c}\gamma^T\, \mathbf{C}_{11} / c_m \\ \sqrt{\alpha_s}\mathbf{s}^T / s_m \end{pmatrix} \tag{11.93}$$

$$\mathbf{u}_{0e} = \begin{pmatrix} \mathbf{u}_e / \sqrt{\left[\mathrm{vec}\left(\mathbf{C}_e^c\right)\right]^T \mathrm{vec}\left(\mathbf{C}_e^c\right)} \\ \left(\mathbf{r}_m - \mathbf{r}_0\right) / \sqrt{\mathbf{r}_m^T\mathbf{r}_m} \\ \left(\bar{\lambda}_m - \mathbf{g}_{00}\right) / \sqrt{\bar{\lambda}_m^T\bar{\lambda}_m} \\ \left(\mathbf{c}_m - \gamma^T\mathbf{c}_{00}\right) / c_m \\ \left(s_m - s_0\right) / s_m \end{pmatrix} \tag{11.94}$$

This model has a unique minimizing solution if and only if the matrix $\mathbf{H}_{0e}^{T}\mathbf{H}_{0e}$ is positive definite. For practical applications, some of the quality criteria may be omitted in the objective function. In this case the model tries to best approximate the ones that appear in it.

Finally, similar to the case of geodetic positioning networks, if there is a case where all the design criteria of precision, reliability, and sensitivity must be satisfied whatever the total expenditure of the project is, there are two ways to accomplish this goal. One way is to add the following constraints to the above mathematical model so that the design criteria are approximated "from one side," i.e.,

$$\mathbf{A}_{1e}\mathbf{w} \le \mathbf{b}_{1e} \tag{11.95}$$

where

$$\mathbf{A}_{1e} = \begin{pmatrix} \mathbf{H}_{1e} \\ -\mathbf{R}_{11} \\ \mathbf{G}_{11} \\ -\mathbf{s}^{T} \end{pmatrix} \tag{11.96}$$

$$\mathbf{b}_{1e} = \begin{pmatrix} \mathbf{u}_{1e} \\ \mathbf{r}_{00} - \mathbf{r}_{m} \\ \bar{\lambda}_{m} - \mathbf{g}_{00} \\ s_{0} - s_{m} \end{pmatrix} \tag{11.97}$$

where $\mathbf{H}_{1e}$ and $\mathbf{u}_{1e}$ have been defined in (11.32) and (11.33), respectively. The inclusion of these constraints forces the optimization model to produce a monitoring network in which the achievable precision, reliability, and sensitivity will be better or at least equal to the specified values. Due to the potential inconsistency among the constraints in (11.95), however, the resulting optimization model may not be solvable. In this case, one may construct a new "ideal point" $\tilde{\mathbf{C}}_{e}^{c}$, $\tilde{\mathbf{r}}_{m}$, $\tilde{\bar{\boldsymbol{\lambda}}}_{m}$, and $\tilde{s}_{m}$ for precision, internal and external reliability, and sensitivity, respectively, such that

$$\operatorname{vecdiag}\left(\tilde{\mathbf{C}}_{e}^{c}\right) < \operatorname{vecdiag}\left(\mathbf{C}_{e}^{c}\right) \tag{11.98}$$

$$\tilde{\mathbf{r}}_{m} > \mathbf{r}_{m} \tag{11.99}$$

$$\tilde{\bar{\lambda}}_{m} < \bar{\lambda}_{m} \tag{11.100}$$

$$\tilde{s}_{m} > s_{m} \tag{11.101}$$

These new ideal points are then used in the optimization procedure. Equations (11.98) to (11.101) mean that the precision, reliability, and sensitivity criteria used for optimization are better than what is actually required. By adjusting the values of the ideal points, the optimization algorithm may produce a network in which the achievable network quality will be better or at least equal to the required values. However, here the upper limits regarding the weights in vector $\mathbf{b}_{00}$ have to be put large enough so that it is physically possible to achieve the required network quality. It should be noted that all the solutions to the above mathematical models should be iterative since Taylor series of linear form is used. A FORTRAN-77 computer program using Quadratic Programming to solve the above optimization model has been completed by the author (Kuang, 1991), and its applications have been presented, for instance, in Kuang (1991, 1993h), Kuang and Chrzanowski (1994), among others.

11.4 Summary of the Optimization Procedures

The optimization procedures for monitoring networks as shown by Figure 11.1 share many features of that for positioning networks (cf. Fig. 10.2). First of all, to establish a monitoring scheme, one has to clearly know the purpose it is to serve, i.e., what kind of deformation parameters are to be detected (i.e., the deformation model **Be** is assumed to be known). At the design stage, the deformation model is usually not fully understood. Therefore, except for the case where the deformation model has been specified by the users or derived from previous observations, the construction of the deformation model may be based on a study of the relevant physical properties of the object and on the knowledge of the acting forces (deterministic modeling). The network quality criteria then have to be defined, which include precision, reliability, sensitivity, and cost. The optimization computation starts with an initial design characterized by initially determined station coordinates $\mathbf{x}^0$, and geodetic observables with initial observation weights $\mathbf{P}^0$ through field reconnaissance and/or other means. One may include all the possible geodetic observables in the initial design, and the selection of the initial weights may be arbitrary. The optimal coordinate changes to be introduced to the initially selected survey points and optimal weight changes to be introduced to the initially adopted approximate weights are then solved iteratively by the optimization model discussed in this chapter. After the solution converges, the optimal values of both coordinates and weights are obtained by adding the solved for optimal changes to their corresponding approximate values. Finally, the survey marks can then be monumented according to the solved optimal coordinates of all the initially selected points. As for the observing scheme, all the observables that obtained zero or insignificant weights can be eliminated from the initial design. For observables that obtained significant optimal weights, instrumentation needed to take the measurements can be chosen according to the required precision calculated from the optimal weights. The optimization examples given in the following chapter demonstrate in detail how the optimization methodology is used in practice.

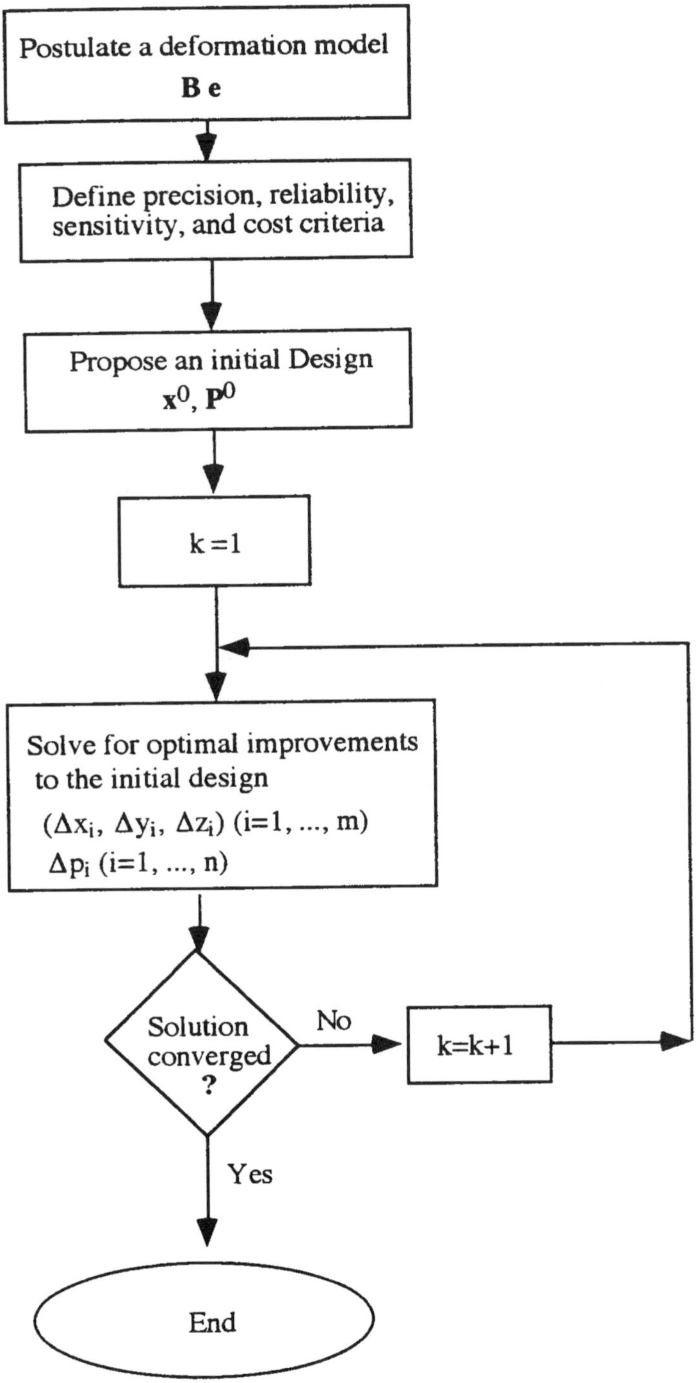

Figure 11.1. Optimization procedures of monitoring networks

12 Application Examples

The theory and procedures for geodetic and deformation network analysis and optimal design have been discussed in Chapters 1 to 11. The purpose of this chapter is to give a few application examples for network optimal design in order to demonstrate how the methodology is applied in actual surveying projects. Application examples for network analysis are omitted here since they are more easily available from textbooks in surveying and geodesy. As mentioned before, a computer software package using FORTRAN 77 has been written by the author to implement in practice all the features of the developed network optimization methodology. The input information includes the design criteria, amount of the maximum possible coordinate shifts for each network point, and all the possible observables together with the maximum available observational precisions (or observational weights). The output information includes the optimal coordinate shifts for each network point and the optimal observational weights for all the observables. Zero or insignificant optimal weights mean that the corresponding observables do not contribute to the accuracy of the network and therefore should be deleted from the final observational scheme. Excellent results have been achieved when it was applied to the planning of both deformation monitoring and geodetic control surveys (cf. Kuang et al., 1991; Kuang, 1992a-f, 1993a-h, 1994; Kuang and Chrzanowski, 1992a,b, 1994).

This chapter starts with §12.1 in which three optimization examples are given in both the two- and three-dimensional space. The specific problems and solutions for the optimal design of geodetic leveling networks and GPS networks are then discussed in §12.2 and §12.3, respectively. Finally, §12.4 summarizes the methodology and gives some suggestions regarding its application.

12.1 Optimal Design of 2D and 3D Networks

Three examples are given in this section. The first example illustrates the application of the optimization modeling developed in Chapter 10 to the optimal design of a three-dimensional engineering network that may be used in industrial metrology, in construction, and deformation surveys, etc. Similar applications in the three-dimensional space can also be made in close range photogrammetry, where an ideal camera position has to be defined. The second example deals with the multi-objective optimal design of a two-dimensional deformation monitoring network shooting for both good precision and high reliability. The deformation parameters to be detected are assumed to be a combination of single point movements and a homogeneous strain field, and the obtained results have shown the advantages of the multi-objective optimization over the single-objective optimization. Finally, the third example demonstrates the optimal re-design of a real-world two-dimensional geodetic network that has been established for monitoring earth deformations. Observations have been made at several epochs separated by several years. The originally planed re-observation scheme was a trilateration scheme that could not satisfy the detection of movements of 2 mm of object points. Therefore, the purpose here is to modify the previous scheme to obtain an optimum triangulateration observation scheme to satisfy this precision requirement with the available instrumentation and with the minimum effort. The network configuration is not supposed to change.

12.1.1 Optimal Design of a 3D Network

Optimal design of three-dimensional engineering networks has been discussed in Kuang (1993f). The concepts of three-dimensional geodetic networks have been well known for centuries. With the development of EDM in geodetic surveys, and with the development of advanced extraterrestrial positioning techniques (e.g., VLBI, SLR, GPS, etc.) in the last two decades, the interest in establishing both global and local three-dimensional geodetic networks has grown. To facilitate the discussion, here we concentrate on the small scale local three-dimensional geodetic networks established for engineering purposes, e.g., the micronetworks used in industrial metrology (Wilkins, 1989), and networks for monitoring local crustal deformations, etc. For large scale (e.g., regional, continental, or global scale) three-dimensional geodetic networks, the influences of gravity field and the earth's geometry on the geodetic observations have to be modeled, and the adjustment of the network should be done in the global geodetic coordinate system. Furthermore, in addition to the conventional geodetic observations, one may also include astronomical azimuth, astronomical latitude and longitude, gravity, and potential differences, etc. (Dermanis, 1985). Nevertheless, the optimization design for such large scale networks can be approached using the same methodology as elaborated on in Chapter 10. Although only precision criteria are considered here, other types of design criteria such as reliability and economy can be easily incorporated to obtain a multi-objective optimal design as illustrated by the following example (cf. also Kuang and Chrzanowski, 1992a).

Suppose a small-scale three-dimensional network is to be established for engineering purposes in the RLG coordinate system (cf. §4.1.1). Figure 12.1 shows a vertical view of the approximate relative locations of the points and the intervisibility

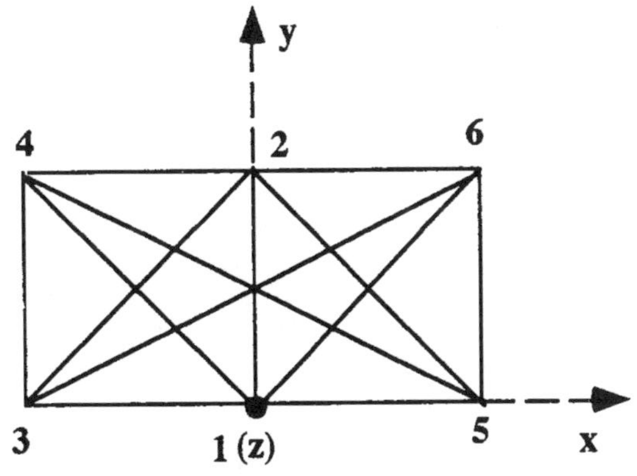

Figure 12.1. The three-dimensional geodetic network

lines as obtained from a field reconnaissance. The minimum and maximum side lengths of the network are approximately 1.0 km and 2.236 km, respectively.

In order to show the reader how the technique of network optimization discussed in Chapter 10 is applied in practice, let us follow the procedures outlined by Figure 10.2 below:

Step 1. Defining network quality criteria

We confine ourselves to precision criteria here. Assume that the RLG coordinate system of Station 1 will be used for the network positioning and the datum of the network provided by fixing point #1, fixing the azimuth from point #1 to #2, as well as fixing two zenith distances from point #1 to #2 and from point #1 to #3. In this case, the datum matrix **D** is expressed as:

$$\mathbf{D}^{\mathrm{T}} = \begin{pmatrix} \mathbf{I} & \mathbf{0}_1 & \mathbf{0}_2 \\ -\mathbf{a}_{\alpha 12}^{\mathrm{T}} & \mathbf{a}_{\alpha 12}^{\mathrm{T}} & \mathbf{0}_3 \\ -\mathbf{a}_{Z12}^{\mathrm{T}} & \mathbf{a}_{Z12}^{\mathrm{T}} & \mathbf{0}_3 \\ -\mathbf{a}_{Z13}^{\mathrm{T}} & \mathbf{a}_{Z13}^{\mathrm{T}} & \mathbf{0}_3 \end{pmatrix} \tag{12.1}$$

where **I** is a 3 by 3 identity matrix; $\mathbf{0}_1$, $\mathbf{0}_2$, and $\mathbf{0}_3$ are respectively 3 by 3, 3 by (3m-6), and 1 by (3m-6) matrices of zeros; vectors $a_{\alpha_{12}}$, $a_{Z_{12}}$, and $a_{Z_{13}}$ have been defined by (4.34) and (4.36), respectively.

As for the precision criteria, we assume that all the standard deviations of the coordinate components with respect to the chosen datum should be less than 2 mm, and use the following 15 by 15 diagonal matrix $\mathbf{C}_s^c$ with diagonal elements being 4 mm^2 as the criterion matrix; i.e.,

Table 12.1. The approximate coordinates of network points

Points	x (m)	y (m)	z (m)
1	0.0	0.0	100.0
2	0.0	1000.0	200.0
3	−1000.0	0.0	300.0
4	−1000.0	1000.0	400.0
5	1000.0	0.0	500.0
6	1000.0	1000.0	600.0

$$C_s^c = \text{diag}\left\{2^2 \cdots 2^2\right\} \left(\text{mm}^2\right) \tag{12.2}$$

Step 2. Specifying an initial design

Let us perform a combined First-Order and Second-Order Design here. In this case, both the network configuration and the observational plan have to be optimally determined in order to achieve the above set precision criteria. As discussed in Chapter 10, the optimization algorithm starts with an initial design that proposes both the approximate locations of network stations as well as the types and precisions of observations to be made. An initial design is usually determined through a combination of the office research and field reconnaissance. In this example, we assume that the network stations have been chosen as shown in Figure 12.1 and Table 12.1 gives the approximate coordinates of the network stations with respect to a selected coordinate system. As for the observing plan, we propose, for the purpose of generality, to measure all the possible spatial distances, azimuths, directions, and zenith distances. The initial precisions of the proposed observations may be given arbitrarily.

Step 3. Solving for an optimal design

The optimization of the network configuration is done by introducing optimal coordinate shifts to the initially given approximate coordinates. We assume that the possible coordinate shifts for all the points #2 to #6 may range from –200 m to +200 m in x-, y-, and z-directions respectively (point #1 is fixed). Furthermore, to solve for an optimal observing plan, we assume that the maximum achievable accuracy for distance is 0.1 parts per million, i.e., $\sigma_s^2 = (0.1 \bullet 10^{-6}\ s)^2$, while for azimuth, directions, and zenith distances the minimum standard deviations attainable are 0.7" (i.e., 0.2 mgon). Group weights are assumed for direction observations, i.e., directions to be observed at the same station are assumed to have the same weights. The initial weights needed to initialize the optimization process are assumed to be 0.1, which is arbitrarily chosen here, for all the proposed observations. The optimization Model I developed in Chapter 10 is now executed to solve for the optimal coordinate shifts and the weight improvements to be introduced to

Table 12.2. Optimal design of the network configuration (unit: meter)

Points		1	2	3	4	5	6
Optimal	Δx	0.0	0.0	200.0	200.0	–200.0	–200.0
Coordinate	Δy	0.0	–58.1	–134.3	–200.0	41.2	–200.0
Shifts	Δz	0.0	–5.8	–38.5	–200.0	–200.0	–200.0
Optimal	x	0.0	0.0	–800.0	–800.0	800.0	800.0
Station	y	0.0	941.9	–134.3	800.0	41.2	800.0
Coordinates	z	100.0	194.2	261.5	200.0	300.0	400.0

the initially given approximate coordinates for points #2 to #6, and to all the initially given weights of observables, respectively.

Step 4. Analyzing the optimization results

After the optimization solution process converged, the obtained optimization results are listed in Tables 12.2 to 12.4. First of all, Table 12.2 shows the optimal coordinate shifts solved by the optimization algorithm, and the optimal station coordinates that are obtained by adding the solved for optimal coordinate shifts to their corresponding approximate coordinates, respectively. Monumentation of the network stations can now be made according to the optimal station coordinates listed in Table 12.2.

In Table 12.3, p(Opti.) and σ(Opti.) are respectively the solved optimal observational weights and the standard deviations calculated from the optimal weights. From Table 12.3 one can see that, in order to achieve the standard deviation of 2 mm for coordinate components, the optimal weights for eight distances, i.e., distances 2-3, 2-5, 3-4, 3-5, 3-6, 4-5, 4-6, 5-6, are zeros or insignificant as compared with the optimal weights of the rest distances. That means that these distances should be deleted in the final observational plan. The azimuths that could be deleted include azimuths 2-3, 2-5, 3-4, and 5-6 since they have also obtained optimal weights of zeros. Similarly, the zenith distance 2-3 could be deleted since it obtained zero optimal weight, and finally all the directions at five stations, i.e., stations 2 to 6 could be deleted since the optimization procedure gave zero or insignificant weights for these five stations. These results show clearly that, in order to satisfy a certain precision criterion, not every distance in the network has to be measured; and/or not every azimuth in the network has to be measured; and/or not every zenith distance in the network has to be measured; and/or not every station has to be occupied to measure horizontal directions. Furthermore, the distribution of observational weight among the observables is heterogeneous. The obtained results are an excellent demonstration of the benefits of network optimization.

Finally, a comparison between the required standard deviations of the coordinate components and the actually achieved ones is given in Table 12.4, which shows that the precision criteria have been satisfied, i.e., all the actual standard

Table 12.3. The optimally solved observational weights

Observations		P(Op.)	σ(Opt.)
Distances	1 - 2	0.21914	2.1 mm
	1 - 3	0.23163	2.1
	1 - 4	0.20910	2.1
	1 - 5	0.24681	2.0
	1 - 6	0.20756	2.2
	2 - 3	0.00141	26.6
	2 - 4	0.09278	3.3
	2 - 5	0.00000	∞
	2 - 6	0.08982	3.3
	3 - 4	0.00677	12.1
	3 - 5	0.00214	21.6
	3 - 6	0.00000	∞
	4 - 5	0.00000	∞
	4 - 6	0.00000	∞
	5 - 6	0.00534	13.7
Azimuths	1 - 3	1.54198	0.8 sec
	1 - 4	2.04082	0.7
	1 - 5	1.46713	0.8
	1 - 6	2.04082	0.7
	2 - 3	0.00000	∞
	2 - 4	0.17930	2.4
	2 - 5	0.00000	∞
	2 - 6	0.34396	1.7
	3 - 4	0.00000	∞
	3 - 5	1.38334	0.8
	3 - 6	2.04082	0.7
	4 - 5	2.04082	0.7
	4 - 6	2.04082	0.7
	5 - 6	0.00000	∞
Zenith Angles	1 - 4	2.04082	0.7 sec
	1 - 5	2.04082	0.7
	1 - 6	2.04082	0.7
	2 - 3	0.00000	∞
	2 - 4	2.04082	0.7
	2 - 5	2.04082	0.7
	2 - 6	2.04082	0.7
	3 - 4	0.75656	1.1
	3 - 5	2.04082	0.7
	3 - 6	2.04082	0.7
	4 - 5	0.4298	1.5
	4 - 6	0.64517	1.2
	5 - 6	0.25267	2.0
Directions	Stn. 1	2.04082	0.7 sec
	Stn. 2	0.00000	∞
	Stn. 3	0.00000	∞
	Stn. 4	0.00000	∞
	Stn. 5	0.05059	4.4
	Stn. 6	0.00000	∞

Table 12.4. Goodness of fit of the precision criteria through the optimization

Points		Precision: σ(Required) [mm]	Precision: σ(Achieved) [mm]
2	σ_x	2.0	0.05
	σ_y	2.0	1.97
	σ_z	2.0	0.20
3	σ_x	2.0	1.99
	σ_y	2.0	1.99
	σ_z	2.0	0.40
4	σ_x	2.0	1.97
	σ_y	2.0	2.07
	σ_z	2.0	2.00
5	σ_x	2.0	2.01
	σ_y	2.0	1.98
	σ_z	2.0	2.02
6	σ_x	2.0	1.98
	σ_y	2.0	2.08
	σ_z	2.0	2.03

deviations of the coordinate components are less than or practically equal to the required ones.

Step 5. Finalizing the observing plan

One may now finalize the observing plan according to the solved optimal weights listed in Table 12.3 and use it as a guideline in choosing the survey instrumentations and field operational procedures. This is the last step in network design.

As discussed above, the proposed observations that obtained zero or insignificant optimal weights should be deleted. And the precisions required for observables that obtained significant optimal weights can be calculated from the optimal weights as follows:

$$\sigma_i = \frac{\sigma_0}{\sqrt{p_i}} \tag{12.3}$$

where σ_i is the required standard deviation for observable l_i, σ_0 the a priori error factor (i.e., the square root of the a priori variance factor) used, and p_i the obtained optimal weight for observable l_i. Survey instruments are then chosen according to the required standard deviations for the observables. It should be noted, however, that during the execution of actual surveys, one may, from the practical point of view, use the same instrument for a certain type of observable, e.g., spatial distances or angles, regardless of the inhomogeneity of the precision requirements of the observables. In this case, one should choose a survey instrument according to the maximum precision required for this type of observable in order to meet the preset network quality criteria from the safe side. For angular measurements, e.g., directions, zenith distances, and azimuth, one may have to calculate the necessary number of observational sets according to the required observational precision for a certain instrumentation chosen. Although it is practical to carry out the actual survey according to the required observational precision as obtained from the optimization since the lower the precision required, the fewer the observational sets one has to measure, one may certainly elect to measure all the angular observables with the maximum precision required for each type of observable, depending on the specific problem under study. In this case, the resulting precision of the station coordinates should be better than the preset criteria.

In this example, according to the precision requirements for distance and angular observables listed in Table 12.3, an EDM instrument with a precision of 2 ppm (parts per million) should be used to measure spatial distances, while for angular measurements, the chosen instrument should be able to produce a precision of 0.7 arc-seconds. After deleting the observations that obtained zero or insignificant optimal weights, the finalized optimal observing plan is given in the following Table 12.5, in which all the angular measurements have been assumed to be measured with the maximum precision required, i.e., 0.7 arc-seconds. Finally, according to this observing plan a comparison between the required precisions and the achievable precisions of coordinate components is shown in Table 12.6 that shows that the precision criteria have been satisfied; that is, all the standard deviations of the coordinate components are less than and close to the required values.

12.1.2 Multi-Objective Optimal Design of a Deformation Monitoring Network

This example shows the reader the advantages of multi-objective optimal design. A two-dimensional deformation monitoring network is to be established to determine deformation parameters that are assumed to be a combination of single point movements and a homogeneous strain field. As discussed in Chapters 10 and 11, the advantage of the multi-objective optimization model is that all the different types of design criteria such as precision, internal- and external-reliability, sensitivity, and economy, which may be contradictory, are considered simultaneously in constructing a global objective function to avoid biasing to any particular criterion. To facilitate the discussion, only two network quality criteria, i.e., the precision and internal reliability, will be considered in this example, and solutions of both single-objective and multi-objective optimization models have been given. Analysis of the obtained results have shown the advantages of the multi-objective optimization over the single-objective optimization.

Let us assume that a trilateration geodetic monitoring network is to be established to detect possible crustal deformations in the concerned area as outlined in Figure 12.2, which also shows the approximate relative locations of the selected network points and the intervisibility lines. Following the optimization procedures

Table 12.5. The finalized optimal observing plan

Observations		Side length	σ(required)
Distances to be observed	1 - 2	0.946 km	1.8 mm
	1 - 3	0.827	1.6
	1 - 4	1.135	2.2
	1 - 5	0.825	1.6
	1 - 6	1.170	2.2
	2 - 4	0.812	1.6
	2 - 6	0.838	1.6
Azimuths to be observed	1 - 3		0.7 sec.
	1 - 4		0.7
	1 - 5		0.7
	1 - 6		0.7
	2 - 4		0.7
	2 - 6		0.7
	3 - 5		0.7
	3 - 6		0.7
	4 - 5		0.7
	4 - 6		0.7
Zenith distances to be observed	1 - 4		0.7 sec.
	1 - 5		0.7
	1 - 6		0.7
	2 - 4		0.7
	2 - 5		0.7
	2 - 6		0.7
	3 - 4		0.7
	3 - 5		0.7
	3 - 6		0.7
	4 - 5		0.7
	4 - 6		0.7
	5 - 6		0.7
Directions to be observed	Stn. 1		0.7 sec.

outlined in Figure 11.1 for deformation monitoring networks, the optimization of the monitoring network is performed as follows:

Step 1. Postulating a deformation model

Assume that the deformation model to be detected includes a homogeneous strain field over the whole area plus single point movements of points 1, 2, 3, and 4 (see Figure 12.2). That is, the vector of deformation parameters to be detected can be expressed as:

$$\mathbf{e} = \left(dx_1\ dy_1\ dx_2\ dy_2\ dx_3\ dy_3\ dx_4\ dy_4\ \varepsilon_x\ \varepsilon_{xy}\ \varepsilon_y\right)^T \tag{12.4}$$

Table 12.6. A comparison between the required precisions and the achievable precisions of coordinate components from the finalized observing plan

Points		Precision σ(Required) [mm]	Precision σ(Achieved) [mm]
2	σ_x	2.0	0.1
	σ_y	2.0	1.5
	σ_z	2.0	0.2
3	σ_x	2.0	1.6
	σ_y	2.0	1.8
	σ_z	2.0	0.3
4	σ_x	2.0	1.4
	σ_y	2.0	1.8
	σ_z	2.0	1.7
5	σ_x	2.0	1.7
	σ_y	2.0	1.8
	σ_z	2.0	1.7
6	σ_x	2.0	1.4
	σ_y	2.0	1.8
	σ_z	2.0	1.7

where dx_i, dy_i (i=1, 2, 3, 4) represent the displacements of points 1, 2, 3, and 4 in x- and y-directions, respectively, and ε_x, ε_y and ε_{xy} the normal strain and shear strain parameters respectively. The deformation model can be expressed as:

$$\begin{aligned} u_i &= \varepsilon_x x_i + \varepsilon_{xy} y_i \\ v_i &= \varepsilon_{xy} x_i + \varepsilon_y y_i \end{aligned} \quad \text{for } i = 5, 6, 7, 8; \text{ and} \tag{12.5}$$

$$\begin{aligned} u_j &= dx_j \\ v_j &= dy_j \end{aligned} \quad \text{for } j = 1, 2, 3, 4 \tag{12.6}$$

Step 2. Defining network quality criteria

As discussed above, in order to facilitate the discussion, only two network quality criteria, i.e., the precision and internal reliability, will be considered in this example. First, we assume that the displacements have to be determined with a standard deviation of 2 mm, while the strains with a standard deviation of 1 ppm. The following diagonal matrix will be used as the precision criterion matrix, i.e.,

$$C_e^c = \mathrm{Diag}\left\{(2.0\ \mathrm{mm})^2 \quad \cdots \quad (2.0\ \mathrm{mm})^2 \quad (1.0\ \mathrm{ppm})^2 \quad \cdots \quad (1.0\ \mathrm{ppm})^2\right\} \tag{12.7}$$

The target function for precision is then used to best approximate the above criterion matrix, i.e.,

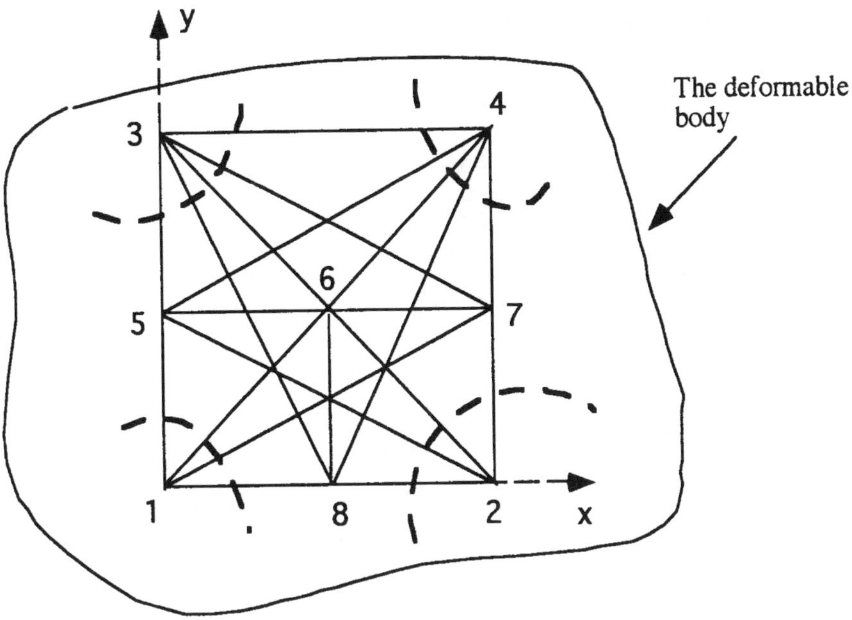

Figure 12.2. The monitoring network

$$f_p = \left\| \mathbf{C}_e - \mathbf{C}_e^c \right\|_2 = \min \tag{12.8}$$

where $\|\cdot\|$ represents the L_2 norm of a matrix as defined by (10.139).

Second, the internal reliability criterion is provided such as to maximize the minimum redundancy numbers of observables [cf. (9.36)], i.e.,

$$f_r = \min\left(r_i,\ i = 1,\ \ldots,\ n\right) = \max \tag{12.9}$$

For the calculation of redundancy numbers of observations, inner constraints are used here. In this case, the datum matrix $\mathbf{D} = \mathbf{H}$, with

$$\mathbf{H} = \begin{pmatrix} 1 & 0 & -y_1^0 \\ 0 & 1 & x_1^0 \\ \vdots & \vdots & \vdots \\ 1 & 0 & -y_8^0 \\ 0 & 1 & x_8^0 \end{pmatrix} \tag{12.10}$$

where (x_i^0, y_i^0) (i=1, ..., 8) are the approximate values of the point coordinates.

Table 12.7. Approximate coordinates of the monitoring network stations

Point	x(m)	y(m)
1	0.0	0.0
2	2000.0	0.0
3	0.0	2000.0
4	2000.0	2000.0
5	0.0	1000.0
6	1000.0	1000.0
7	2000.0	1000.0
8	1000.0	0.0

Step 3. Specifying an initial design

As with the above example, we consider a combined First-Order and Second-Order Design here. The approximate coordinates of the network stations are given in Table 12.7.

As for the initial observing plan, we assume to use only an EDM instrument to measure all the possible distances.

Step 4. Solving for an optimal design

To achieve the above set design criteria, both an optimum network configuration and an observing plan have to be determined. To optimize the network configuration, we assume that the possible coordinate shifts to be introduced for all points could range from –150.0 m to +150.0 m in both x- and y-directions, respectively. To determine an optimum observing plan, we assume that we can have a choice of an EDM instrument with accuracies ranging from $\sigma_s^2 = (2\text{ mm})^2 + (2 \cdot 10^{-6}\text{ s})^2$ to $\sigma_s^2 = (10^{-6}\text{ s})^2$ (where s represents the side length). The input approximate weights of observables are calculated from the minimum available accuracy of $\sigma_s^2 = (2\text{ mm})^2 + (2 \cdot 10^{-6}\text{ s})^2$.

Three mathematical models have been executed in this example. Models I and II use (12.8) and (12.9) as objective function, respectively, while in Model III a multi-objective target function as described by (11.88) is used. As discussed in §10.4 to §10.6 of Chapter 10 and §11.3 of Chapter 11, Models I and II aim at a best approximation of the given precision criterion matrix and maximizing the network internal reliability, respectively, while Model III is a compromise of the two (cf. also Kuang, 1993h).

Step 5. Analyzing the optimization results

After the optimization solution process converged, the optimization results obtained from optimization Models I to III are listed in Tables 12.8 and 12.9. At first, Table 12.8 lists the optimal coordinate shifts solved for each point by optimi-

Table 12.8. Optimal coordinate shifts of netpoints resulting from optimization Models I to III

	Optimal Coordinate Shifts					
	Model I		Model II		Model III	
Point	Δx(m)	Δy(m)	Δx(m)	Δy(m)	Δx(m)	Δy(m)
1	125.4	−150.0	−10.8	−66.9	150.0	−96.7
2	−150.0	150.0	47.3	89.6	−121.8	96.7
3	150.0	−150.0	150.0	−93.3	150.0	−150.0
4	−125.4	−150.0	−150.0	−150.0	−150.0	−150.0
5	−150.0	150.0	−150.0	150.0	−150.0	150.0
6	150.0	150.0	150.0	81.8	121.8	150.0
7	150.0	150.0	82.7	138.8	150.0	150.0
8	−150.0	−150.0	−119.2	−150.0	−150.0	−150.0

zation Models I to III, respectively. Different results were obtained by different models. The optimal observational weights solved by different optimization models are listed in Table 12.9, where p(i) (i=I, II, III) and σ(i) (i=I, II, III) are the solved optimal weights of observables and their corresponding optimal standard deviations calculated from the optimal weights for optimization Models I to III, respectively. σ(Min.) represents the minimum achievable standard deviations for each line calculated according to the maximum available accuracy of the instrumentation.

From Table 12.9, one can see that optimization Model I gave zero weights for six distances, i.e., distances 1-2, 1-3, 1-4, 1-7, 2-3, and 3-4. It means that these distances should be deleted in the final observing plan. The optimal weights solved by Models II and III are more uniform. The optimal standard deviations required for each observable are then used as a guide in choosing an EDM instrument with an adequate accuracy as discussed in Step 5 of the above example.

The goodness of fit of the preset precision criterion is shown in Table 12.10, from which one can see that the precision criterion has been achieved by all the three optimization models. That is, all the obtained standard deviations of deformation parameters are less than or equal to the required ones.

Table 12.11 lists the redundancy numbers of observations corresponding to optimization Models I to III, respectively. From Table 12.11, one can see that the minimum redundancy numbers as given by Models I to III are 0.0648, 0.4324, and 0.3477, respectively. Both Models II and III have obtained relatively uniform redundancy numbers as expected.

Finally, Table 12.12 lists the values of the objective functions f_p and f_r resulting from optimization Models I to III, respectively. From Table 12.12, one can see that Model I achieved the best fitting of the criterion matrix and Model II obtained the maximum reliability of the network, while the requirements of best fitting of the criterion matrix and maximum reliability are balanced by Model III. These results obviously comply with the motive of single-objective and multi-objective optimization.

Table 12.9. Optimal weights and standard deviations of observables resulting from optimization Models I to III

Distances		Optimal Weights			Optimal Standard Deviations (mm)			σ(Min.)
From	To	p(I)	p(II)	p(III)	σ(I)	σ(II)	σ(III)	(mm)
1	2	0.00000	0.25000	0.22947	∞	2.0	2.1	2.0
1	3	0.00000	0.25000	0.24580	∞	2.0	2.0	2.0
1	4	0.00000	0.12500	0.12500	∞	2.8	2.8	2.8
1	5	0.62710	0.29515	0.29872	1.3	1.8	1.8	1.0
1	6	0.13988	0.50000	0.28602	2.7	1.4	1.8	1.4
1	7	0.00000	0.20000	0.20000	∞	2.2	2.2	2.2
1	8	0.47542	0.37552	0.30541	1.4	1.6	1.8	1.0
2	3	0.00000	0.12500	0.12500	∞	2.8	2.8	2.8
2	4	0.00000	0.25000	0.25000	∞	2.0	2.0	2.0
2	5	0.15952	0.20000	0.20000	2.5	2.2	2.2	2.2
2	6	0.50000	0.50000	0.28113	1.4	1.4	1.9	1.4
2	7	0.29198	0.34212	0.31361	1.8	1.7	1.8	1.0
2	8	0.48941	0.37886	0.34227	1.4	1.6	1.7	1.0
3	4	0.02519	0.25000	0.21850	6.3	2.0	2.1	2.0
3	5	0.53981	0.26626	0.33969	1.4	1.9	1.7	1.0
3	6	0.44353	0.49809	0.34436	1.5	1.4	1.7	1.4
3	7	0.20000	0.20000	0.20000	2.2	2.2	2.2	2.2
3	8	0.20000	0.20000	0.20000	2.2	2.2	2.2	2.2
4	5	0.20000	0.20000	0.20000	2.2	2.2	2.2	2.2
4	6	0.39565	0.50000	0.35352	1.6	1.4	1.7	1.4
4	7	0.85475	0.29523	0.34211	1.1	1.8	1.7	1.0
4	8	0.20000	0.20000	0.20000	2.2	2.2	2.2	2.2
5	6	1.00000	0.61060	0.34321	1.0	1.3	1.7	1.0
5	7	0.25000	0.25000	0.25000	2.0	2.0	2.0	2.0
5	8	0.36476	0.38358	0.27765	1.7	1.6	1.9	1.4
6	7	1.00000	0.65089	0.35929	1.0	1.2	1.7	1.0
6	8	1.00000	0.50514	0.65733	1.0	1.4	1.2	1.0
7	8	0.50000	0.50000	0.50000	1.4	1.4	1.4	1.4
Sum		9.85701	9.30146	7.98810				

Table 12.10. Goodness of fit of the precision criterion

Parameters	σ(Required)	σ(Model I)	σ(Model II)	σ(Model III)
dx_1	2.00 mm	1.98 mm	1.77 mm	1.91 mm
dy_1	2.00	2.00	2.00	1.96
dx_2	2.00	1.77	1.68	1.78
dy_2	2.00	2.00	2.00	2.00
dx_3	2.00	2.00	2.00	2.00
dy_3	2.00	1.80	1.91	1.84
dx_4	2.00	2.00	2.00	2.00
dy_4	2.00	1.85	2.00	2.00
ε_x	1.00 ppm	0.62 ppm	0.71 ppm	0.76 ppm
ε_{xy}	1.00	0.70	0.66	0.70
ε_y	1.00	0.89	1.00	0.99

Table 12.11. The redundancy numbers of observations resulting from optimization Models I to III (Note: N/A = not applicable)

Distances		Redundancy Numbers of Observations		
From	To	(Model I)	(Model II)	(Model III)
1	2	N/A	0.5647	0.5592
1	3	N/A	0.5036	0.5173
1	4	N/A	0.8110	0.7753
1	5	0.0869	0.4324	0.4234
1	6	0.6043	0.4584	0.5576
1	7	N/A	0.7338	0.6858
1	8	0.1097	0.4324	0.4610
2	3	N/A	0.8109	0.7632
2	4	N/A	0.5392	0.5278
2	5	0.7018	0.7169	0.6492
2	6	0.2561	0.4324	0.5236
2	7	0.3541	0.4324	0.4577
2	8	0.2155	0.4324	0.4540
3	4	0.8951	0.4343	0.4761
3	5	0.1168	0.4324	0.3542
3	6	0.3613	0.4324	0.4871
3	7	0.5880	0.6566	0.6156
3	8	0.6215	0.6656	0.6363
4	5	0.5487	0.6318	0.6042
4	6	0.4278	0.4544	0.5227
4	7	***0.0648***	***0.4324***	***0.3477***
4	8	0.6994	0.7214	0.7058
5	6	0.3034	0.4324	0.5259
5	7	0.7343	0.6765	0.6139
5	8	0.3921	0.4324	0.4954
6	7	0.2963	0.4324	0.5278
6	8	0.2175	0.4324	0.3160
7	8	0.4047	0.4324	0.4162

Table 12.12. The objective functions after optimization

Objective Function		f_p	f_r
Model	I	12.48949	0.0648
	II	53.20174	0.4324
	III	46.36257	0.3477

In summary, let us examine which of the three optimization Models I to III has given the best results. Table 12.12 shows that Model I has achieved the best fit of the precision criterion matrix, Model II has produced the highest network internal reliability, while Model III is a balance of the two. On the other hand, one can see, from Table 12.9, that Model III has required the minimum total weight, implying a more economical observation campaign. From this analysis, one can conclude that the multi-objective optimization Model III has given the best results in this example.

Step 6. Finalizing the observing plan

As with the above example, the last step of network design is to finalize the observing plan and choose a proper instrument. Monumentation of the network stations can be made according to the optimal station coordinates that are obtained by adding the solved optimal coordinate shifts to their corresponding approximate coordinates. In this example, if we adopt the optimization results of Model III, from the solved optimal weights and standard deviations, i.e., p(III) and σ(III) in Table 12.9, one can see that in order to produce a network that will achieve the pre-set precision requirements and also have a high internal reliability, one has to measure all the possible distances in the network with an EDM with an accuracy of 1 ppm.

12.1.3 Optimal Design of a Two-Dimensional Geodetic Network for Monitoring Displacements of Object Points

The following example demonstrates the practical application of the network optimization methodology to redesigning a geodetic network that has been established to monitor earth deformations in the vicinity of a hydro-electric power generating station. Observations have been made at several epochs separated by several years, and some new object points were constantly added almost each year to determine the stability of some newly suspected unstable objects. The present design refers to the survey campaign conducted in 1991 in which the network consists of 17 reference points and 15 object points. The relative locations of the network stations are shown by Figure 12.3. The originally planned re-observation scheme was a trilateration scheme that could not satisfy the detection of movements of 2 mm of object points. Also, it was felt that too many redundant distance observations were included in the observation scheme. Therefore, the purpose here is to modify the previous scheme to obtain an optimum triangulateration observation scheme to satisfy this precision requirement with the available instrumentation and with the minimum effort. The network configuration is not supposed to change. With the optimization algorithm discussed in Chapters 10 and 11, the following procedures have been followed:

Step 1. Selection of the network datum

According to the results of deformation analysis from the previous observation epochs, it was found that points p3 and p14 are the most stable points. Thus, the datum of the monitoring network is provided by minimum constraints with the coordinates of point p3 and the azimuth from p3 to p14 being fixed.

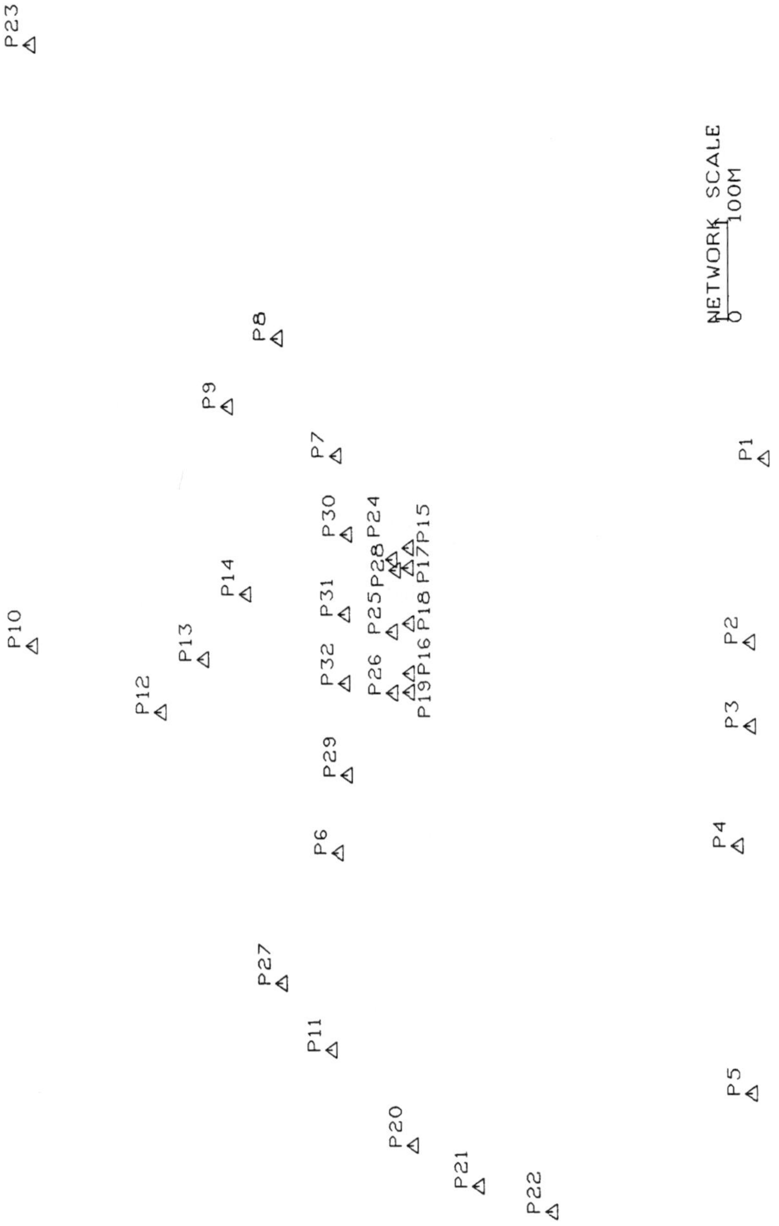

Figure 12.3. The monitoring network: distribution of network points

Step 2. Definition of the criterion matrix

For the purpose of network optimization, the precision criterion is set as such to detect displacements of a magnitude of 2.0 mm of the object points at 95% confidence level from two independently repeated surveys with respect to the chosen network datum.

Denote the vector of displacements at point i as

$$\mathbf{d}_i = \left(d_{xi}\ d_{yi}\right)^T \tag{12.11}$$

The statistical significance of an estimated value $\hat{\mathbf{d}}_i$ of $\mathbf{d}_i$ may be tested using the following statistic T, i.e., [cf. (11.60)]

$$T = \left[\hat{\mathbf{d}}_i - E\left(\hat{\mathbf{d}}_i\right)\right]^T \mathbf{C}_{\hat{d}_i}^{-1}\left[\hat{\mathbf{d}}_i - E\left(\hat{\mathbf{d}}_i\right)\right] \tag{12.12}$$

where $E\{\cdot\}$ represents the statistical expectation operator. The matrix $\mathbf{C}_{\hat{d}_i}$ in Equation (12.12) is the a priori variance-covariance matrix of $\hat{\mathbf{d}}_i$. Referring to Equation (11.4), one has

$$\mathbf{C}_{\hat{d}_i} = 2\,\mathbf{C}_{xi} \tag{12.13}$$

with $\mathbf{C}_{xi}$ being the variance-covariance matrix for coordinate components (x_i, y_i) of point i.

The null hypothesis

$$H_0\colon E\left\{\hat{\mathbf{d}}_i\right\} = \mathbf{0} \tag{12.14}$$

is tested against the alternative hypothesis

$$H_a\colon E\left\{\hat{\mathbf{d}}_i\right\} \neq \mathbf{0} \tag{12.15}$$

It can be shown that under the null hypothesis H_0, the statistic T follows the chi-square distribution with two degrees of freedom, i.e.,

$$T\big|H_0 \in \chi^2(2) \tag{12.16}$$

Given a certain significance level α, if

$$\begin{aligned} T &= \hat{\mathbf{d}}_i^T \mathbf{C}_{\hat{d}_i}^{-1} \hat{\mathbf{d}}_i \\ &= \frac{1}{2}\hat{\mathbf{d}}_i^T \mathbf{C}_{xi}^{-1} \hat{\mathbf{d}}_i > \chi^2_{1-\alpha}(2) \end{aligned} \tag{12.17}$$

one rejects the null hypothesis.

Geometrically speaking, Equation (12.17) represents an error ellipse with semi-major and semi-minor axes and their orientations being defined by the square root of the eigenvalues and their corresponding eigenvectors of matrix $2\chi^2_{1-\alpha}(2)\ C_{xi}$, respectively. Any estimated displacement vector $\hat{\mathbf{d}}_i$ that falls within the ellipse cannot be detected, otherwise it is detectable. Therefore, in order to detect the displacements of 2 mm at 95% confidence level, the maximum semi-major axis a_{max} of the station standard error ellipses for point coordinates determination should satisfy the following relation:

$$a_{max} \leq \frac{2}{\sqrt{2} \cdot \sqrt{\chi^2_{0.95}(2)}} = \frac{2}{\sqrt{2} \cdot 2.4484} = 0.6(\text{mm}) \tag{12.18}$$

To satisfy this condition, a precision criterion matrix for point coordinates determination can be established by using a diagonal matrix with all its diagonal elements being 0.6^2 (mm^2).

Step 3. Deletion of unnecessary distance observations

The originally planned re-observation scheme was a trilateration scheme (using EDM ME3000 with $\sigma_s^2 = (0.3\ \text{mm})^2 + (3 \bullet 10^{-6} \bullet s)^2$) that consisted of measuring all 206 observable distances (Figure 12.4). The station 95% error ellipses for this observation scheme are shown in Figure 12.8. The major semi-major axes of the 95% station error ellipses corresponding to this case are listed in column 3; i.e., a(I), of Table 12.13, from which one can see that the values of the semi-major axes of object points reach up to 2.7 mm for point p15. Displacements less than 3.8 mm will not be detectable for this point.

The optimization algorithm found and deleted 26 distances that do not have significant contributions to the improvement of the network accuracy. These distances obtained zero values for their optimal weights. The deleted distances are shown in Figure 12.5, and the optimized trilateration network is shown in Figure 12.6. The station 95% error ellipses after deleting these distances are shown in Figure 12.9, and values of the semi-major axes are listed under a(II) of Table 12.13. Numerically, from a(II) in Table 12.13, one can see that deleting these 26 distances only causes the semi-major axis of station 95% error ellipses of some points to increase by 0.5 mm at the maximum. This result shows that measuring these 26 distances could not significantly add to the improvement of accuracies of displacement detection for both the reference and object points.

Step 4. Addition of optimum directions

From the *optimized trilateration scheme* the accuracies of most object points are not satisfactory, the maximum semi-major axis of station 95% error ellipses of object points is 3.0 mm. Thus, roughly in the worst case, displacements less than 4.2 (i.e., $\sqrt{2} \bullet 3.0$) mm cannot be detected. To improve the accuracies, the optimiza-

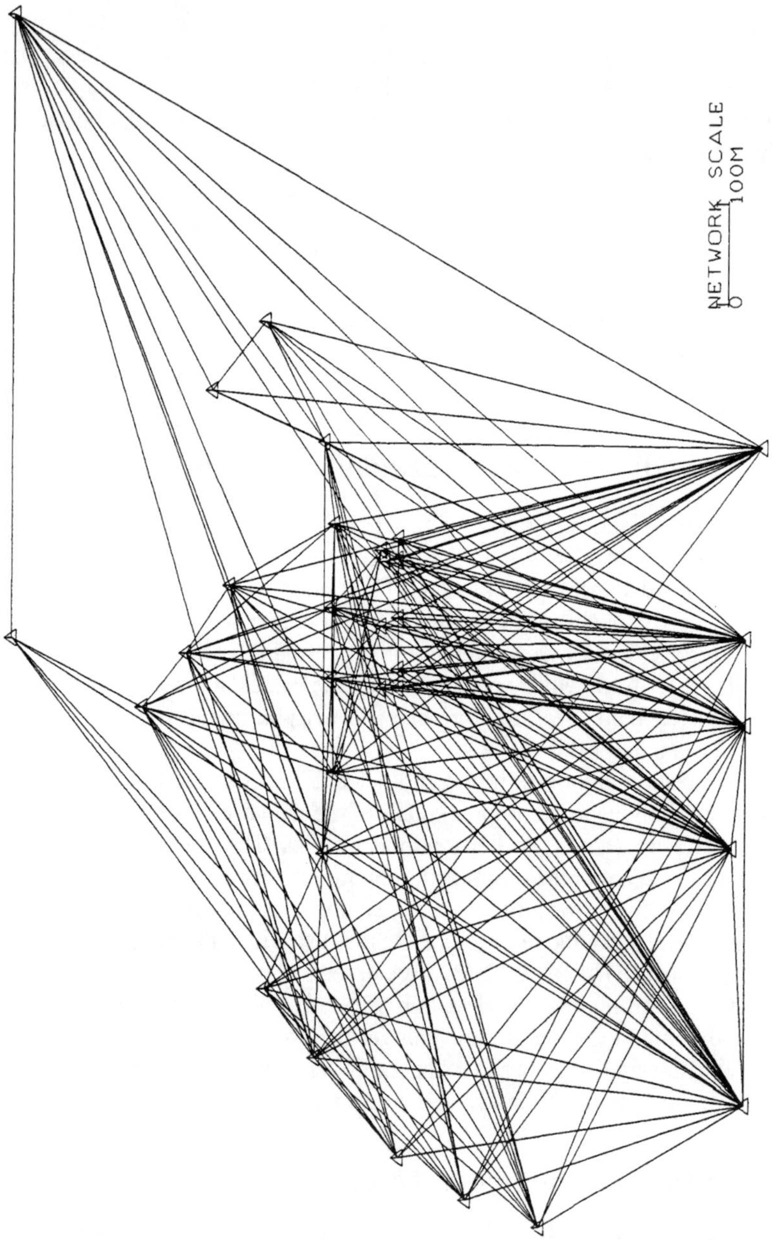

Figure 12.4. The monitoring network—the original trilateration (total: 206 distances)

Table 12.13. Semi-major axes of 95% station error ellipses corresponding to different observing schemes

Point Name		a(I) (mm)	a(II) (mm)	a(III) (mm)	a(IV) (mm)
Reference points	P1	5.0	5.3	3.7	2.1
	P2	1.2	1.4	1.3	1.2
	P3	0.0	0.0	0.0	0.0
	P4	1.4	1.5	1.3	1.1
	P5	2.0	2.0	1.8	1.1
	P8	4.4	4.6	2.0	1.3
	P9	4.5	4.7	1.9	1.9
	P10	6.7	6.7	2.3	2.3
	P11	2.0	2.0	1.9	1.8
	P12	1.6	1.7	1.6	1.4
	P13	1.3	1.3	1.3	1.2
	P14	1.3	1.5	1.3	1.3
	P20	2.4	2.4	2.3	1.7
	P21	2.2	2.2	2.1	1.8
	P22	2.4	2.5	2.3	2.0
	P23	8.7	8.9	4.3	2.9
	P27	1.8	1.8	1.7	1.4
Object points	P6	1.6	1.7	1.6	1.2
	P7	2.5	3.0	1.4	1.3
	P15	2.7	3.0	1.5	1.3
	P16	2.5	2.7	1.4	1.2
	P17	2.6	2.9	1.4	1.2
	P18	2.5	2.8	1.4	1.2
	P19	2.4	2.7	1.4	1.3
	P24	1.7	1.7	1.7	1.5
	P25	1.5	1.7	1.5	1.4
	P26	1.4	1.6	1.5	1.4
	P28	2.6	2.9	1.5	1.3
	P29	1.4	1.5	1.4	1.2
	P30	1.4	1.8	1.4	1.2
	P31	1.2	1.3	1.2	1.2
	P32	1.2	1.2	1.2	1.2

tion algorithm was applied to figure out optimal directions, i.e., those directions that have maximum contribution to the improvement of accuracies of these points.

Assuming the use of a Kern E-2 with minimum achievable standard deviation of direction observation 0.7", we have input all the 412 directions at 32 stations, but the optimization algorithm gives only ten stations with 81 directions that are significant for the improvement of accuracies of the object points. The stations to be occupied by theodolites are P1, P2, P3, P4, P5, P7, P10, P15, P23, and P28, and the standard deviations of direction observation given by the optimization algorithm for all these stations are 0.7". These directions are represented by Figure 12.7. This would be the optimal solution for the given configuration of the network

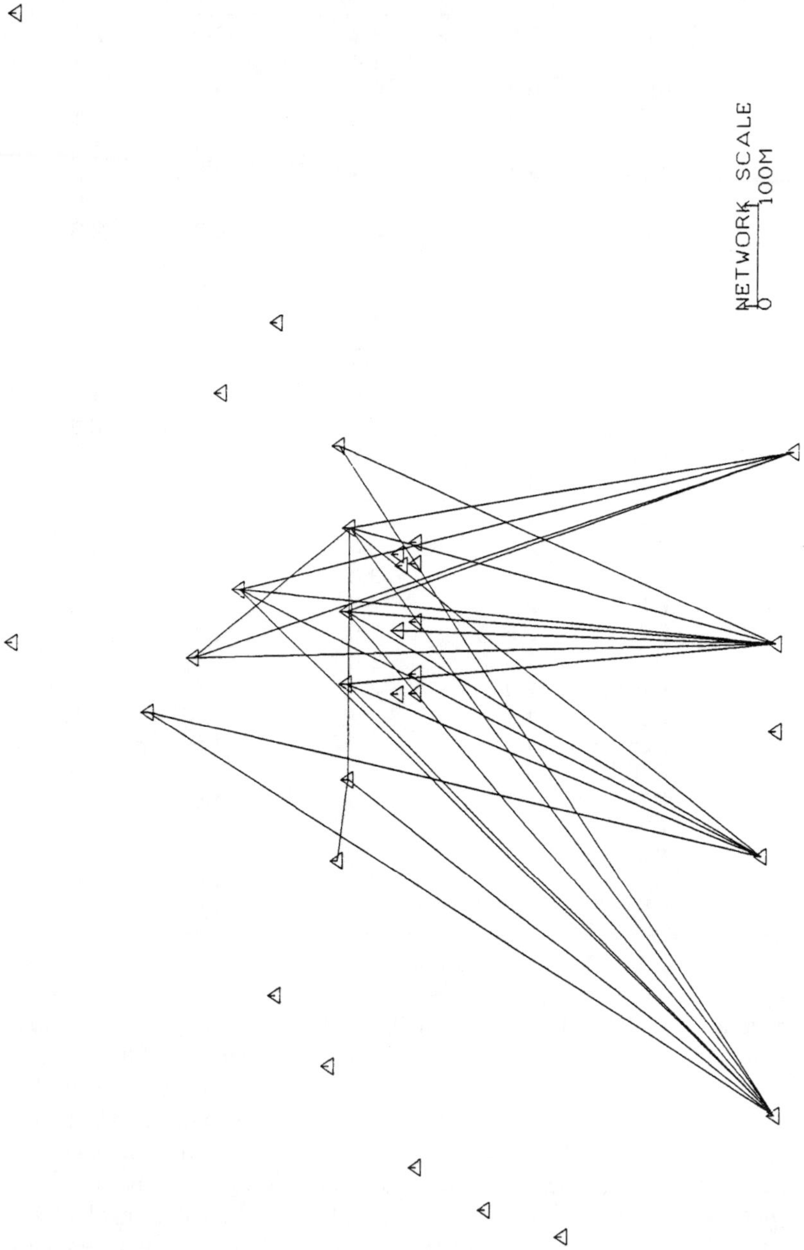

Figure 12.5. The monitoring network—the deleted distances by optimization (total: 26)

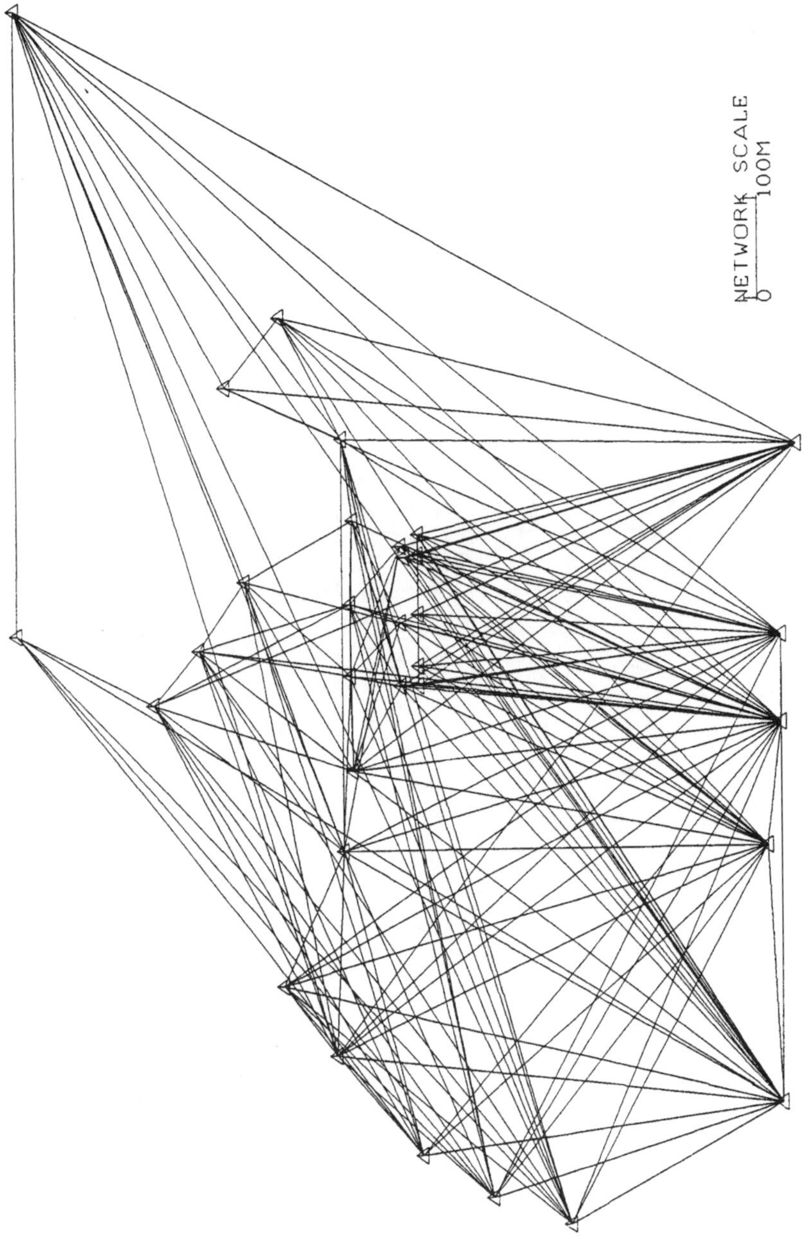

Figure 12.6. The monitoring network—the optimized trilateration (total: 180 distances)

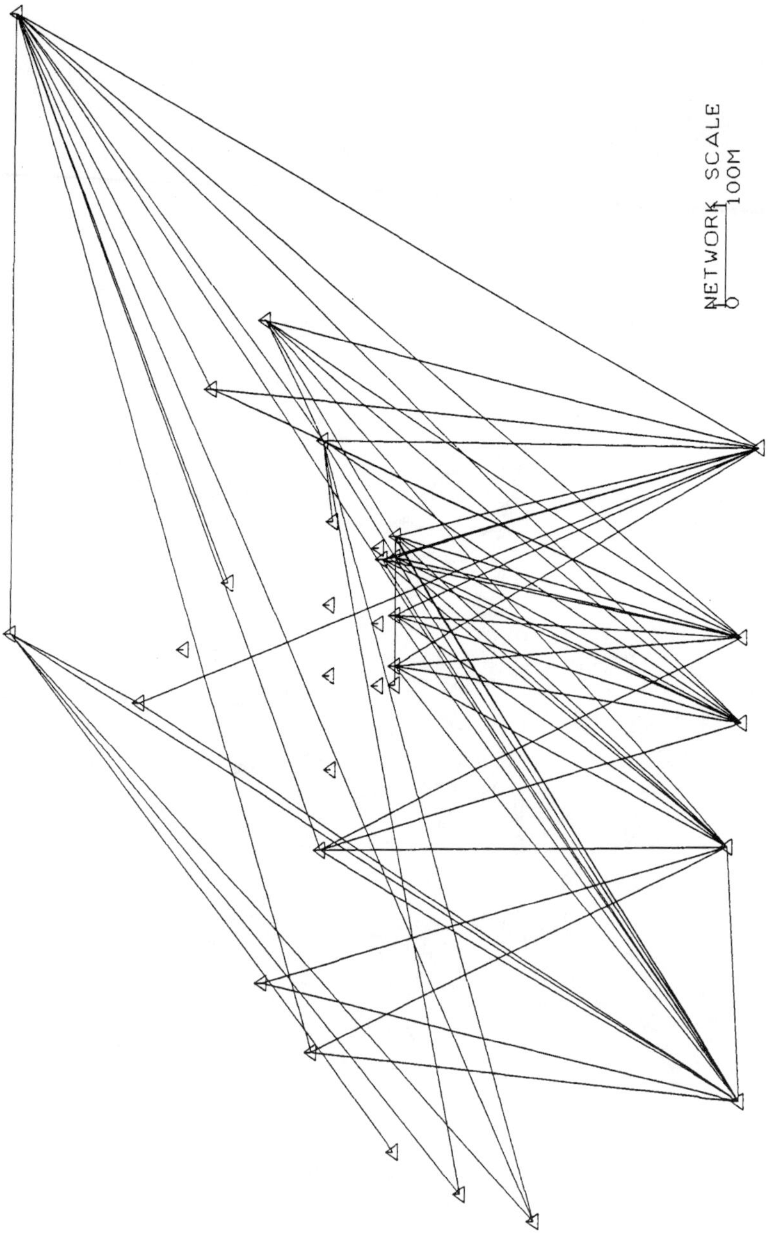

Figure 12.7. The monitoring network—the proposed directions by optimization (total: 81)

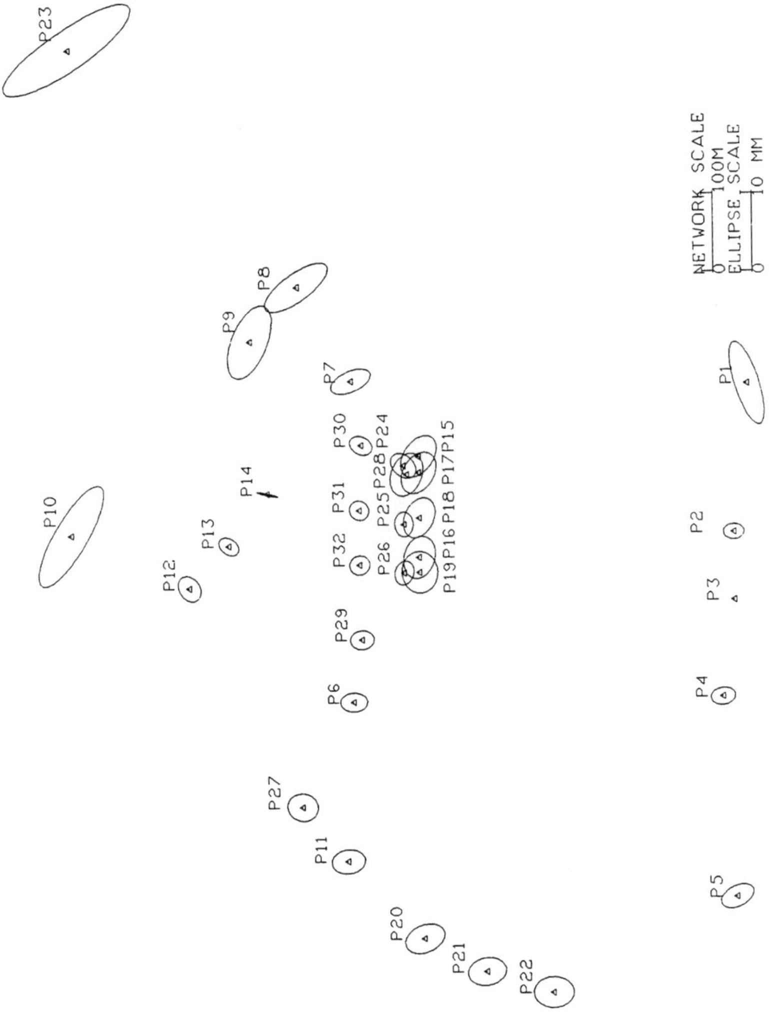

Figure 12.8. 95% station error ellipses for the original trilateration scheme

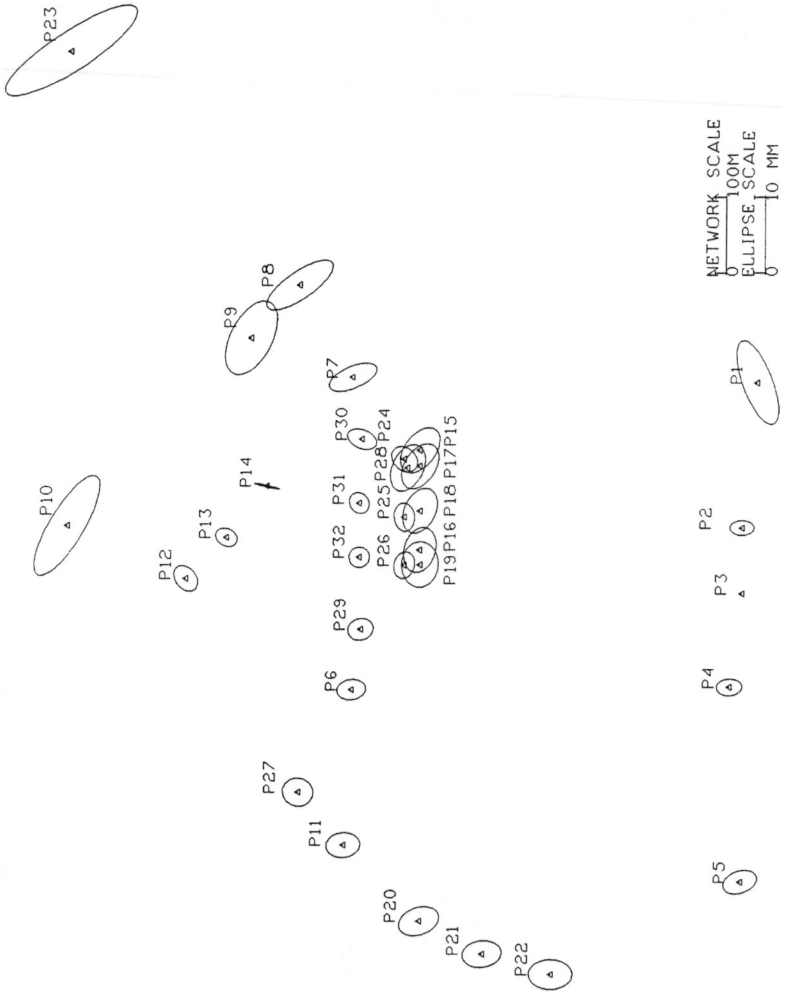

Figure 12.9. 95% station error ellipses—after deleting 26 distances

and for the available instrumentation. The station 95% error ellipses after adding these directions are shown in Figure 12.10, and values of the semi-major axes are listed under a(III) of Table 12.13. This optimal design has satisfied the above given requirement of 2.0 mm in detecting the displacements of all object points except for points P6 and P24, for which the maximum semi-major axes of station 95% error ellipses are 1.6 mm and 1.7 mm, respectively, allowing for detecting displacements larger than 2.4 mm.

To test the optimization results obtained from Step 3, the author has also tried to re-input the above deleted 26 distances into the optimized triangulateration observation scheme. It was found that the semi-major axes of station 95% error ellipses would decrease by only 0.1 mm at the maximum. That once again verifies the optimization results: these 26 distances are really useless for the accuracy improvement of the network.

Step 5. Optimum configuration design

Finally, although only the Second-Order Design can be considered in this example from the practical point of view, an optimum solution for the network configuration has also been given under the assumption that the locations of the reference stations are allowed to change in a certain extent. If one could allow for the change of the locations of the reference points (except for the datum points P3 and P14) by a maximum of ±150.0 meters in both x- and y-directions, respectively, the optimal solution would be obtained for the coordinate shifts as shown in Table 12.14. The semi-major axes of the station error ellipses after optimizing the configuration are shown as a(IV) in Table 12.13, from which one can see that this optimal design satisfies the above given requirement of 2.0 mm in detecting the displacements of all the object points.

12.2 Optimal Design of Geodetic Leveling Networks

The problems and solutions of the optimal design of leveling networks have been studied in Kuang (1993a). In contrast with the optimal design of two- or three-dimensional geodetic networks, there are some special features concerning the optimal design of geodetic leveling networks. First, in the case of leveling networks, the network configuration (i.e., the point location) is considered to be given; therefore only the optimum set of observations have to be chosen (Niemeier and Rohde, 1981). Second, concerning the observations, in leveling networks only height differences between neighboring points can be measured. Important for the precision are the chosen instruments, the measuring techniques, and the surveying instructions (Niemeier and Rohde, 1981). In contrast with the two- or three-dimensional networks, where there is a large range of distance, direction, and azimuth measuring instruments all with different precisions, we rarely have a choice as to the precision of the measuring process and hence cannot tailor our field methods and instrumentation to fit the required weight matrix (Cross and Fagir, 1982). Thus

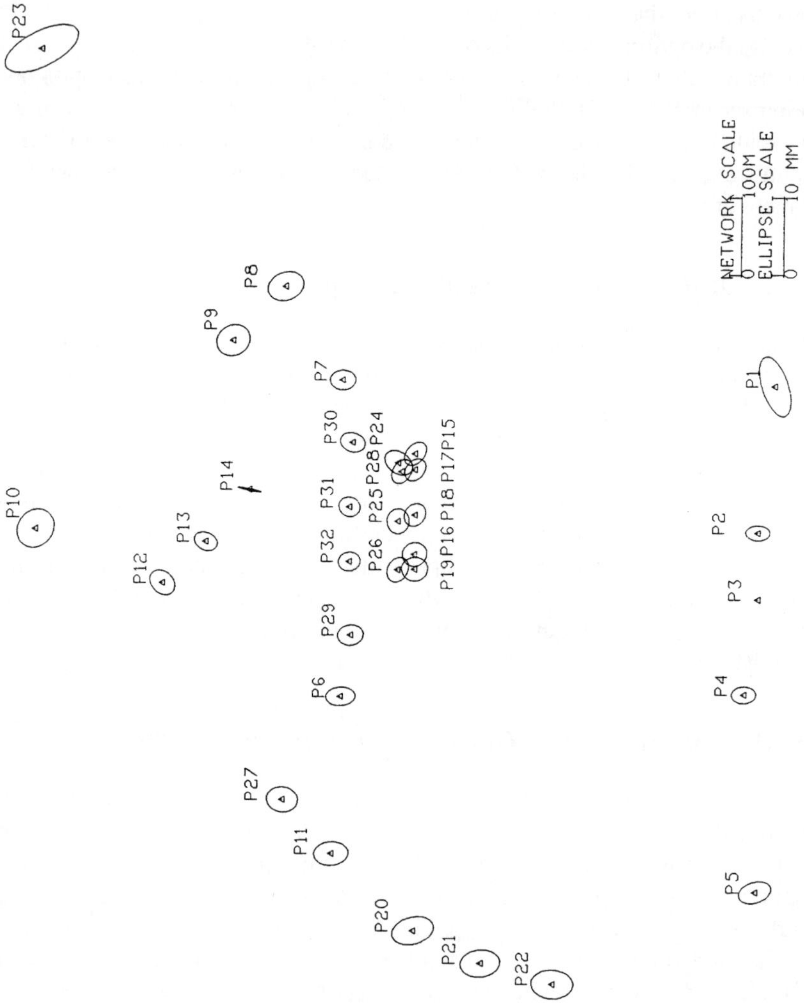

Figure 12.10. 95% station error ellipses—for the optimized triangulateration scheme

Table 12.14. Optimal coordinate shifts of the reference points

Point	x	y
P1	−125.1	150.0
P2	53.4	150.0
P4	−69.6	150.0
P5	150.0	150.0
P8	−150.0	−150.0
P9	85.4	−150.0
P10	−56.8	−150.0
P11	−86.1	−118.0
P12	−25.6	−143.1
P13	−150.0	−114.2
P20	150.0	113.2
P21	150.0	−150.0
P22	150.0	−35.4
P23	−150.0	−150.0
P27	150.0	−150.0

if the configuration is once given, there only remains the 0 or 1 (i.e., yes or no) decision for the observation of leveling lines (Niemeier and Rohde, 1981). The design problem of leveling networks is usually posed as (cf. Cross and Fagir, 1982; Kuang, 1993a):

> *Given a set of stations whose heights are to be found, a list of all possible level routes, and the quality criteria for the estimated heights, find the optimal subset of level routes that will satisfy the design criteria with minimum cost.*

Since in the case of leveling, networks costs are proportional to the length of level routes, one advocates that the optimal subset of level routes should have the minimum total length.

Effort was made in Cross and Thapa, 1979; Cross and Whiting, 1981; and Cross and Fagir, 1982, 1983 to study the optimal design of leveling networks, and some iterative solution procedures were proposed based on Linear Programming. One, however, can see that the procedures are not always successful and therefore not universally applicable. The optimization modeling given in Chapter 10 is evaluated here for the optimal design of leveling networks based on Quadratic Programming. Only the precision criteria will be considered here. Once given a set of stations whose heights are to be found, a list of all possible level routes and the precision criterion matrix, the optimization modeling solves for an optimal set of observational weights. The level routes that obtained zero or insignificant optimal weights are then deleted, and those that obtained significant optimal weights constitute the optimal subset of level routes to be observed.

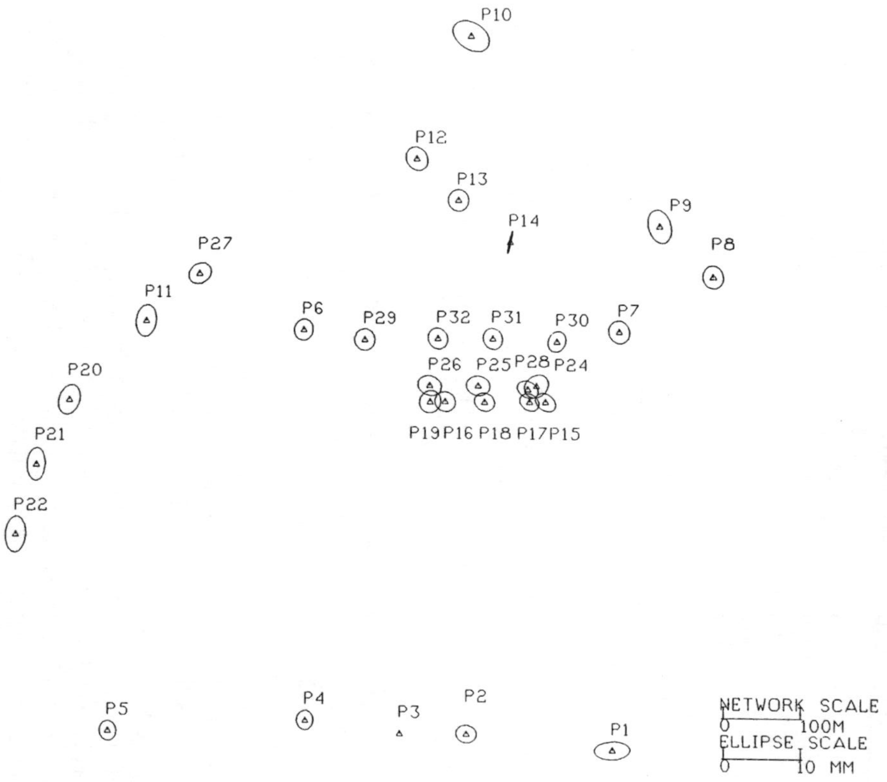

Figure 12.11. 95% station error ellipses—after optimizing the configuration

Precision Criterion Matrix for Leveling Networks

The structure of the criterion matrix for two- and three-dimensional networks has been discussed in §9.1.2. It may take the form of Taylor-Karman structure (abbreviated as TK-structure), or constructed by modifying the present variance-covariance matrix according to the special purpose of the network. However, little effort was made to study the structure of criterion matrices for leveling networks. Cross and Thapa (1979) proposed a structure of the criterion matrices for leveling networks based on the fact that one is usually interested in the height differences derived from the least squares estimates of the station heights. Usually, one would demand that the variance of a height difference be less than some specified function of the distance between the points, i.e.,

$$\sigma^2_{\Delta h_{ij}} = f(s_{ij}) \tag{12.19}$$

where $\sigma^2_{\Delta h_{ij}}$ is the variance of the height difference between two stations i and j separated by a distance of s_{ij}; $f(s_{ij})$ usually takes the form of a linear function, i.e.,

$$f(s_{ij}) = a \cdot s_{ij} \tag{12.20}$$

with a being a properly selected positive constant.

Once Equations (12.19) and (12.20) are given, elements of the criterion matrix for the station heights are calculated as follows:

$$\sigma^2_i = \sigma^2_{ori.} + f(s_i) \tag{12.21}$$

$$\sigma_{ij} = \frac{1}{2}\left(\sigma^2_i + \sigma^2_j - \sigma^2_{\Delta h_{ij}}\right) \quad (i \neq j) \tag{12.22}$$

where $\sigma^2_{ori.}$ and σ^2_i are the variances of the origin and station i, respectively, s_i the distance of the i-th station from the origin, and σ_{ij} is the covariance of every pair of station. The variance of the origin can be assigned either zero if it is considered to be fixed or nonzero if it is considered to be a weighted station. No correlation is considered between the origin and the rest stations in the network.

Modeling of Optimization

Note that for a leveling network, the matrices **A**, **H**, and **D** involved in the calculation of the variance-covariance matrix for station heights are all constant matrices. Therefore, from (4.137), the variance-covariance matrix $\mathbf{C}_x$ of station heights consists of nonlinear functions of only observational weights p_i (i=1, ..., n), that are, therefore, the basic unknown parameters to be optimally solved for at the design stage of the network. Given a set of approximate values for observational weights, matrix $\mathbf{C}_x$ can be approximated by Taylor series of linear form as (cf. Kuang, 1993a)

$$\mathbf{C}_x = \mathbf{C}_x^0 + \sum_1^n \frac{\partial \mathbf{C}_x}{\partial p_i} \Delta p_i \tag{12.23}$$

where

$$\mathbf{C}_x^0 = \sigma_0^2 \left[\left(\mathbf{A}^T \mathbf{P} \mathbf{A} + \mathbf{D} \mathbf{D}^T \right)^{-1} - \mathbf{H} \left(\mathbf{H}^T \mathbf{D} \mathbf{D}^T \mathbf{H} \right)^{-1} \mathbf{H}^T \right] \Big|_{\mathbf{P}^0} \tag{12.24}$$

$$\frac{\partial \mathbf{C}x}{\partial p_i} = \sigma_0^2 \left\{ -\left(\mathbf{A}^T \mathbf{P} \mathbf{A} + \mathbf{D} \mathbf{D}^T \right)^{-1} \left[\mathbf{A}^T \frac{\partial \mathbf{P}}{\partial p_i} \mathbf{A} \right] \left(\mathbf{A}^T \mathbf{P} \mathbf{A} + \mathbf{D} \mathbf{D}^T \right)^{-1} \right\} \Big|_{\mathbf{P}^0} \tag{12.25}$$

$\mathbf{P}^0$ is the initial observation weight matrix, and Δp_i (i=1, ..., n) are the ***improvements***, which are to be optimally solved for, to the initially given weights. Assuming that no correlation between observations exists, the partial derivatives of the weight matrix $\mathbf{P}$ with respect to the individual observational weight p_i are calculated by (10.118), and the final optimal values of observational weights are obtained by adding the solved weight improvements to their corresponding approximate weights. For leveling networks the maximum possible number of datum defects is equal to one, and the datum matrix $\mathbf{D}$ and matrix $\mathbf{H}$ in (12.24) and (12.25) are given by (4.98) and (4.113), respectively. Equation (12.23) is then used to establish optimization modeling for geodetic leveling networks as discussed in §10.4 of Chapter 10.

Optimization Examples

In order to demonstrate the efficiency of the optimization modeling given in Chapter 10, two examples are given below. The first example deals with the optimal design of a simulated network. Given the precision requirement for the estimated height differences between any two stations in the network, the optimization procedure searches for the optimal observational weights (or optimal observational precision) for each of the proposed level route. The level routes that obtained zero or insignificant optimal weights should be deleted from the final observing scheme, while those that obtained significant optimal weights are to be observed according to the solved for optimal weights. The practical significance of the optimization modeling is shown by the second example that deals with the optimal design of an actual leveling network established to provide vertical control for a large engineering project. In this example, the optimization problem is posed slightly differently. Since in practice it is usually impractical to observe each level route using a different precision function (cf. Equation 12.19), therefore, starting with a "maximum" network that includes all the possible level routes with a given observational precision, instead of solving for different optimal observing precision for each individual level route, *the optimization modeling is applied only to determine which of the level routes could be deleted with an allowable decrease in the precision of the network.* That has resulted in significant savings on the cost of the project.

Example No. 1. Optimal design of a simulated leveling network

Suppose that we are given a leveling network to be optimized as shown by Figure 12.12. There are six stations and a total of 11 possible level routes. The length of each level route is listed in Table 12.15. The precision requirement is given such that the variance of the estimated height difference between any two stations should satisfy

$$\sigma^2_{\Delta h_{ij}} \leq 4 \cdot s_{ij} \left(mm^2\right) \tag{12.26}$$

where Δh_{ij} is the height difference between stations i and j, and s_{ij} the distance in kilometers between stations i and j, respectively. In order to satisfy the above precision requirement, one assumes that point #1 is fixed to provide the datum of the network, and the criterion matrix for the optimization can be established according to Equations (12.21) and (12.22) as follows

$$\mathbf{C}^c = \begin{pmatrix} 0.000 & 0.000 & 0.000 & 0.000 & 0.000 & 0.000 \\ & 8.000 & 2.600 & 7.000 & 1.600 & 0.000 \\ & & 12.000 & 3.000 & 7.800 & 4.000 \\ & & & 10.000 & -1.200 & 0.000 \\ & \text{Symmetric} & & & 23.600 & 13.800 \\ & & & & & 28.000 \end{pmatrix} \left(mm^2\right) \tag{12.27}$$

The task of optimal design is to find the optimal weights for the above proposed level routes to satisfy the above set precision criterion. One assumes that each level route may be surveyed using different precision, but the maximum achievable precision could be 4·s mm^2, i.e., the field leveling process could not be carried out with a variance of better than 4·s mm^2 (with s being the length of a leveled line). Taking all the initial weights as being 0.001, which is arbitrarily chosen, the unknown parameters to be optimally solved for are then the weight improvements to all the initial weights. After applying the optimization modeling, the optimization results are given in Table 12.15, where p(appro.), p(opti.), p(max.) represent the initially input approximate weights, the optimally solved for weights, and the maximum achievable weights that are calculated from the maximum available precision of the field surveys, respectively; σ(opti.) and σ(min.) represent the optimal standard deviations calculated from the optimal observational weights and the mini-

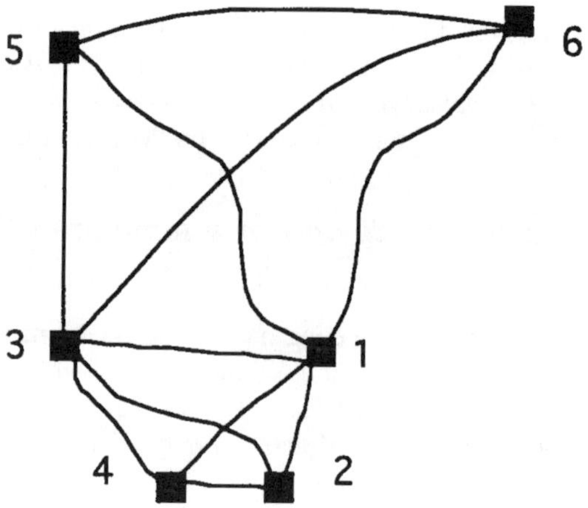

Figure 12.12. The proposed leveling network

Table 12.15. Optimization results for the simulated level network

From	To	Distance (km)	p(Appro.)	p(opti.)	p(max.)	σ(opti.) (mm)	σ(min.) (mm)
4	2	1.0	0.001	0.22568	0.25000	2.1	2.0
4	3	4.0	0.001	0.01255	0.06250	8.9	4.0
4	1	2.5	0.001	0.02169	0.10000	6.8	3.2
2	1	2.0	0.001	0.09605	0.12500	3.2	2.8
3	1	3.0	0.001	0.05917	0.08333	4.1	3.5
3	5	5.0	0.001	0.03087	0.05000	5.7	4.5
1	5	5.9	0.001	0.00871	0.04237	10.7	4.9
1	6	7.0	0.001	0.02091	0.03571	6.9	5.3
5	6	6.0	0.001	0.02903	0.04167	5.9	4.9
2	3	3.7	0.001	0.00086	0.06757	34.1	3.8
3	6	8.0	0.001	0.00000	0.03125	∞	5.7

mum achievable standard deviations calculated from the maximum achievable weights of observables, respectively.

From Table 12.15, one can see that two level routes, i.e., (2-3) and (3-6) obtained zero or insignificant (as compared with the rest of the optimal weights) optimal weights. It means that these level routes can be deleted from the final observing plan. After removing these two level routes with a total length of 11.7 km the optimized network is shown in Figure 12.13, and the actual variance-covariance matrix of the station heights from the optimally designed weights is given as follows:

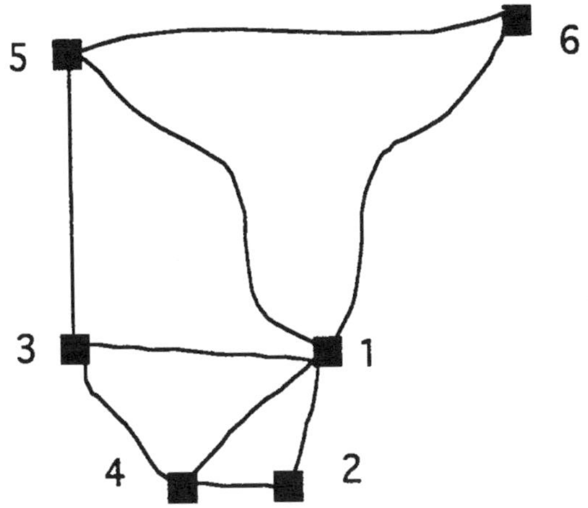

Figure 12.13. The optimized leveling network

$$
\mathbf{C}_X = \begin{pmatrix} 0.000 & 0.000 & 0.000 & 0.000 & 0.000 & 0.000 \\ & 8.000 & 1.114 & 7.000 & 0.665 & 0.387 \\ & & 12.000 & 1.547 & 7.160 & 4.162 \\ & & & 10.000 & 0.923 & 0.537 \\ & \text{Symmetric} & & & 23.600 & 13.719 \\ & & & & & 28.000 \end{pmatrix} \left(\text{mm}^2\right) \qquad (12.28)
$$

By comparing $\mathbf{C}^c$ with $\mathbf{C}_x$, one can see that the preset precision criterion has been satisfied.

Example No. 2. Optimal design of a practical leveling network

The second example demonstrates the practical significance of the optimization modeling that has been applied to the optimal design of an actual leveling network established to provide vertical control for a large engineering project. The original design of the network consisted of a total of 87 junction points and 139 leveling sections between the junction points with a total leveling distance of 686.3 km as shown by Figure 12.14, in which the junction points are numbered from #2 to #88. The center point #1 is fixed to provide the datum of the network.

Due to the special purpose of the network, one expects the benchmarks in the network to have high precision. Therefore all the possible level routes are pro-

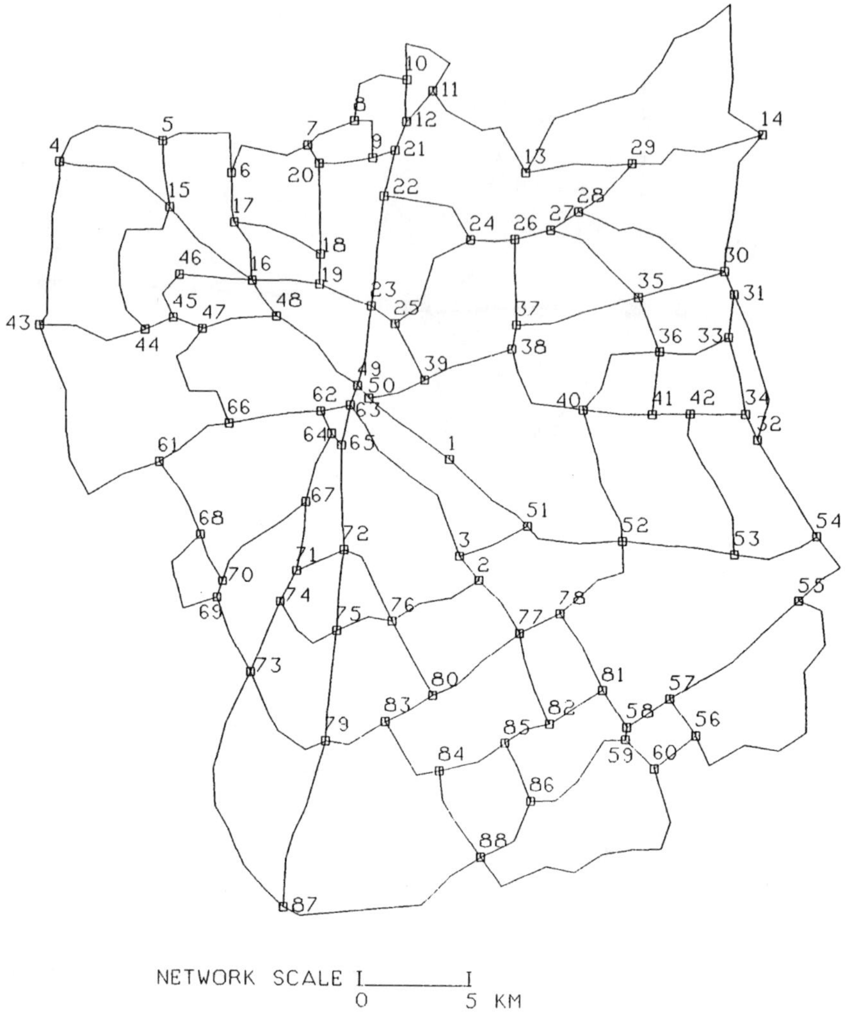

Figure 12.14. The originally planned leveling scheme

posed as shown in Figure 12.14 and the network is to be observed with a precision as follows

$$\sigma^2_{\Delta h_{ij}} = 2 \cdot s_{ij} \left(mm^2\right) \tag{12.29}$$

where Δh_{ij} and s_{ij} are the leveled height differences and distances in kilometers between points i and j, respectively. Since it is impractical to observe each level route with a different precision, here all the level routes are to be observed with the same precision as given in Equation (12.29).

According to the originally planned leveling scheme (Figure 12.14) and the observational precision (Equation 12.29), the standard deviations of the junction points of the network as obtained from the originally planned observing scheme are listed under σ_0 in Columns 2 and 6 of Table 12.17. The results indicate that the originally designed network would give a maximum standard deviation of 4.8 mm in the determined heights of benchmarks with respect to the center point #1. This precision is found to be a little higher than what is actually required. Therefore, in order to save time and effort in the field, i.e., to save the cost of the project, instead of solving for different optimal observing precision for each individual level route, *the purpose of the optimization here is to determine which of the leveling sections between junction points could be deleted from the originally planned observing scheme without a significant decrease in the precision of the network.*

Depending on the allowable decrease in the precision of the network, the optimization procedure will find out which of the level sections could be deleted. Let us accept 1.0 mm as the allowable decrease in the precision of the network points (junction benchmarks) at the 1σ level, and use the following diagonal matrix $\mathbf{C}_s^c$ as the precision criterion matrix for the optimization, i.e.,

$$\mathbf{C}_s^c = \text{diag}\left\{0.0 \quad \sigma_2^2 \quad \sigma_3^2 \quad \cdots \quad \sigma_{88}^2\right\} \tag{12.30}$$

where

$$\sigma_i^2 = \left(\sigma_{0i} + 1.0\right)^2 \quad \left(\text{mm}^2\right) \quad (i = 2, \ldots, 88) \tag{12.31}$$

σ_{0i} is the standard deviation for point i (i=2, ..., 88) from the original observing scheme, point #1 obtained zero variance since it is fixed.

Taking the maximum achievable weights calculated from the maximum achievable precision expressed by Equation (12.29) as the initial approximate weights of observations, and after only one iteration the optimization procedure gives the optimal observational weights that can be grouped into nonzero and zero or insignificant weights. The nonzero weights, some of which may be smaller than the maximum achievable weights, are then all replaced by their corresponding maximum achievable weights. The level routes that obtained zero or insignificant observational weights represent those that should be deleted from the final observing plan. By following this procedure, not more than one iteration is necessary.

The optimization results are listed in Tables 12.16 to 12.17, from which one can see that if we allow for a decrease of 1.0 mm in the precision of the network, a total of 27 leveling sections between junction points with a total length of 164.0 km could be deleted from the originally designed leveling scheme. The deleted sections are listed in Table 12.16, and graphically represented by Figure 12.15. The optimized leveling scheme in this case is shown by Figure 12.16. The standard deviations of the junction points as obtained from the optimized leveling scheme is listed under σ_h in Columns 3 and 7 of Table 12.17. Columns 4 and 8 of Table 12.17 lists the differences between σ_0 and σ_h, from which one can see that the preset optimization criterion of the 1.0 mm decrease in the precision has been satisfied for

Table 12.16. The deleted leveling lines by optimization

From	To	Length (km)
5	6	5.2
8	10	5.4
29	14	7.2
17	18	5.1
16	19	3.6
20	9	3.6
23	25	1.8
24	26	3.6
28	30	10.8
31	32	7.2
35	36	3.6
38	40	5.4
36	41	1.8
42	53	10.8
57	56	1.8
63	3	10.8
47	66	8.4
68	69	9.6
67	71	3.4
74	75	5.4
78	52	7.2
73	79	6.3
72	76	5.4
84	85	3.6
86	59	7.2
88	60	16.2
82	81	3.6
Total length of deleted level lines = 164.0 km		

all the junction points; i.e., all the differences are less than or practically equal to 1.0 mm. Deleting these 164.0 km of leveling routes out of a total of 686.3 km leveling routes without a significant decrease of precision of the network means a saving of 23.9% of field work, implying a significant saving on the cost of the project.

The examples given above have demonstrated the efficiency and practicability of the optimization modeling also for geodetic leveling networks. From the optimization results, one can see clearly that once a list of all possible level routes have been given, not all the level routes can significantly contribute to the improvement of the precision of the network. Therefore the subset of level routes that do not significantly contribute to the improvement of the precision of the network should be deleted in order to save time and effort, i.e., the cost of the survey campaign. For instance, this example shows that deleting these 164.0 km of leveling routes out of a total of 686.3 km leveling routes while satisfying an allowable

Table 12.17. Comparison between the original and optimized network precision

Point	σ_0 (mm)	σ_h (mm)	σ_h-σ_0 (mm)	Point	σ_0 (mm)	σ_h (mm)	σ_h-σ_0 (mm)
1	0.00	0.00	0.00	45	4.01	4.65	0.64
2	3.38	3.67	0.29	46	4.17	4.79	0.62
3	3.22	3.52	0.30	47	3.94	4.65	0.71
4	4.60	5.30	0.70	48	3.77	4.28	0.51
5	4.39	5.41	1.03	49	3.02	3.25	0.22
6	4.20	4.87	0.66	50	2.78	2.92	0.14
7	4.15	4.65	0.49	51	2.77	2.90	0.13
8	4.31	4.80	0.49	52	3.53	4.09	0.56
9	4.10	4.60	0.51	53	4.22	4.83	0.62
10	4.44	5.08	0.65	54	4.21	4.64	0.42
11	4.37	4.77	0.41	55	4.42	4.88	0.46
12	4.21	4.63	0.42	56	4.24	5.23	0.99
13	4.72	5.05	0.33	57	4.15	4.91	0.75
14	4.78	5.58	0.80	58	3.99	4.75	0.76
15	4.25	4.99	0.74	59	4.01	4.78	0.77
16	3.78	4.49	0.71	60	4.19	5.04	0.85
17	4.04	4.76	0.72	61	4.11	4.46	0.36
18	3.92	4.67	0.75	62	3.43	3.75	0.31
19	3.77	4.55	0.78	63	3.18	3.49	0.31
20	4.07	4.72	0.65	64	3.48	3.79	0.31
21	4.01	4.41	0.41	65	3.38	3.65	0.27
22	3.93	4.32	0.38	66	3.92	4.44	0.52
23	3.48	4.08	0.61	67	3.75	4.30	0.55
24	3.86	4.54	0.68	68	4.09	4.47	0.37
25	3.52	4.22	0.70	69	3.99	4.36	0.37
26	3.90	4.82	0.91	70	3.98	4.31	0.33
27	4.00	4.78	0.79	71	3.67	4.17	0.50
28	4.16	4.95	0.79	72	3.59	3.94	0.35
29	4.54	5.18	0.64	73	3.86	4.31	0.46
30	4.09	4.60	0.51	74	3.77	4.29	0.53
31	4.10	4.60	0.50	75	3.71	4.02	0.31
32	4.16	4.68	0.52	76	3.61	3.98	0.37
33	4.10	4.55	0.45	77	3.57	3.97	0.40
34	4.11	4.59	0.48	78	3.67	4.24	0.57
35	3.96	4.62	0.67	79	3.86	4.31	0.45
36	3.90	4.89	1.00	80	3.78	4.12	0.34
37	3.86	4.43	0.57	81	3.89	4.63	0.74
38	3.74	4.33	0.59	82	3.96	4.75	0.79
39	3.35	3.65	0.30	83	3.89	4.32	0.43
40	3.79	4.61	0.82	84	4.09	4.76	0.67
41	3.92	4.83	0.91	85	4.08	4.97	0.89
42	4.04	4.83	0.79	86	4.20	5.00	0.81
43	4.44	4.90	0.46	87	4.46	4.73	0.27
44	4.15	4.74	0.59	88	4.23	4.84	0.61

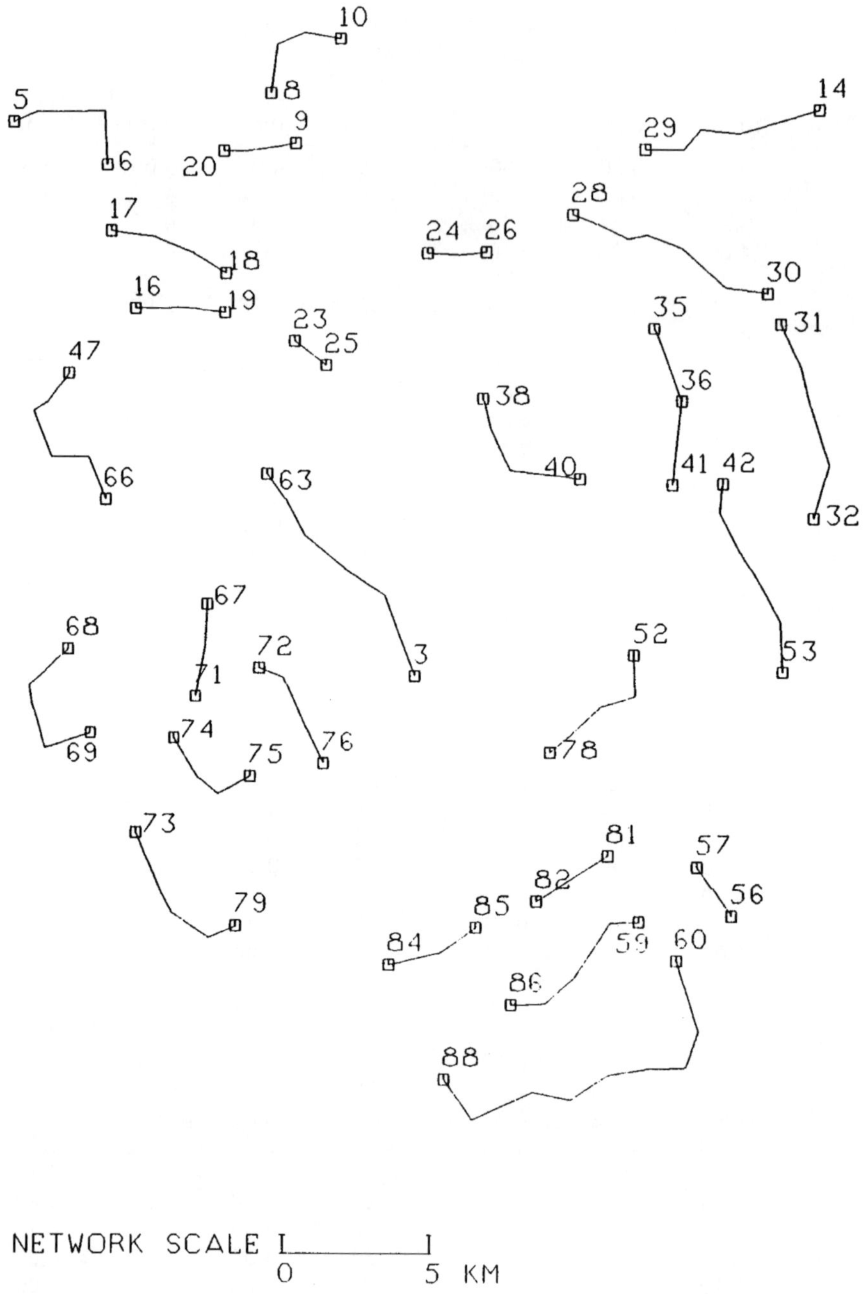

Figure 12.15. The deleted leveling lines by optimization

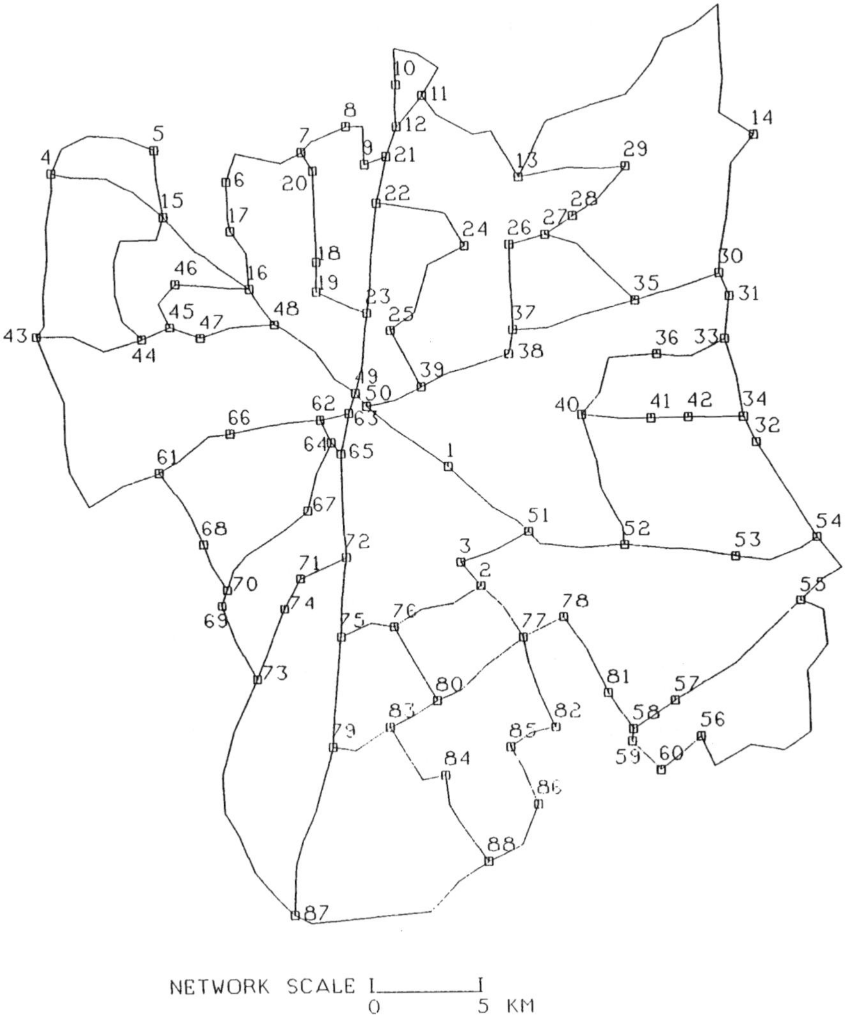

Figure 12.16. The optimized leveling scheme

decrease of 1.0 mm in the original precision of the network means a saving of 23.9% on the cost of the survey campaign. Especially for leveling networks that are to be repeatedly observed on a regular basis such as encountered in networks established to detect vertical crustal movements and to monitor the relationship between mean sea level and geodetic height datums, the accumulated savings as brought out by network optimization can be very significant. Although only the use of criterion matrices are discussed here, optimization models including other types of design criteria such as reliability and/or sensitivity can also be incorporated elegantly to obtain a multi-objective optimal design as discussed in Chapter 10.

12.3 GPS Survey Planning: Choice of Optimum Baselines

The optimal design problem of the GPS network positioning for engineering applications has been discussed in Kuang (1994a). Assuming that the configuration of the satellites, the locations of the ground network stations, and the precision of the GPS observations are known, a methodology has been developed for the selection of optimum baselines to be measured in the field that will satisfy the prescribed network precision with the minimum effort. As is obvious, the highest precision and reliability of a GPS network are expected if all the possible combinations of baselines in the network would be measured. Since in practice that will rarely happen due to the limitations of time and expenses, an optimum survey planning has to be made to achieve some prescribed design criteria with the minimum effort. The optimization modeling developed in Chapter 10 is also used for this purpose. For practical applications, once given a set of stations whose relative positions are to be found, a list of all possible baselines (i.e., starting with the "maximum" network), and the criterion matrix of the estimated coordinates, the optimization modeling solves for an optimal subset of observational weights that will satisfy the prescribed precision criteria with the minimum cost. The baselines that obtained zero or insignificant optimal weights are then deleted, and those that obtained significant optimal weights constitute the optimal observing scheme. After the set of optimum baselines are chosen, observation session planning can then be done easily by the GPS field crew according to the available receivers to find out the total number of observation sessions needed and the ground stations to be occupied in each observation session in order to produce the required quasi-independent optimum baselines.

The General Optimal Design Problem of GPS Networks

The general optimal design problem of GPS networks has been discussed in, among others, the works of Lindohr and Wells (1985), Wells et al. (1986, 1987), and Delikaraoglou and Lahaye (1989). Some critical points are addressed here. For a more detailed discussion, the interested readers are referred to the references.

As with the conventional terrestrial geodetic networks, one may define the general optimal design problem of a GPS network as designing an optimum combined configuration of both the satellites to be tracked and the network of ground stations and designing an optimum observing accuracy to achieve the prescribed network quality criteria, which can be precision, reliability and/or sensitivity, with the minimum cost.

If we consider the entire set of ground and satellite points to be analogous to the set of terrestrial points in a classical design problem, a detailed breakdown of the general design problem for GPS networks can be obtained by following the classical approach to the optimization of terrestrial geodetic networks as follows (cf. Grafarend, 1974, Grafarend et al., 1979; 1985, Delikaraoglou and Lahaye, 1989):

1. Zero-Order Design (ZOD): Optimum datum definition
2. First-Order Design (FOD): Design an optimum network configuration
3. Second-Order Design (SOD): Specification of observation weights; and
4. Third-Order Design (THOD): Improvement of the existing network, e.g., integration with existing terrestrial networks such as triangulation, trilateration, leveling, etc.

Nevertheless, as compared with the conventional geodetic networks, there are some special features concerning the specific problems to be solved at each order's design stage for GPS networks:

First, the datum definition for GPS networks (ZOD) is usually more complex. The reference system is defined by the set of global tracking stations used to generate the broadcast or precise ephemerides;

Second, GPS First Order Design (FOD) has to deal with the following three parts: *GPS redundancy design*, *GPS configuration design*, and *logistics design*. GPS redundancy design is to establish a minimum network configuration that allows a number of unknown parameters to be uniquely defined using the principle that the number of observations must be equal to or greater than the number of unknowns. When such minimal designs are satisfactory, increasing the redundancy appropriately will improve the reliability, otherwise, it may not generally result in a more satisfactory design (Tsimis, 1973; Grafarend et al., 1979; Delikaraglou, 1985). GPS configuration design is to establish a strongest combined geometry of satellites and ground stations that results in the optimum dilution of precision over the ground and satellite points that comprise the entire network. Finally, GPS logistics design is concerned with the selection of a set of mappings of available receivers onto the ground stations selected in the configuration design with minimum cost. In contrast to the design of terrestrial geodetic networks where the selection of stations is limited by intervisibility between stations, the selection of ground points in GPS configuration design has no such limitations. However, since the configuration of the satellite constellation is determined by designers of the GPS system, we have no control over the satellite constellation and we are only free to select the times and time span for which to make observations, and select which satellites to track if we are unable to track all available satellites;

Third, in GPS networks the precision of the observed ranges is one of the most crucial parameters which control the overall precision of the determined positions. In contrast to the Second-Order Design (SOD) of terrestrial geodetic networks where one usually searches for the optimum distribution of observational weights in a fixed network configuration, GPS Second-Order Design is considered in a much broader sense here. In practice, GPS random errors are usually small and have little effect on the adjustment solution due to the high measurement redundancy. Thus the quality of GPS surveying is largely influenced by unmodeled systematic errors that are difficult to quantify. The key to precise GPS surveying is therefore the reliable detection of systematic errors (e.g., the resolution of the cycle ambiguities, clock errors, ionosphere and troposphere delays) (i.e., with high internal reliability), and minimization of the effect of the unmodeled systematic errors on the estimated coordinates (i.e., with high external reliability). Therefore, in addition to the choice of instruments, as in the case of Second-Order Design for terrestrial geodetic networks, the GPS Second-Order Design is also concerned with: the observational bias modeling, modeling of the effects of orbit errors, modeling of the clock behavior, atmospheric influences on the propagation of the GPS signals, and the influence of any unmodeled or improperly modeled biases. For instance, the selection of the measurement modes (carrier range or code range); the time span in which to make observations (the longer we track the selected set of satellites, the more accurate the resulting position differences); selection of the mask angle (vertical angle cutoff) during tracking (usually from 8° to 15°) to reduce tropospheric modeling; selection of the appropriate time of the day for the observations to reduce the effect of the ionosphere; etc.; are all the concerns of the GPS SOD. Although with the full GPS coverage available today observations made for equal periods, but at different times of the day, are unlikely to provide significantly different results (apart perhaps from refraction considerations), the observing time span may still be an important design selection.

In mathematical terms, one can summarize that the general task in the planning stage of a GPS survey is to design such a combined geometry of both the satellites to be tracked and the network of ground stations together with the choice of instruments and bias modeling that are optimal in the sense of maximizing the precision of the obtained coordinates (high precision), maximizing the chances of the detection of systematic errors (high internal reliability), and minimizing the effect of the unmodeled systematic errors on the estimated coordinates (high external reliability) and in the meantime with minimum cost, as expressed by the following objective function:

$$\alpha_p \cdot \text{precision} + \alpha_i \cdot (\text{internal reliability}) + \alpha_e \cdot (\text{external reliability}) + \\ + \alpha_c \cdot (\text{cost})^{-1} = \max \tag{12.32}$$

where α_p, α_i, α_e and α_c are the appropriately selected weight coefficients for the corresponding precision, internal reliability, external reliability, and cost criteria included in the global objective function, respectively.

The research on the optimal design of GPS networks using the above multi-objective function has not yet been done. In the past years, great effort has been made by geodesists to work on the selection of the geometry of the satellites to be tracked, and the bias modeling, e.g., the tropospheric and ionospheric delay modeling, of GPS observations. Nevertheless, the conventional DOP (dilution of precision) factors (Wells et al., 1986) are inappropriate for GPS network design since they were developed for the purpose of GPS navigation (point positioning) using pseudoranges. Effort was made in, among others, Vanicek and Craymer (1986), and Ashtech (1991) to develop methods for satellite selection for differential positioning with multiple receivers. In Merminod et al. (1990), the BDOP (bias dilution of precision) factors were proposed for satellite selection for differential positioning based on carrier beat phase observation with the intent of high reliable resolution of phase ambiguities. Even with the full GPS coverage available today, the optimal design of a GPS network in the sense of both high precision and reliability will still be the first step towards the success of the survey.

Choice of Optimum Baselines

From the above discussions, one can say that the quality of a GPS network generally depends on the following factors:

1. The geometry of the satellite configuration and the network stations on the ground;
2. The precision of the original observations, such as pseudoranges and carrier phases; and
3. The total number and the distribution of quasi-independent baselines measured.

Theoretically, the best precision and reliability of the relative positions of a GPS network can be obtained if all visible satellites are tracked as long as possible and all possible baselines in the network are measured. Due to the limitations of time and expenses, however, that will rarely happen in practice, and therefore an optimum survey design has to be made in order to achieve some prescribed design criteria with the minimum effort.

Here we consider the special case in which the geometry of the satellite configuration, the locations of the ground network points, and the precision of the original GPS observations are known. We then search for the optimum set of quasi-independent baselines to be measured in order to achieve the prescribed precision criteria of the network. This is of profound practical significance for practitioners of GPS. Since in practice a surveyor has no control over the configuration of the satellite constellation, he usually tries to obtain an optimum configuration of the satellites to be tracked by selecting the optimum observation window(s) under the existing satellite constellation using a computer software that is developed according to some quality criteria (e.g., precision criteria), and the visibility and availability of the satellites for a certain location. This kind of software is usually available from the manufacturers of GPS receivers, e.g., Ashtech's MSMP (Multi-Site Mission Planning); for engineering applications the locations of the ground network stations are selected to accommodate the purpose of the network; and the precision of GPS observations is dictated by the chosen instruments (GPS receivers). Therefore, the only variable left for a surveyor to decide with freedom is the total number and the distribution of the quasi-independent baselines to be measured in the field. As is well known, the baseline length and orientation affect the precision of the estimated baseline components, i.e., relative positions. The selection of baselines within the network can either reduce or enhance geometrical redundancy (Wells et al., 1986). A set of properly chosen baselines may also result in a great amount of savings of field work while adhering to the required precision.

As with the optimal design of leveling networks, there is only the 0 or 1 (i.e., yes or no) decision as to the choice of baselines. The optimal design problem to be considered here for a GPS relative survey campaign can thus be posed as:

> *Given a set of stations whose relative positions are to be found, the available instruments, a list of all possible quasi-independent baselines that could be measured, and the quality criteria for the estimated relative coordinates, find the optimal subset of baselines that will satisfy the design criteria with the minimum cost.*

Since, in the case of relative GPS networks, costs of field survey are proportional to the total number of quasi-independent baselines observed, one advocates that an optimal survey scheme should have the minimum total number of baselines.

A "Stepwise" design approach to the selection of optimum baselines was discussed in, among others, Janes et al. (1986) and Wells et al. (1986). This approach is laborious, and since it is a "trial and error" approach, it may never find the theoretical optimum set of baselines. A fully analytical solution approach to this optimization problem has therefore been needed.

Gauss-Markov Models for GPS Network Adjustment

In the surveying community, any network optimization problem is always related to the adjustment of the relevant observations, and is in fact formulated with the help of the same terms (i.e., design matrix, weight matrix, and variance-covariance matrix of unknown parameters) that ultimately appear in the mathematical model established for the adjustment and evaluation of the collected observations. In the case of GPS network positioning, the unknown parameters are in general the coordinates of both the satellites and the ground points that make up the network, and other nuisance parameters (e.g., clock errors, atmospheric delays, carrier-phase ambiguities, etc.) that describe the inconsistencies between the available observations and the corresponding observables. Establishment of the relation between the GPS measurements and the unknown parameters to be estimated is more involved here than in the case of terrestrial geodetic observations. Thus, a good understanding of the GPS adjustment models along with the assessment of the likely influence of systematic errors is a prerequisite to any successful GPS survey design (cf. also Delikaraoglou and Lahaye, 1989).

Suppose a general network of ground stations is to be positioned by the NAVSTAR GPS system. After the field survey campaign is completed, the Gauss-Markov Model for the network adjustment may take different forms depending on the processing mode used. The following Gauss-Markov Model I deals with the case that the GPS observations observed simultaneously at all stations are taken directly into a network adjustment with the unknown parameters to be solved being all the coordinates of the network stations and some nuisance parameters used for orbit improvement and/or other systematic biases. Gauss-Markov Model II is used when the observation campaign is first reduced to simultaneous data observed from pairs of stations (baselines), and then combine these baselines into a network to perform network adjustment (cf. Wells et al., 1986). The Gauss-Markov Model I is generally used by GPS researchers for data analysis and software development, while the Gauss-Markov Model II is widely used in practice by surveyors for production purposes.

Gauss-Markov Model I

The GPS system has a number of measurement modes (Wells et al., 1986; Leick, 1990), but the one relevant to high-precision surveying is the carrier beat phase measurement. The general Gauss-Markov Model of the undifferenced carrier beat phase observations of a GPS network can be written as follows (Wells et al., 1987):

$$E\{\varphi\} = \mathbf{A}_1\mathbf{x} + \mathbf{A}_2\xi = (\mathbf{A}_1 \quad \mathbf{A}_2)\begin{pmatrix}\mathbf{x}\\ \xi\end{pmatrix} \tag{12.33}$$

$$D\{\varphi\} = \mathbf{P}_\varphi^{-1}\sigma_0^2 \tag{12.34}$$

where it is assumed that receiver Re_i at station i is simultaneously tracking a number of satellites S_i (i=1, ..., N_g); $E\{\cdot\}$ and $D\{\cdot\}$ are the statistical expectation and dispersion operators, respectively; φ is a misclosure vector computed from the observed phases ϕ and the a priori approximate coordinates (which are assumed to be non-stochastic here) of both the satellites and ground stations; $\mathbf{x}$ is the vector of corrections to the approximate coordinates of both satellites and ground stations; ξ is the vector of nuisance parameters consisting of the biases (e.g., clock, ambiguities, atmospheric errors, etc.); $\mathbf{A}_1$ and $\mathbf{A}_2$ are called the parameter design matrix and the bias design matrix, respectively; $\mathbf{P}_\varphi$ is the observational weight matrix; and σ_0^2 is the a priori variance factor.

As with the conventional terrestrial geodetic networks, the parameter design matrix $\mathbf{A}_1$ is singular if the datum defects of the network are not resolved. A three-dimensional GPS network datum is defined geometrically by three translation, three orientation, and one scale parameters. The distance content of the GPS measurements (e.g., pseudoranges or carrier beat phases) may provide the datum scale to the extent that the contributing biases can be sufficiently accounted for. The missing translation and orientation parameters must be provided by some constraints on the coordinates. As discussed in, among others, Delikaraoglou and Lahaye (1989), the datum definition for a GPS network may usually be complex and overconstrained. In contrast to the terrestrial networks where the origin and orientation of a network are usually defined by holding fixed the coordinates of only one point, two zenith distances and one azimuth observation, the constraints used to define a GPS network datum can be in the form of fixed constraints (i.e., specified "known" positions), conditional constraints (i.e., through additional relationships between certain points), or weighted constraints [i.e., weights assigned to some or all (station or satellite) coordinate values involved in the adjustment]. For most surveying applications, the satellite position vectors are generally assumed to be known as defined by the adopted broadcast or precise ephemerides of the GPS satellites used in the data reduction. The GPS ephemerides are generated by a set of global tracking stations whose coordinates are determined by some method independent of GPS; e.g., VLBI (Very Long Baseline Interferometry) or SLR (Satellite Laser Ranging). The GPS datum is then transferred to the stations of the survey network via these ephemerides without the need of direct observations between the survey and the global tracking stations. For a detailed discussion of the datum problem, the reader is referred to Delikaraoglou and Lahaye (1989) and Wells et al. (1987). Here let us assume that the datum of the GPS network has been defined, and that reduces the total number of unknown parameters by six, leaving the parameter design matrix $\mathbf{A}_1$ nonsingular. In this case, elimination of the bias parameters ξ allows for the least squares solution of $\mathbf{x}$ and its associated variance-covariance matrix given by (cf. also Wells et al., 1987)

$$\mathbf{x} = \left(\mathbf{A}_1^T \mathbf{P}_\varphi \mathbf{R}_1 \mathbf{A}_1\right)^{-1} \mathbf{A}_1^T \mathbf{P}_\varphi \mathbf{R}_1 \varphi \tag{12.35}$$

$$\mathbf{C}_x = \sigma_0^2 \left(\mathbf{A}_1^T \mathbf{P}_\varphi \mathbf{R}_1 \mathbf{A}_1\right)^{-1} \tag{12.36}$$

where

$$\mathbf{R}_1 = \mathbf{I} - \mathbf{A}_2 \left(\mathbf{A}_2^T \mathbf{P}_\varphi \mathbf{A}_2\right)^{-1} \mathbf{A}_2^T \mathbf{P}_\varphi \tag{12.37}$$

with $\mathbf{I}$ being an identity matrix.

If the differencing techniques (e.g., single differencing, double differencing, triple differencing) are used, Equation (12.33) is modified by linear combinations of the original phase observations, and Equation (12.34) by applying the variance-covariance propagation technique. The undifferenced carrier beat phase observables offer the flexibility of modeling rigorously the clock errors while still allowing the adjustment of the observations in a manner equivalent to double differencing. Furthermore, the simultaneous observations are independent of each other and thus no correlation modeling is necessary as in differencing schemes. But, the differencing technique removes or greatly reduces errors related to the satellites, the propagation medium, and station instrumental effects. Theoretically, both approaches give identical results (Schaffrin and Grafarend, 1986; Lindohr and Wells, 1985).

Gauss-Markov Model II

The second approach for GPS network processing is first to reduce the observation campaign to simultaneous data observed from pairs of stations (baselines), and then combine these baselines into a network to perform network adjustment. Figure 12.17 shows a GPS network consisting of interconnected baselines.

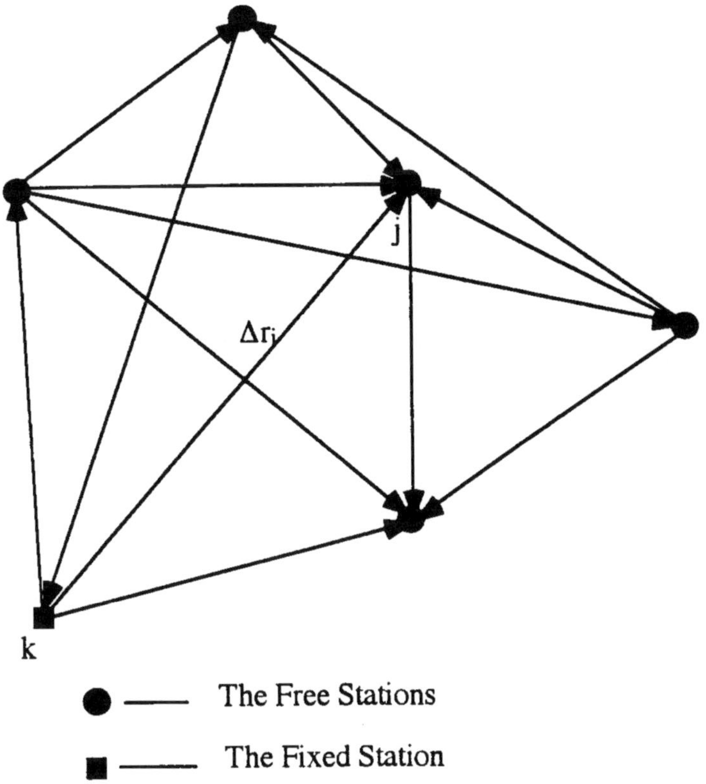

Figure 12.17. A GPS network

This approach is called the "Baseline Approach," and it is recommended for GPS networks of local extent (not larger than 100 km) and is widely used in practice for production purposes. As is clear, the "Baseline Approach" consists of two consecutive stages of data processing. In the first stage, for each observation session the baseline vectors are computed. The Gauss-Markov Model as expressed by Equations (12.33) and (12.34) can be used for the computation of the baselines observed in different observation sessions, separately. In practice, some commercial software, e.g., Ashtech's GPPS, Terrasat's TOPAS, and Trimble Navigation's TRIMVEC, are used by surveyors for this purpose. *It should be pointed out that in the computation of baselines observed in a certain observation session, only a set of quasi-independent baselines should be selected; that is, if there are three receivers in one observation session, only two baselines should be used. The third baseline is a trivial baseline. However, if the third baseline is observed again in a different observation session, it becomes a quasi-independent baseline, and should be used. Similarly, if there are n receivers in a certain observation session, only (n-1) baselines from that session are considered as being quasi-independent and should be used* (cf. Leick, 1990; Wells et al., 1986).

In the second stage, the obtained baselines are combined to form a network to perform network adjustment. Considering only the ground station coordinates as the unknown parameters to be solved, one obtains the following Gauss-Markov Model for the three-dimensional adjustment of the network:

$$E\{\Delta\mathbf{r}\} = \mathbf{A}\mathbf{x} \tag{12.38}$$

$$D\{\Delta\mathbf{r}\} = \mathbf{C}_{\Delta r} \tag{12.39}$$

where **A** is the parameter configuration matrix with its elements consisting of zeros and plus/minus ones, **x** the vector of corrections to the approximate coordinates of ground stations, and the observational vector $\Delta\mathbf{r}$ is defined as follows:

$$\Delta\mathbf{r} = \begin{pmatrix} \Delta\mathbf{r}_1^T & \Delta\mathbf{r}_2^T & \cdots & \Delta\mathbf{r}_n^T \end{pmatrix}^T \tag{12.40}$$

with $\Delta\mathbf{r}_i$ being the quasi-observed vector of the coordinate differences between stations j and k (see Figure 12.17), i.e.,

$$\Delta\mathbf{r}_i = \mathbf{r}_j - \mathbf{r}_k \quad (i = 1, \ldots, n) \tag{12.41}$$

where $\mathbf{r}_j$, and $\mathbf{r}_k$ are the vectors of unknown coordinates to be solved for points j and k, respectively. In the case that the geodetic Cartesian (x, y, z) coordinate system is used, a vector of the baseline components in x-, y- and z-directions is:

$$\Delta\mathbf{r}_i = \mathbf{r}_j - \mathbf{r}_k = \begin{pmatrix} x_j \\ y_j \\ z_j \end{pmatrix} - \begin{pmatrix} x_k \\ y_k \\ z_k \end{pmatrix} = \begin{pmatrix} \Delta x_i \\ \Delta y_i \\ \Delta z_i \end{pmatrix} \tag{12.42}$$

where (x_j, y_j, z_j) and (x_k, y_k, z_k) are the three-dimensional geodetic coordinates for the two terminal points j, k of the baseline, respectively.

There are two basic approaches to obtain the stochastic part (i.e., Equation 12.39) of Gauss-Markov Model II. One way is to take the a posterior variance-covariance matrix of the baseline components obtained from the data processing of stage one. Another way is to estimate the variance-covariance matrix $\mathbf{C}_{\Delta r}$ using the method of Minimum Norm Quadratic Unbiased Estimation (MINQUE) based on the following empirical model (cf. Lichten, 1989; Wells et al., 1986):

$$\sigma^2(\text{baseline}) = a^2 + b^2\, s^2 \tag{12.43}$$

with a and b being some constants and s the baseline length. Some studies using MINQUE have shown that the error structure of a GPS survey should always be modeled for each particular application rather than blindly use a model suggested by others (Gokalp, 1991; Kuang, 1993d).

At this stage of the survey design, let us use the above empirical model and extend the error model (12.43) to express the precision of the baseline components. One therefore obtains

$$\begin{aligned}\sigma^2_{\Delta x_i} &= \sigma^2_{\Delta y_i} = \sigma^2_{\Delta z_i} = a^2 + b^2 s_i^2 \\ &= \sigma_i^2\end{aligned} \tag{12.44}$$

For a GPS network with n baselines, if the correlations among all the GPS baseline components are neglected, the variance-covariance matrix of the total vector of observations is denoted by

$$\begin{aligned}\mathbf{C}_{\Delta r} &= \text{diag}\left(\mathbf{C}_{\Delta r i}\right) \\ &= \text{diag}\left(\sigma_i^2 \cdot \mathbf{I}_i\right) \quad (i = 1,\ \ldots,\ n)\end{aligned} \tag{12.45}$$

where diag($\sigma_i^2 \bullet \mathbf{I}_i$) represents a block diagonal matrix consisting of n submatrices ($\sigma_i^2 \bullet \mathbf{I}_i$) (i=1, ..., n) with $\mathbf{I}_i$ all being 3 by 3 unit matrices.

Once the network datum is defined, the adjusted coordinates and the variance-covariance matrix of the estimated coordinates can be expressed as follows:

$$\mathbf{x} = \left(\mathbf{A}^T\mathbf{P}\mathbf{A}\right)^{-1}\mathbf{A}^T\mathbf{P}\Delta\mathbf{r} \tag{12.46}$$

$$\mathbf{C}_x = \sigma_0^2 \left(\mathbf{A}^T\mathbf{P}\mathbf{A}\right)^{-1} \tag{12.47}$$

where

$$\begin{aligned}\mathbf{P} &= \sigma_0^2\, \mathbf{C}_{\Delta r}^{-1} \\ &= \text{diag}\left(p_i \cdot \mathbf{I}_i\right) \quad (i = 1,\ \ldots,\ n)\end{aligned} \tag{12.48}$$

with

$$p_i = \frac{\sigma_0^2}{\sigma_i^2} \quad (i = 1,\ \ldots,\ n) \tag{12.49}$$

In practice, Equations (12.47) to (12.49) are in fact widely used by surveyors for the purpose of accuracy pre-analysis of a GPS network.

Modeling of Optimization

As discussed before, any design optimization problem is related to the adjustment of the relevant observations. Since we confine ourselves to GPS networks of local extent established for engineering applications, the Gauss-Markov Model II as described in the above section can be used as the base of the optimization modeling. From Equation (12.47), the variance-covariance matrix of station coordinates is written as

$$\mathbf{C}_x = \sigma_0^2 \left(\mathbf{A}^T\mathbf{P}\mathbf{A}\right)^{-1}$$

Since in the above equation matrix **A** is a constant matrix, matrix $\mathbf{C}_x$ consists of nonlinear functions of only observational weights p_i (i=1, ..., n), which are, therefore, the basic unknown parameters to be optimally solved for at the design stage of the network. Given a set of approximate values for observational weights, matrix $\mathbf{C}_x$ can be approximated by Taylor series of linear form as (cf. Kuang, 1994a)

$$\mathbf{C}_x = \mathbf{C}_x^0 + \sum_{i=1}^{n} \frac{\partial \mathbf{C}_x}{\partial p_i} \Delta p_i \tag{12.50}$$

where

$$\mathbf{C}_x^0 = \sigma_0^2 \left(\mathbf{A}^T\mathbf{P}\mathbf{A}\right)^{-1}\Big|_{\mathbf{P}^0} \tag{12.51}$$

$$\frac{\partial \mathbf{C}_x}{\partial p_i} = \sigma_0^2 \left\{ -\left(\mathbf{A}^T\mathbf{P}\mathbf{A}\right)^{-1} \left[\mathbf{A}^T \frac{\partial \mathbf{P}}{\partial p_i} \mathbf{A}\right] \left(\mathbf{A}^T\mathbf{P}\mathbf{A}\right)^{-1} \right\}\Bigg|_{\mathbf{P}^0} \tag{12.52}$$

$\mathbf{P}^0$ is the initial observation weight matrix, and Δp_i (i=1, ..., n) are the ***improvements***, which are to be optimally solved for, to the initially given weights. Assuming that no correlation between observations exists, the partial derivatives of the weight matrix **P** with respect to the individual observational weight p_i are given below [cf. (12.48)]:

$$\frac{\partial \mathbf{P}}{\partial p_i} = \text{diag}\left(\mathbf{E}_j\right) \text{ with } \mathbf{E}_j = \begin{cases} \mathbf{I}_3 & \text{if } j = i \\ \mathbf{0}_3 & \text{if } j \neq i \end{cases} \quad (j = 1, \ldots, n) \tag{12.53}$$

where $\mathbf{I}_3$ and $\mathbf{0}_3$ are the 3 by 3 identity and zero matrices, respectively. Equation (12.50) then serves as the basis in formulating the optimization modeling as discussed in Chapter 10.

The Example

In order to demonstrate the efficiency of the optimization modeling, an example is given below. In principle, the optimization modeling developed above solves for the optimal observational weights for each of the proposed baselines given the precision requirement for the estimated coordinates of stations in the network. The baselines that obtained zero or insignificant optimal weights should be deleted from the final observing scheme, while those that obtained significant optimal weights are to be observed according to the precisions that can be calculated according to the solved for optimal weights. However, as with the choice of optimal level routes in geodetic leveling networks, in practice it is impractical to observe each individual baseline using different precision function (cf. Equations 12.43 and 12.44), there is only the 0 or 1 (i.e., yes or no) decision as to the choice of a baseline. Therefore, for practical applications, we consider the case to start with a "maximum" network that includes all the possible baselines with a given observational precision, instead of solving for different optimal observing precision for each individual baseline, *the optimization modeling is applied only to determine which of the baselines could be deleted while adhering to the prescribed precision of the network.* That may result in significant savings on the cost of the project.

Let us assume that we have a network of 18 stations to be positioned using the GPS relative positioning technique. Suppose that the selected GPS receivers can allow for a baseline to be determined with the following precision:

$$\sigma_s^2 = 2^2 \text{ mm}^2 + (1 \text{ ppm} \cdot s)^2 \tag{12.54}$$

where s is the length of a baseline. Since it is impractical to observe each individual baseline with a different precision function, here all the baselines are to be observed with the precision function given in Equation (12.54).

According to the proposed configuration of the ground stations (see Figure 12.18), the maximum number of quasi-independent baselines is 153. If all the 153 baselines are measured with the observational precision expressed by Equation (12.54), the standard deviations of the coordinate components of each station, in the case that station #1 is chosen as the fixed datum point, are listed under σ_0 in Columns 3 and 8 of Table 12.19. This is the maximum achievable precision of the network if all the possible combinations of baselines would be measured. Since, in practice, one in general, except for some special cases, does not expect the network to have the highest precision due to the limitations of time and expenses, an optimum survey plan has to be made to achieve a certain desired precision, which is lower than the maximum achievable precision, in order to save time and effort in the field, i.e., to save on the cost of the project. The required precision can be defined by introducing a certain amount of decrease in the maximum achievable precision of the network. In order to show the significance of the optimization methodology through the present example, let us determine which of the baselines could be deleted by allowing for a decrease of 2.0 mm in the precision of the network stations at the 1σ level.

The optimization modeling has been executed by using the following diagonal matrix $\mathbf{C}_s^c$ as the criterion matrix; i.e.,

$$\mathbf{C}_s^c = \text{diag}\{0.0 \quad 0.0 \quad 0.0 \quad \sigma_{2x}^2 \quad \sigma_{2y}^2 \quad \sigma_{2z}^2 \quad \cdots \quad \sigma_{18x}^2 \quad \sigma_{18y}^2 \quad \sigma_{18z}^2\} \tag{12.55}$$

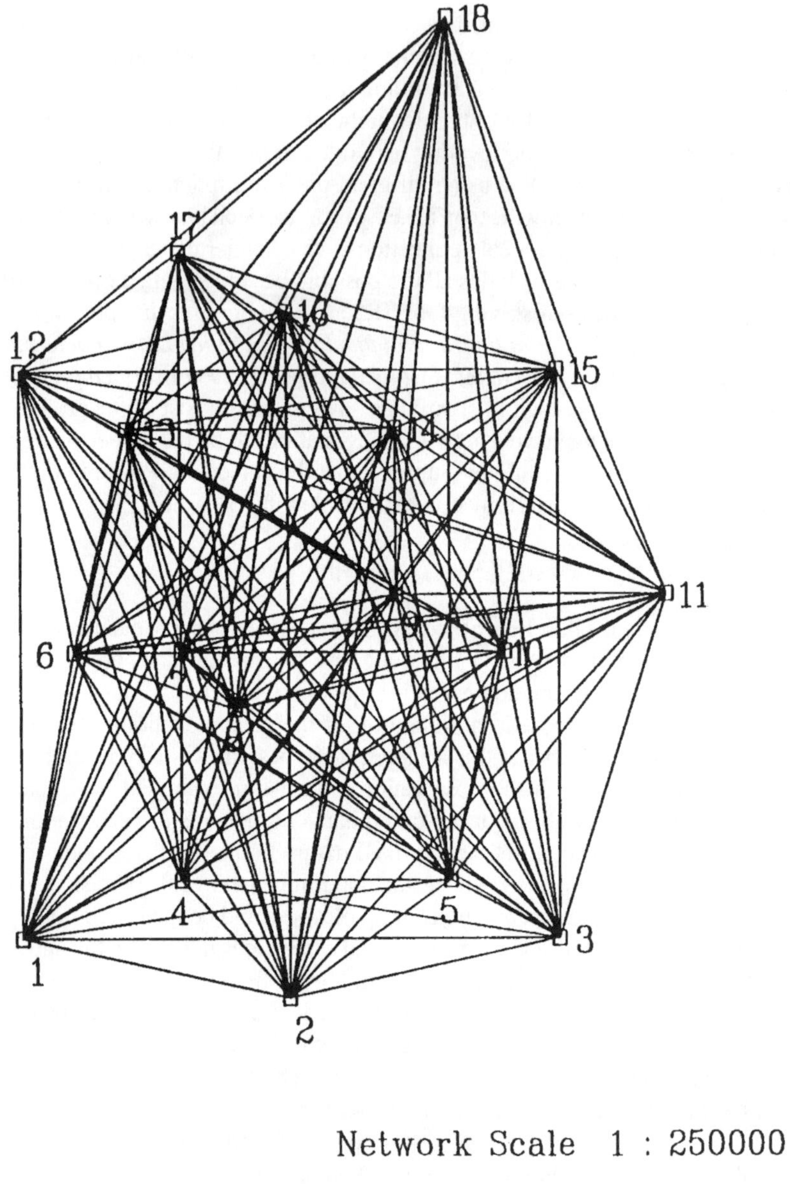

Figure 12.18. All possible combinations of baselines (total:153)

where

$$\sigma_{ij}^2 = \left(\sigma_{0ij} + 2.0\right)^2 \quad \left(mm^2\right) \quad \left(i = 2, \ldots, 18; \ j = x, y, z\right) \qquad (12.56)$$

σ_{0ij} (i=2, ..., 18; j=x, y, z) are the minimum attainable standard deviations of the coordinate components of the network stations by observing all the possible baselines. Point #1 obtained zero variance since it is fixed.

Taking the maximum achievable weights calculated from the maximum achievable precision expressed by Equation (12.54) as the initial approximate weights of observations, and after convergence, the optimization procedure gives the optimal observational weights that can be grouped into significant and zero or insignificant weights. The significant weights, some of which may be smaller than the maximum achievable weights, are then all replaced by their corresponding maximum achievable weights. The baselines that obtained zero or insignificant weights represent those that should be deleted from the final observing plan. In the latter case, in addition to the solved for optimum observational weights with zeros, a solved for optimum observational weight is considered insignificant if it is significantly less than its corresponding maximum achievable observational weight. This judgment is to some extent subjective. For instance, in this example if a solved for optimum observational weight is five or more times smaller than its corresponding maximum achievable observational weight, it is considered insignificant. The criterion may vary depending on the specific problems being solved. Nevertheless, an objective rule of thumb for this is that if the optimal precision of an observation calculated from the corresponding solved for optimal weight is well below the precision that could be realized using the available instruments, this solved for optimal weight can be considered insignificant. The optimization results are listed in Tables 12.18 and 12.19.

At first, Table 12.18 shows that if we allow for a decrease of 2.0 mm in the precision of the network, a total of 73 baselines could be deleted. The deleted baselines are also graphically represented by Figure 12.19. The optimized observing scheme is shown by Figure 12.20. The standard deviations of the station coordinates as obtained from the optimized observing scheme is listed under σ_p in Columns 4 and 9 of Table 12.19. Columns 5 and 10 of Table 12.19 list the differences between σ_0 and σ_p, from which one can see that the prescribed optimization criterion of the 2.0 mm decrease in the original precision has been satisfied for all the station coordinates, i.e., all the differences are less than or practically equal to 2.0 mm. Since deleting 73 baselines out of a total of 153 baselines while adhering to the required precision means a saving of 47.7% of field work, the optimization example has again demonstrated the benefit of the network optimization theory. Especially for GPS networks that are to be repeatedly observed on a regular basis such as met in networks established for deformation monitoring purposes, the accumulated savings as brought out by network optimization can be very significant.

Choice of optimum baselines may be seen only as part of the GPS configuration design as discussed before. Mathematically, however, it may be considered as the classical Second-Order Design problem since in the optimization modeling only the observational weight for each proposed baseline is solved for. The developed methodology is of profound practical significance for practitioners of GPS. The assumptions made with the optimization modeling are applicable in practice. For instance, Gauss-Markov Model II and the stochastic model (12.43) and (12.44)

Table 12.18. The deleted baselines by optimization

From	To	Distance (km)	From	To	Distance (km)
2	11	19.799	2	6	14.422
2	12	24.166	2	7	12.649
2	13	20.880	2	8	10.198
2	15	24.166	3	5	4.472
4	9	12.806	3	10	10.198
4	10	14.422	4	6	8.944
4	11	20.591	4	7	8.000
4	12	18.973	4	8	6.324
4	13	16.124	5	6	16.124
4	14	17.888	5	7	12.806
4	15	22.803	5	8	10.000
4	16	20.396	5	9	10.198
4	17	22.000	5	10	8.246
4	18	31.623	6	7	4.000
11	15	8.944	6	8	6.324
11	17	21.633	6	9	12.165
11	18	21.540	7	8	2.828
12	14	14.142	7	9	8.246
12	15	20.000	8	9	7.211
12	18	20.000	9	10	4.472
13	14	10.000	9	14	6.000
13	15	16.124	10	11	6.324
15	18	12.649	11	14	11.661
3	6	20.591	11	16	17.204
3	7	17.204	12	13	4.472
3	8	14.422	12	16	10.198
3	13	24.083	12	17	7.211
3	15	20.000	13	16	7.211
8	10	10.198	13	17	6.324
7	10	12.000	14	15	6.324
9	13	11.661	14	16	5.656
10	13	16.124	14	17	10.000
11	13	20.880	14	18	14.142
6	15	20.591	15	16	10.198
6	11	22.090	16	17	4.472
2	4	5.656	16	18	11.661
2	5	7.211			

are in fact used by surveyors at the design stage to predict the expected precision of a GPS network (i.e., accuracy preanalysis). The manufacturer of GPS receivers provides nominal values of a and b in Equation (12.43). These values can also be estimated a posterior by MINQUE from the previous GPS measurements if available (cf. Kuang, 1993d). Further research can be done to approach the general multi-objective optimization problem as expressed by Equation (12.32) starting with the Gauss-Markov Model I, where the basic unknown parameters to be opti-

Table 12.19. A comparison of the standard deviations of the coordinates before and after optimization

Point		σ_0 (mm)	σ_p (mm)	$\sigma_p - \sigma_0$ (mm)	Point		σ_0 (mm)	σ_p (mm)	$\sigma_p - \sigma_0$ (mm)
	σ_x	0.00	0.00	0.00		σ_x	4.44	6.30	1.86
1	σ_y	0.00	0.00	0.00	10	σ_y	4.44	6.30	1.86
	σ_z	0.00	0.00	0.00		σ_z	4.44	6.30	1.86
	σ_x	4.45	6.34	1.89		σ_x	4.96	6.97	2.01
2	σ_y	4.45	6.34	1.89	11	σ_y	4.96	6.97	2.01
	σ_z	4.45	6.34	1.89		σ_z	4.96	6.97	2.01
	σ_x	4.81	6.47	1.66		σ_x	4.71	6.60	1.88
3	σ_y	4.81	6.47	1.66	12	σ_y	4.71	6.60	1.88
	σ_z	4.81	6.47	1.66		σ_z	4.71	6.60	1.88
	σ_x	4.01	5.57	1.55		σ_x	4.44	6.23	1.78
4	σ_y	4.01	5.57	1.55	13	σ_y	4.44	6.23	1.78
	σ_z	4.01	5.57	1.55		σ_z	4.44	6.23	1.78
	σ_x	4.49	6.31	1.82		σ_x	4.46	6.26	1.79
5	σ_y	4.49	6.31	1.82	14	σ_y	4.46	6.26	1.79
	σ_z	4.49	6.31	1.82		σ_z	4.46	6.26	1.79
	σ_x	4.23	5.71	1.48		σ_x	4.88	6.84	1.96
6	σ_y	4.23	5.71	1.48	15	σ_y	4.88	6.84	1.96
	σ_z	4.23	5.71	1.48		σ_z	4.88	6.84	1.96
	σ_x	4.06	5.69	1.63		σ_x	4.53	6.52	1.98
7	σ_y	4.06	5.69	1.63	16	σ_y	4.53	6.52	1.98
	σ_z	4.06	5.69	1.63		σ_z	4.53	6.52	1.98
	σ_x	4.07	5.90	1.84		σ_x	4.66	6.52	1.86
8	σ_y	4.07	5.90	1.84	17	σ_y	4.66	6.52	1.86
	σ_z	4.07	5.90	1.84		σ_z	4.66	6.52	1.86
	σ_x	4.32	6.03	1.71		σ_x	5.96	7.91	1.95
9	σ_y	4.32	6.03	1.71	18	σ_y	5.96	7.91	1.95
	σ_z	4.32	6.03	1.71		σ_z	5.96	7.91	1.95

mized are both the positions of the satellites to be tracked and ground network points and the observational weights simultaneously. The theories applicable to this problem can be obtained from Chapter 10. Interested readers are encouraged to work on the research and applications of this interesting field of study.

12.4 Summary and Discussions

The essential purpose of geodetic network optimization is to decide on a most economical survey campaign that will satisfy the desired network quality determined by the purpose of the network. As discussed before, in the design stage of a geodetic positioning and/or network, the fundamental problem that a surveying engineer faces is how to decide on its configuration, i.e., the point location, and the

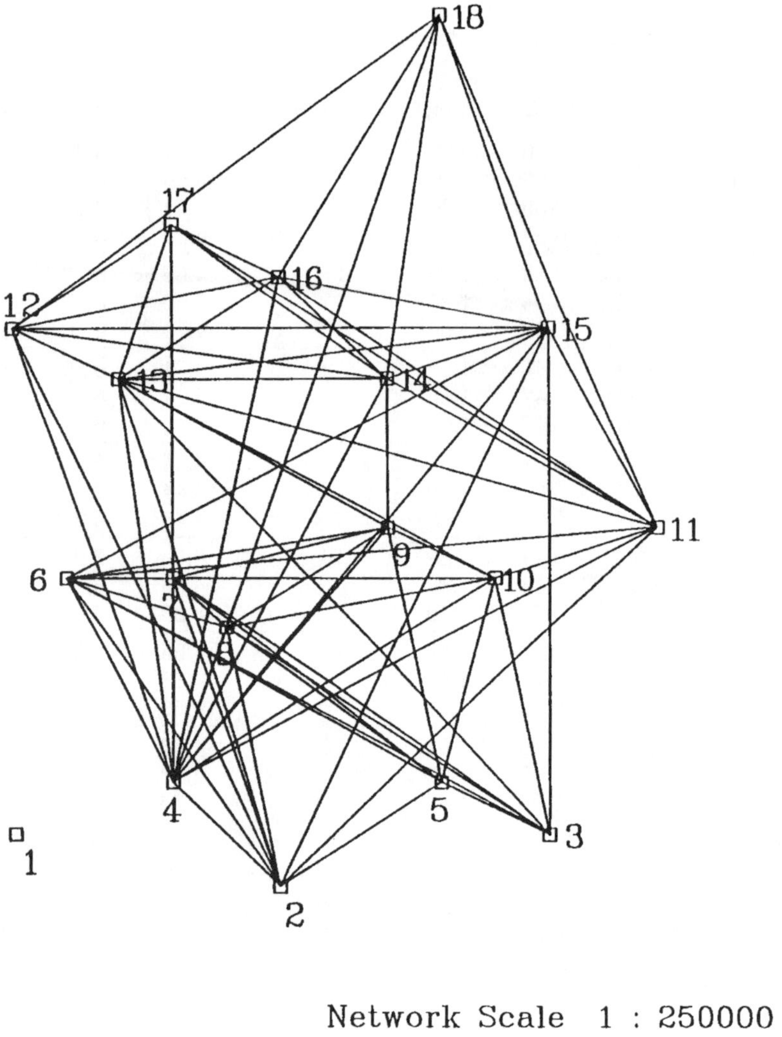

Figure 12.19. The deleted baselines by optimization (total: 73)

types of observations together with the distribution of observational work among them; i.e., the precisions of the measurements to be made. This leads to the need for the optimal design of geodetic networks. It allows for the geodesists to know, before any geodetic measurement campaign is started, about the result of their work according to the set objectives, and therefore to prevent the measurement campaigns from failing with reasonably low cost. The results of optimization enable us to make decisions on which instruments should be selected from the hundreds of available models of various geodetic instruments and where they should be located

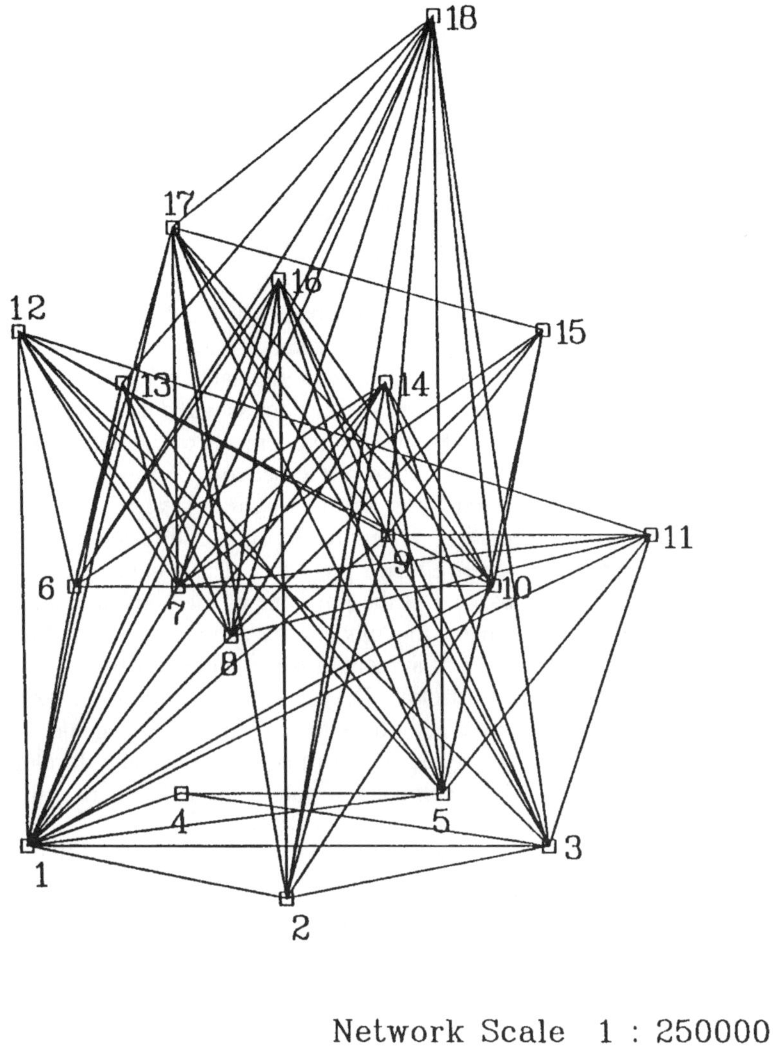

Figure 12.20. The optimized observing scheme (80 baselines)

in order to estimate the unknown parameters and achieve the desired criteria derived from and determined by the purpose of the network. Of interest to surveying engineers is that it enables surveyors to avoid unnecessary observations, and therefore may result in saving considerable time and effort in the field.

Having recognized the benefits of network optimization, geodesists have taken a century to find the most appropriate methods for its solution, and the recently developed methodology (Kuang, 1991), which has been discussed in Chapters 10 and 11, has proved to be the most efficient method up to date. One can now obtain a fully analytical solution for both the optimum network configuration and the optimum observing plan, and the different network quality criteria of precision, reliability, sensitivity, and economy can be considered simultaneously. Based on the theory of the multi-objective optimization, a suitable target function that includes different types of design criteria can be formulated under a common scale. The possibly contradictory requirements of different criteria are, therefore, balanced. The resulting optimization model is solved by linear or quadratic programming that allows for an easy implementation of the methodology and proves its practical significance.

The presented mathematical models in Chapters 10 and 11 deal with the simultaneous optimization of the network configuration and observational weights. For practical applications, after appropriate network quality criteria are given, this optimization procedure gives the optimal weights or standard deviations for each observable as well as the optimal position improvements of the initially selected points to obtain an optimal configuration of the network. All the aspects of the conventional First-Order, Second-Order and Third-Order Designs are, therefore, embedded in a single mathematical procedure. Separate optimization of the configuration and weights can be performed simply by deleting unnecessary parameters from the model. For instance, if the positions of the netpoints selected in reconnaissance cannot be changed, this model reduces to the Second-Order Design. On the other hand, if the measurement instrumentation for this job is fixed, then the model reduces to the First-Order Design. And finally, if position of some points and weights of some observables have to be optimized, this model reduces to the Third-Order Design. Therefore, the optimization procedure developed is very flexible and practical. All the solutions are automatic, and it removes the need for the method of "trial and error."

The network configuration is optimized by introducing coordinate shifts to the initially given approximate coordinates, which may be determined by reconnaissance in the field, of network points. This is practical, especially for engineering networks, since the general locations of the network points are fixed by the topography of the network area. Changes of the locations of network points can be made to a limited extent. The boundary conditions on the coordinate shifts to be optimized are dictated by the topography of the network area and/or other preferences. The optimization procedure starts with a set of approximate observational weights; whatever approximate values for observation weights are used, the optimization procedure will give practically the same results. The optimally solved for observational weights are then used as a guideline for the choice of instrumentation and field operational procedures. Observables that obtained zero or insignificant weights should not be observed. One can calculate the minimum observational precision required for each observable that obtained significant optimal weights, and may be different for different observables. During the execution of actual surveys, however, one may, for the purpose of convenience, use the same precision for a certain type of observables (e.g., spatial distances). In this case, one should carry out the actual survey using the maximum observational precision required for this type of observable in order to meet the preset network quality criteria on the safe side. For angular measurements, e.g., directions, zenith distances, and azimuth, it is practical to carry out the actual survey according to the required observational precision as obtained from the optimization since the lower the precision required, the fewer the observational sets one has to measure; i.e., one can calculate the necessary number of observational sets from the required observational precision for a certain instrumentation chosen. Although the methodology has been induced by the concept of introducing relatively small position changes, relatively large changes can actually be accommodated simply by increasing the number of iterations in the solution process as discussed in Kuang (1991).

References

Alberda, J.E. (1974). "Planning and optimization of networks: Some general considerations." *Boll, Geod. Sci. Aff.* No. 33, pp. 209–240.

Alberda, J.E. (1980). A review of analysis techniques for engineering survey control schemes. Paper presented at the Industrial and Engineering Survey Conferences, London, UK.

Ashtech (1991). Ashtech XII GPPS GPS Multi-Site Mission Planning System. Ashtech. 1170 Kifer Road, Sunnyvale, CA 94086, U.S.A.

Asplund, L. (1963). "Specifications for fundamental geodetic networks." *Travaux de L'Association Internationale de Geodesie,* 21:53–57.

Ashkenazi, V. (1965). "Strength of a triangulation layout." *Bulletin Geodesique,* 76:125–134.

Baarda, W. (1962). "A generalization of the concept Strength of Figure." Publ. of the Computing Center of the Geodetic Inst., Delft.

Baarda, W. (1967). "Statistical concepts in geodesy." *Neth. Geod. Com., publ. on Geodesy,* New Series 2, No. 4, Delft, Netherlands.

Baarda, W. (1968). "A testing procedure for use in geodetic networks." *Neth. Geod. Com., publ. on Geodesy,* New Series 2, No. 5, Delft, Netherlands.

Baarda, W. (1973). "S-Transformations and Criterion matrices." *Neth. Geod. Com., publ. on Geodesy,* New Series 5, No. 1, Delft, Netherlands.

Baarda, W. (1976). Reliability and precision of networks. VII International Course for Engineering Surveys of High Precision, Darmstadt, pp. 17–27.

Baran, W. (1982). Some new procedures of sequential adjustment. Int. symp. on geodetic networks and computations (1981), *Deutsche Geodätische Kommission,* Munchen, VIII: 69–79.

Blazs, E.I. and G.M. Young (1982). Corrections applied by the National Geodetic Survey to precise leveling observations. NOAA Technical Memorandum NOS NGS 34, U.S. Department of Commerce, Rockville, MD.

Boedecker, G. (1977). The development of an observation scheme from the variance covariance matrix of the unknowns. *Proc. Int. Symposium on optimization of design and computation of control networks,* Sopron.

Bomford, G. (1980). *Geodesy.* Clarendon Press, Oxford, 855 pp.

Boot, J.C.G. (1964). *"Quadratic Programming,"* North-Holland, Amsterdam, 1964.

Bossler, J., E. Grafarend, and R. Kelm (1973). "Optimal design of geodetic nets 2." *Journal of Geophysical Research,* Vol. 78, No. 26.

Caspary, W.F. (1987). Concepts of network and deformation analysis. Monograph 11, School of Surveying, The University of New South Wales, Kensington, N.S.W., Australia.

Chen, Y.Q. (1983). Analysis of Deformation Surveys—A generalized method. Technical Report No. 94, Dept. of Surveying Engineering, University of New Brunswick, Canada.

Chen, Y.Q., M. Kavouras, and J.M. Secord (1983). Design considerations in deformation monitoring. FIG XVII International Congress, Sofia, Bulgaria, Paper No. 608.2.

Chen, Y.Q. and A. Chrzanowski (1986). An overview of the physical interpretation of deformation measurements, at the Deformation Measurements Workshop, M.I.T., Cambridge, Mass. 31 October–1 November 1986.

Chen Y.Q., A. Chrzanowski, and M. Kavouras (1990). "Assessment of observations using Minimum Norm Quadratic Unbiased Estimation (MINQUE)." CISM Journal ACSGS, Vol. 44, No. 4, pp. 39–46.

Chrzanowski, A. (1977). Design and error analysis of surveying projects. Lecture Notes No. 47, Dept. of Surveying Engineering, University of New Brunswick, Canada.

Chrzanowski, A. (1981). A comparison of different approaches into the analysis of deformation measurements. Proc. of FIG XVI International Congress, Montreux, Paper No. 602.3, August 9–18.

Chrzanowski, A. (1983). Economization of vertical control surveys in hilly areas by using modified trigonometric levelling. Presented at the 1983 ACSM-ASP Convention, Washington, D.C., March.

Chrzanowski, A. (1986). Geotechnical and other non-geodetic methods in deformation measurements at the Deformation Measurements Workshop, M.I.T., Cambridge, Mass. 31 October–1 November 1986.

Chrzanowski, A., Y.Q. Chen, J.M. Secord (1983). "On the strain analysis of tectonic movements using fault crossing geodetic surveys." *Tectonophysics,* 97, pp. 295–315.

Chrzanowski, A. and J.M. Secord (1983). Report of the 'ad hoc' Committee on the Analysis of Deformation Surveys. Presented at the FIG-17th International Congress, Paper 605.2, Sofia, June 19–28.

Chrzanowski, A., Y.Q. Chen, P. Romero, and J.M. Secord (1986). "Integration of geodetic and geotechnical deformation surveys in geoscience." *Tectonophysics,* 130, pp. 369–383.

Chrzanowski, A., Y.Q. Chen, J.M. Secord, G.A. Thompson, and Z. Wroblewicz (1988). Integration of geotechnical and geodetic observations in the geometrical analysis of deformations at the Mactaquac Generating Station. Conference 1988 on deformation surveys, Fredericton, Canada.

Conzett, R., A. Frank, and C. Misslin (1980). Interactive triangulation. VIII Int. Kurs für Ingenieurermessung, Zürich.

Cooper, M.A.R. (1982). Modern theodelites and levels. 2nd edition. London: Granada.

Cooper, M.A.R. (1987). Control surveys in civil engineering. Nichols Publishing Company.

Cottle, R.W. and G.B. Dantzig (1968). "Complementary pivot theory of mathematical programming." *Linear Algebra Application* No. 1 (1968), 103–125.

Crosilla, F. (1982). "Geodetic network optimization for the detection of crustal movements using a mekometer." *Bollettino di Geodesia e Sc. Affini.*

Crosilla, F. (1983a). "A criterion matrix for the second order design of control networks." *Bulletin Geodesique* 57:226–239.

Crosilla, F. (1983b). "Procrustean transformation as a tool for the construction of a criterion matrix for control networks." *Manuscripta Geodetica* 8: 343–370.

Crosilla, F. (1985). "A criterion matrix for deforming networks by multifactorial analysis" in *Optimization and design of geodetic networks* edited by Grafarend and Sanso. Springer: Berlin etc. pp. 429–435.

Cross, P.A. and B.M. Whiting (1980). The design of national vertical control networks. *Proc. 2nd Int. Symposium On Problems Related to the Redefinition of North American Vertical Geodetic Networks,* Ottawa, Canada.

Cross, P.A. and K. Thapa (1979). "The optimal design of levelling networks." *Survey Review,* vol. XXV, No. 192, pp. 68–79.

Cross, P.A. and B.M. Whiting (1981). On the design of vertical geodetic control networks by iterative methods. VI International Symposium On Geodetic Computations, Munich.

Cross, P.A. (1981). The geodetic suite. Working paper 4, North East London Polytechnic, Department of Land Surveying, 92 pp.

Cross, P. and A. Fagir (1982). Procedures for the First and Second Order Design of vertical control networks. *Proc. Survey Control Networks,* Heft 7, Munchen.

Cross, P.A. and A. Fagir (1983). "On the application of linear programming to the design of one and two dimensional geodetic networks." Proceedings of the FIG XVII International Congress, Sofia, Bulgaria, June 19–28, 1983, pp. 506.2/1–506.2/13.

Cross, P.A. (1985). "Numerical methods in network design." in *Optimization and design of geodetic networks* edited by Grafarend and Sanso. Springer: Berlin etc. pp. 429–435.

Dantzig, G.B. (1963). *Linear programming and extensions.* Princeton University Press, Princeton, NJ, 1963.

Delikaraoglou, D. (1985). "Estimability analyses of the free networks of differential range observations to GPS satellites." In *Optimization and design of geodetic networks,* edited by E.W. Grafarend and F. Sanso. Springer: Berlin etc., pp. 196–220.

Delikaraoglou, D. and F. Lahaye (1989). "Optimization of GPS Theory, Techniques and Operational Systems: Progress and Prospects." Global Positioning System: An Overview. International Association of Geodesy Symposia, symposium 102, Springer-Verlag, pp. 218–239.

Dermanis, A. (1985). "Optimization problems in geodetic networks with signals," in *Optimization and design of geodetic networks* edited by Grafarend and Sanso. Springer: Berlin etc. pp. 221–256.

Dinkelbach, W. and H. Isermann (1973). "On decision making under multiple criteria and under incomplete information," in *Multiple Criteria Decision Making,* J.L. Cochrane and M. Zeleny, (eds.), University of South Carolina Press, Columbia, SC.

Dunnicliff, J. (1988). *Geotechnical instrumentation for monitoring field performance.* John Wiley & Sons, New York/Chichester/Brisbane/Toronto/Singapore.

Dyer, J.S. (1972). "Interactive goal programming." *Management Science,* No. 19, pp. 62–70.

Faig, W. (1978). The utilization of photogrammetry of deformations of structural parts in the shipping building industry. *Proc. of the II international Symposium of deformation measurements by geodetic methods,* Bonn, West Germany, Konrad Wittwer, Stuttgart.

Faig, W. and C. Armenakis (1982). Subsidence monitoring by photogrammetry. *Proceedings of the Fourth Canadian Symposium On Mining surveying and deformation measurements,* Bannf, Alberta, June, The Canadian Institute of Surveying, 197–208.

FGCS (Federal Geodetic Control Subcommittee) (1984). "Standards and Specifications for Geodetic Control Networks."

Förstner, W. (1986). "Reliability, error detection, and self calibration." Program of the ISPRS Commission III Tutorial on Statistical Concepts for Quality Control.

Frank, C.F. (1966). "Deduction of earth strain from survey data." *Bulletin Seismological Society of America,* 56 (1):35–42.

Frank, A. and C. Misslin (1980). INTRA: a programme for interactive design and adjustment of geodetic networks. Institute für Geodasie und Photogrammetrie, Eidgenssische Technische Hochschule Zürich, Bericht 43, 11 pp.

Gokalp, E. (1991). Evaluation of GPS measurements by MINQUE and integration of GPS and levelling in subsidence monitoring. M.Eng. Report, Department of Surveying Engineering, University of New Brunswick, Fredericton, N.B., Canada.

Grafarend, E. (1972). "Genauigkeitsmasse geodatischer Netze." *Deutsche Geodätische Kommission*-publ. A-73, Munchen.

Grafarend, E.W. (1974). "Optimization of geodetic networks." *Bolletino di Geodesia a Science Affini,* Vol. 33, No. 4, pp. 351–406.

Grafarend, E.W. (1985). "Criterion matrices for deforming networks," in *Optimization and design of geodetic networks* edited by Grafarend and Sanso. Springer: Berlin etc. pp. 363–428.

Grafarend, E.W. and B. Schaffrin (1979). "Kriterion-matrizen I-zweidimensional homogene und isotrope geodatische netze." *Zeitschrift Für Vermessungswesen,* No. 4, pp. 133–149.

Grafarend, E.W., H. Heister, R. Kelm, H. Kropff, and B. Schaffrin (1979). *Optimierung geodetischer messoperationen.* Karlsruhe.

Grafarend, E.W., A. Kleusberg, and B. Richter (1979). "Free Doppler network adjustment." Proceedings of the Second International Geodetic Symposium on Satellite Doppler Positioning, Austin, Texas, Vol. II, pp. 1053–1069.

Grafarend E., A. Kleusberg, and B. Schaffrin (1980). "An introduction to the variance-covariance components estimation of Helmert type." ZFV, 105, pp. 129–137.

Grafarend, E.W., F. Krumm, and B. Schaffrin (1986). "Kriterion-matrizen III—zweidimensional homogene und isotrope geodatische netze." *Zeitschrift Für Vermessungswesen,* No. 5, pp. 197–207.

Grafarend, E.W., W. Lindlohr, and A. Stomma (1985). "Improved second order design of the Global Positioning System—ephemeris, clocks and atmospheric influences." Proceedings of the First International Symposium on Precise Positioning with the Global Positioning System, Rockville, MD, April, Vol. I., pp. 273–284.

Hanson, Richard J., R. Lehoucq, J. Stolle, and A. Belmonte (1990). Improved performance of certain matrix eigenvalue computations for the IMSL/MATH Library, IMSL Technical Report 9007, IMSL, Houston.

Heiskanen, W.A. and H. Moritz (1979). *Physical Geodesy.* Institute of Physical Geodesy, Technical University, Graz, Austria, 364 pp.

Helmert, F.R. (1868). "Studien über rationelle Vermessungen im Gebiet der höheren Geodäsie." *Z. Math. Phys. Schlömilch,* Vol. 13 (1868), pp. 73–120, 163–168.

Henderson, H.V. and S.R. Searle (1979). "Vec and Vech operators for matrices with some uses in Jacobians and multivariate statistics." *Canadian Journal of statistics* 7, 65–81.

Hugget, G.R. (1982). A Terrameter for deformation measurements. *Proceedings of the Fourth Canadian Symposium on Mining Surveying and Deformation measurements,* Banff, Alberta, June 7–9, The Canadian Institute of Surveying, 179.

Illner, M. (1986). "Anlage und Optimierung von Verdichtungsnetzen." *Deutsche Geodätische Kommission,* C-317, Munich.

Illner, M. (1988). "Different models for the sequential optimization of geodetic networks." *Manuscripta Geodaetica* 13 (1988), 306–315.

Jäger, R. (1988). "Analyse und Optimierung geodatischer Netze nach spectralen Kriterien und mechanische Analogien." *Deutsche Geodätische Kommission,* C-342, Munich, 1988.

Jäger, R. (1989). "Optimum positions for GPS-points and supporting fixed points in geodetic networks." Proc. IAG General Meeting (Edinburgh, 1989), Springer Lect. Notes, 1990.

Jäger, R. and H. Kaltenbach (1990). "Spectral analysis and optimization of geodetic networks based on eigenvalues and eigenfunctions." *Manuscripta Geodaetica* 15 (1990), 302–311.

Jäger, R. and M. Vogel (1990a). Theoretisches Konzept zum Design 1., 2., 3., and 0. Ordnung mittels analytischer Differentiation der Kovarianzmatrix und spektralem Ansatz für Kriteriummatrix-Zielfunktionen. *Zeitschrift für Vermessungswesen,* 10:425–435.

Jäger, R. and M. Vogel (1990b). Kriteriummatrix-orientierte Netzoptimierung im Design 0. bis 3. Ordnung—Das Programmsystem CADSOC und einige Optimierungsstudien zu unterschiedlichen Netztypen und Designs. *Zeitschrift für Vermessungswesen,* 11:473–481.

Janes H., K. Doucet, B. Roy, D.E. Wells, R. Langley, P. Vanicek, and M. Craymer (1986). "GPSNET: A program for the interactive design of geodetic networks." Geodetic Survey of Canada CR 86-003.

Jung, I. (1924). "Uber die gunstigste Gewischitsverteilung in Basisnetzen." *Akadem. Abh.* Uppsala 1924.

Kahmen, H. and W. Faig (1988). Surveying. Walter de Gruyter & Co.

Karadaidis, D. (1984). "A substitute matrix for photogrammetrically determined point fields." *ITC-Journal,* 1984–3.

Karman, T. (1937). "On the statistical theory of turbulence." *Proc. Natl. Acad. Sci.* USA, 23.

Kavouras, M. (1982). On the detection of outliers and the determination of reliability in geodetic networks. Technical report No. 87, Department of Surveying Engineering, University of New Brunswick, Fredericton, Canada.

Koch, K.R. (1982). "Optimization of the configuration of the geodetic networks." *Deutsche Geodätische Kommission,* B, 258/III, 82-89, Munchen, 1982.

Koch, K.R. (1985). "First Order Design: Optimization of the configuration of a network by introducing small position changes," in *Optimization and design of geodetic networks* edited by Grafarend and Sanso. Springer: Berlin etc. pp. 56–73.

Koch, K.R. (1987). "Parameter estimation and hypothesis testing in linear models." Springer-Verlag.

Koch, K.R. and D. Fritsch (1981). "Multivariate hypothesis tests for detection of crustal movements." *Tectonophysics,* 71, pp. 301–313.

Krakiwsky, E.J. and D.E. Wells (1971). Coordinate systems in geodesy. Lecture notes No. 16, Department of Surveying Engineering, University of New Brunswick, Fredericton, Canada.

Kuang, S.L. (1994a). "A Strategy for GPS Survey Planning: Choice of Optimum Baselines." Proceedings of the 1994 ASPRS/ACSM Annual Convention & Exposition, Technical Papers, Vol. 2, pp. 42–58, Reno, Nevada, April 25–28.

Kuang, S.L. (1994b). Introducing a new computer program for processing and analysis of geodetic leveling networks.

Kuang, S.L. and A. Chrzanowski (1994). "Optimization of Integrated Survey Scheme for Deformation Monitoring." *Geomatica* Vol. 48, No. 1, Winter 1994, pp. 245–259.

Kuang, S.L. (1993a). "On Optimal Design of Levelling Networks." *The Australian Surveyor.* Vol. 38, No. 4, pp. 257–273.

Kuang, S.L. (1993b). "Optimization of Horizontal Direction Observations in Triangulation or Triangulateration Networks." *Journal of Surveying Engineering,* Vol. 119, No. 4, pp. 156–173.

Kuang, S.L. (1993c). "Quality Control of GPS Surveys for Tunnelling: Precision and Reliability Aspects." *Zeitschrift für Vermessungswesen.* Vol. 118, No. 7, pp. 329–345.

Kuang, S.L. (1993d). "Evaluating the accuracy of GPS baseline measurements using Minimum Norm Quadratic Unbiased Estimation (MINQUE)." *Surveying and Land Information Systems,* Vol. 53, No. 2, pp. 103–110.

Kuang, S.L. (1993e). "Second Order Design: Shooting for Maximum Reliability." *Journal of Surveying Engineering.* Vol. 119, No. 3, pp. 102–110.

Kuang, S.L. (1993f). "On Optimal Design of Three-Dimensional Engineering Networks." *Manuscripta Geodaetica,* Vol. 18, No. 1, pp. 33–45.

Kuang, S.L. (1993g). "A methodology for the accuracy analysis of the Finite Element Computations applied to structural deformation studies." *Allgemeine Vermessungs-Nachrichten,* International Edition, Vol. 10, pp. 1–14.

Kuang, S.L. (1993h). "Bi-objective Optimal design of deformation monitoring schemes." *The Australian Surveyor* Vol. 38, No. 2, pp. 106–119.

Kuang, S.L. (1992a). "Analytical Models of the Optimal First Order Design of Geodetic Networks." *Zeitschrift für Vermessungswesen* Vol. 117, No. 3, pp. 145–155.

Kuang, S.L. (1992b). "A New Approach to the Optimal Second Order Design of Geodetic Networks." *Survey Review,* Vol. 31, No. 243, pp. 279–288.

Kuang, S.L. (1992c). "Improvement of the reliability of geodetic networks by modifying the configuration." *Bollettino di Geodesia e Scienze Affini.* No. 3, pp. 215–226.

Kuang, S.L. (1992d). "Geodetic Network Optimization—A Handy Tool for Surveying Engineers." *Surveying and Land Information Systems.* Vol. 52, No. 3, pp. 169–183.

Kuang, S.L. (1992e). "A Study of the Choice of Norm Problem for Network Optimization." *Allgemeine Vermessungs-Nachrichten,* International Edition, Vol. 9, pp. 22–32.

Kuang, S.L. (1992f). "Theory of Single-objective and Multi-objective optimization design of Integrated Deformation Monitoring Schemes with Geodetic and Non-geodetic Observables." *Bollettino di Geodesia e Scienze Affini.* No. 4, pp. 335–369.

Kuang, S.L. and A. Chrzanowski (1992a). "Multi-objective optimization design of Geodetic Networks." *Manuscripta Geodaetica.* Vol. 17, No. 4, pp. 233–244.

Kuang, S.L. and A. Chrzanowski. (1992b). "Rigourous Combined First Order and Second Order Optimal Design of Geodetic Networks." *Bollettino di Geodesia e Scienze Affini.* No. 2, pp. 141–156.

Kuang, S.L. (1991). "Optimization and Design of Deformation Monitoring Schemes." *Ph.D. dissertation.* Dept. of Surveying Engineering Technical Report No. 157, University of New Brunswick, P.O. Box 4400, Fredericton, Canada, July, 1991, 179 pp.

Kuang, S.L., A. Chrzanowski, and Y.Q. Chen (1991). "A unified Mathematical Modelling for the Optimal Design of Monitoring Networks." *Manuscripta Geodaetica.* Vol. 16, No. 6, pp. 376–383.

Kuang, S.L. (1989). "Using Minimax Principle to Estimate Surveying Parameters and Perform Accuracy Analysis." *Journal of Hohai University,* (in Chinese with English abstract) Vol. 17, No. 3, pp. 16–23.

Kuang, S.L. (1988). "Study of Outlier Detection and Position Estimation for Point Positioning in Marine Surveying." *Journal of Wuhan Technical University of Surveying and Mapping (WTUSM),* (in Chinese with English abstract), Vol. 12, No. 4, pp. 85–97.

Kuang, S.L. (1987). "Application of Kalman Filtering to the Data Post-processing for Marine Geophysical Exploration in East Sea of China." Paper presented at the National Conference on Marine Geophysical Exploration (in Chinese), Tianjing, P.R. China, July, 1987.

Kuang, S.L. (1986). "Study on the Post-Analysis of Positioning Data for Marine Geophysical Exploration." *M.Sc.E thesis,* Wuhan Technical University of Surveying and Mapping (WTUSM) (in Chinese), Wuhan, P.R. China, April, 1986, 128 pp.

Leick, A. (1990). *GPS Satellite Surveying.* John Wiley & Sons, New York/Chichester/Brisbane/Toronto/Singapore, 352 pp.

Lichten, M.S. (1989). "High Accuracy Global Positioning System Orbit Determination: Progress and Prospects." Global Positioning System: An Overview. International Association of Geodesy Symposia, symposium 102, Springer-Verlag, pp. 146–164.

Liew, C.K. and J.K. Shim (1978). "Computer program for inequality constrained least squares estimation." *Econometrica,* vol. 46, pp. 237.

Lindlohr, W. and D. Wells (1985). "GPS design using undifferenced carrier phase observations." *Manuscripta Geodaetica*, Vol. 10, No. 4, pp. 255–295.

Maling, D.H. (1973) Coordinate systems and map projections. Ebenezer Baylis & Son Limited, The Trinity Press, Worcester, and London.

Meissl, P. (1976). " Strength analysis of two-dimensional Angular Anblock networks." *Manuscripta Geodetica,* Vol. 1, No. 4.

Mepham, M.P.A., A.P. Mackenzie, and E.J. Krakiwsky (1982). Design engineering and mining surveys using Interactive Computer Graphics. Presented at the 4th Canadian Symposium on Mining Surveying and Deformation Measurements, Banff, Canada.

Mepham, M.P.A. and E.J. Krakiwsky (1983). Description of CANDSN: a computer aided survey network design and adjustment system. Presented at the 76th Annual meeting of the Canadian Institute of Surveying, Victoria, Canada.

Merminod, B., D.B. Grant, and C. Rizos (1990). "Planning GPS Surveys—Using Appropriate Precision Indicators." *CISM Journal ACSGC* Vol. 44, No. 3, pp. 233–249.

Mierlo, J. Van (1981). "Second Order Design: precision and reliability aspects." *Algemeine Vermessungs-Nachrichten,* pp. 95–101.

Mittermayer, E. (1972). "Adjustment of free networks." *Zeitschrift Für Vermessungswesen,* No. 11, pp. 481–489.

Molenaar, M. (1981). "A further inquiry into the theory of s-transformations and criterion matrices." *Netherlands Geodetic commission, Publications on geodesy,* New series, Vol. 7, No. 1, Delft 1981.

Molenaar, M. (1986a). "Statistical fundamentals and estimation theory, hypotheses testing, and decision theory." Program of the ISPRS Commission III Tutorial on Statistical Concepts for Quality Control.

Molenaar, M. (1986b). "Criterion matrices, S-system, and precision evaluation." Program of the ISPRS Commission III Tutorial on Statistical Concepts for Quality Control.

Moniwa, H. (1977). "Analytical Photogrammetric System with Self Calibration and its Applications," Ph.D. Dissertation, Department of Surveying Engineering, U.N.B., Fredericton, Canada.

Müller, H. (1986). Zur Berucksichtigung der Zuverlassigkeit bei der Gewichtsoptimierung geodatischer Netze." *Zeitschrift Für Vermessungswesen* 111 (1986), 157–169.

Mueller, I. Ivan (1977). *Spherical and Practical Astronomy as applied to Geodesy.* Second edition, Frederick Ungar Publishing Co. New York.

Nickerson, B.G. (1978). A Priori Estimation of Variance for Surveying Observables. Technical Report No. 57, Dept. of Surveying Engineering, University of New Brunswick, Fredericton, Canada.

Nickerson, B.G. (1979). Horizontal network design using computer graphics. M.sc. Eng. Thesis, Dept. of Surveying Engineering, University of New Brunswick, Fredericton, Canada.

Niemeier, W. and G. Rohde (1981). On the optimization of levelling nets with respect to the determination of crust movements. IAG Symposium on Geodetic networks and computations, Munich.

Niemeier, W. (1982). Principal component analysis and geodetic networks-some basic considerations. *Proc. of the meeting of FIG study Group 5B.*

Pearson, E.S. and H.O. Hartley (1951). "Charts of the power function of the analysis of variance tests, derived from the non-central F-distribution." *Biometrika,* Vol. 38, pp. 112–130.

Pelzer, H. (1971). Zur Analyse Geodatischer Deformationsmessungen, *Deutsche Geodätische Kommission,* C 164, Munchen.

Pope, A.J. (1976). The statistics of residuals and the detection of outliers. NOAA Technical Report NOS 65 NGS 1, U.S. Department of Commerce, Rockville, MD.

Ramsay, J.G. (1967). *Folding and Fracturing of Rocks.* McGraw-Hill, New York, 568 pp.

Rao, C.R. (1970). "Estimation of heterogeneous variances in linear models." *Journal of American Statistics Association,* 65, pp. 161–172.

Rao, C.R. (1971). "Estimation of variance and covariance components MINQUE theory." *Journal of Multivariate Analysis,* 65, pp. 161–172.

Rao, C.R. (1973). *Linear statistical Inference and its applications.* John Wiley & Sons, New York.

Rao, C.R. (1979). "MINQUE theory and its relation to ML and MML estimation of variance components." Sankhya, 41, Series B, pp. 138–153.

Richardus, P. (1984). Project surveying. Second Edition, A.A. Balkema/Rotterdam/ Boston.

Rüeger, J.M. (1980). *Introduction to Electronic Distance Measurement.* Second Edition, School of Surveying, The University of New South Wales, Australia.

Sawaragi, Y., H. Nakayama, and T. Tanino (1985). *Theory of multi-objective optimization.* Academic Press, Inc. Orlando, San Diego, New York, London, Toronto, Montreal, Sydney, Tokyo.

Schaffrin, B. (1977). A study of the Second Order Design problem in geodetic networks. Proceedings of the International Symposium on Optimization of Design and Computation of control Networks, Sopron, 1977, pp. 175–177.

Schaffrin, B., E.W. Grafarend, and G. Schmitt (1977). "Kanonisches Design Geodatischer Netze I." *Manuscripta Geodaetica* 2 (1977) 263–306.

Schaffrin, B., F. Krumm, and D. Fritsch (1980). Positive diagonale genauigkeitsoptimierung von realnetzen uber den komplementaritats-algorithms. VIII International Kurs Fur Ingenieurvermessung, Zurich.

Schaffrin, B. (1981). "Some proposals concerning the diagonal Second Order Design of geodetic networks." *Manuscripta Geodetica,* Vol. 6, No. 3, p. 303–326.

Schaffrin, B. and E.W. Grafarend (1982). "Kriterion-Matrizen II-Zweidimensionale homogene und isotrope geodatische Netze." *Zeitschrift Für Vermessungswesen,* Tei IIa, No. 5, pp. 183–194, Tei IIb, No. 11, pp. 485–493, 1982.

Schaffrin, B. (1983). "On some recent modifications regarding the optimal design of geodetic networks." *Geodezja* (Krakow) 79, 43–56.

Schaffrin, B. (1983). Varianz-Kovarianz-Komponenten-Schatzung bei der Ausgleichung heterogener Wiederholungsmessungen, Publ. DGK C-282.

Schaffrin, B. (1985a). "On design problems in geodesy using models with prior information." *Statistics and Decision,* Suppl. Issue 2 (1985), 443–453.

Schaffrin, B. (1985b). "Aspects of network design," in *Optimization and design of geodetic networks* edited by Grafarend and Sanso. Springer: Berlin etc. pp. 548–597.

Schaffrin, B. (1986). "New estimation/prediction techniques for the determination of crustal deformations in the presence of geophysical prior information." *Tectonophysics* 130 (1986), 361–367.

Schaffrin, B. and E. Grafarend (1986). "Generating classes of equivalent linear models by nuisance parameter elimination—Application to GPS observations." *Manuscripta Geodaetica*, Vol. 11, No. 4, pp. 262–271.

Schaffrin, B. (1988). Tests for random effects based on homogeneously linear predictors, in H. Lauter (ed.), Proc. of the Intl. Workshop on "Theory and Practice in Data Analysis," Acad. of Sci. of GDR, Report No. R-MATH-01/89, Berlin (East), pp. 209–227.

Schmitt, G. (1977). Experience with the second order design of geodetic networks. *Proc. International Symposium of Design and Computation of Control Networks,* Sopron, Hungary.

Schmitt, G. (1978). Numerical problems concerning the second order design of geodetic networks. Second International Symposium Related to the Redefinition of North American Geodetic Networks, Washington.

Schmitt, G. (1980). "Second Order Design of free distance networks considering different types of criterion matrices." *Bulletin Geodesique,* 54 pp. 531–543.

Schmitt, G. (1982). "Optimal design of geodetic networks." *Deutsche Geodätische Kommission,* B, 258/III, 82–89, Munchen, 1982.

Schmitt, G. (1985). "A review of network designs, criteria, risk functions and design ordering," in Optimization and design of geodetic networks edited by Grafarend and Sanso. Springer: Berlin etc. pp. 6–10.

Schmitt, G. (1985). "Second Order Design," in *Optimization and design of geodetic networks* edited by Grafarend and Sanso. Springer: Berlin etc. pp. 74–121.

Schmitt, G. (1985). "Third Order Design," in *Optimization and design of geodetic networks* edited by Grafarend and Sanso. Springer: Berlin etc. pp. 122–131.

Schönemann, P.H. (1966). "The generalized solution of the orthogonal procrustes problem." *Psychometrica* 31:1–16.

Schreiber, O. (1882). "Anordnung der winkelbeobachtung im Gottinger Basisnetz." *Zeitschrift Für Vermessungswesen* 11, 129–161.

Searle, S.R. (1982). Matrix algebra useful for statistics. John Wiley & Sons.

Secord, J. M. (1985). Implementation of a generalized method for the analysis of deformation surveys. Technical Report No. 117, Dept. of Surveying Engineering, University of New Brunswick, Fredericton, N.B., Canada.

Secord, J.M. (1986). Terrestrial survey methods for precision deformation measurements. At the Deformation Measurements Workshop, M.I.T., Cambridge, Mass. 31 October–1 November 1986.

Sokolnikoff, I.S. (1956). "Mathematical Theory of Elasticity." McGraw-Hill, New York.

Sprinsky, W.H. (1978). "Improvement of parameter accuracy by choice and quality of observation." *Bulletin Geodesique* 52, 269–279.

Steeves, P.A. (1984). Numerical processing of horizontal control data: Economization by automation. Ph.D. dissertation, Department of Surveying Engineering, University of New Brunswick, Fredericton, Canada.

Stem, J.E. (1991). *State Plane Coordinate System of 1983.* NOAA Manual NOS NGS 5, National Geodetic Survey, Rockville, MD.

Szostak-Chrzanowski, A., A. Chrzanowski, and S.L. Kuang (1993). "Propagation of Random Errors in Finite Element Analysis." Presented at the 1st Canadian Symposium on Numerical Modelling Applications in Mining and Geomechanics. Montreal, Quebec, March 27–30, 1993.

Szostak-Chrzanowski, A., A. Chrzanowski, S.L. Kuang, and A. Lambert (1992). "Finite Element Modelling of Tectonic Movements in Western Canada." Presented at the 6th International (FIG) Symposium on Deformation Measurements. Hanover, Germany, Feb. 24–28, 1992.

Szostak-Chrzanowski, A., A. Chrzanowski, and S.L. Kuang (1991). "Software 'FEMMA' for Modelling Tectonic Movements." Presented at the American Geophysical Union Chapman Conference on Time Dependent Positioning. Annapolis, Maryland, 22–25 Sept., 1991.

Taylor, G.I. (1935). "Statistical theory of turbulence." Proc. R. Soc. London, Ser. A, 151–421.

Teskey, W.F. (1979). Geodetic Aspects of Engineering Surveys Requiring High Accuracy. Department of Surveying Engineering Technical Report #65. University of New Brunswick, Fredericton, Canada, September, 1979.

Teskey, W.F. (1987). Integrated analysis of geodetic, geotechnical, and physical model data to describe the actual deformation behaviour of earthfill dams under static loading. Institute für Anwendungen der geodäsie im Bauwesen. Universität Stuttgart, May, 1987.

Teunissen, P.J.G. (1985). "Zero Order Design: Generalized inverses, Adjustment, the Datum problem and S-Transformations" in *Optimization and design of geodetic networks* edited by Grafarend and Sanso. Springer: Berlin etc. pp. 11–55.

Teunissen, P.J.G. (1985). "Quality control in geodetic networks," in *Optimization and design of geodetic networks* edited by Grafarend and Sanso. Springer: Berlin etc. pp. 526–547.

Thomson, B.D. (1976). Combination of geodetic networks. Technical report No. 30, Department of Surveying Engineering, University of New Brunswick, Fredericton, Canada.

Thomson, B.D., E.J. Krakiwsy, and J.R. Adams (1978). A manual for geodetic position computations in the maritime provinces. Technical report No. 52, Department of Surveying Engineering, University of New Brunswick, Fredericton, Canada.

Thomson, B.D., E.J. Krakiwsy, and B.G. Nickerson (1979). A manual for the establishment and assessment of horizontal survey networks in the maritime provinces. Technical report No. 56, Department of Surveying Engineering, University of New Brunswick, Fredericton, Canada.

Thomson, B.D., M.P. Mepham, and R.R. Steeves (1977). *The Stereographic Double Projection. Technical Report No. 46,* Dept. of Surveying Engineering, University of New Brunswick, Canada.

Torge, W. (1980). *Geodesy.* Walter de Gruyter, Berlin, New York, 254 pp.

Tsimis, E. (1973). "Critical configurations for range and range-difference satellite networks." Department of Geodetic Science and Surveying Report #191, The Ohio State University, Columbus, OH.

Vanicek, P. (1980): Tidal corrections to geodetic quantities. NOAA Technical Report NOS 83 NGS 14, U.S. Department of Commerce, Rockville, MD.

Vanicek, P. and M. Craymer (1986). "Satellite selection methods for differential GPS positioning," Appendix D in Janes H., K. Doucet, B. Roy, D.E. Wells, R. Langley, P. Vanicek, and M. Craymer (1986). "GPSNET: A program for the interactive design of geodetic networks." Geodetic Survey of Canada CR 86-003.

Vanicek, P. and E.J. Krakiwsky (1986). *Geodesy: The Concepts.* Second edition, North-Holland Publishing Company.

Vanicek, P., A. Kleusberg, D. Wells, and R. Langley (1986). "GPS survey design," in *Guide to GPS Positioning,* Chapter 14. Canadian GPS Associates. Printed in Fredericton, N.B., Canada by University of New Brunswick Graphic Services.

Vincenty, T. (1973). Three-Dimensional Adjustment of Geodetic Networks. Internal Report, DMAAC Geodetic Survey Squadron.

Wells, D.E. and E.J. Krakiwsky (1971). The method of least squares. Dept. of Surveying Engineering, University of New Brunswick, Fredericton, N.B., Canada.

Wells, D.E., N. Beck, D. Delikaraoglou, A. Kleusberg, E. Krakiwsky, G. Lachapelle, R. Langley, M. Nakiboglu, K. Schwarz, J. Tranquilla, and P. Vanicek (1986). *Guide to GPS Positioning*. Canadian GPS Associates, Fredericton, N.B., Canada.

Wells, D.E., W. Lindlohr, B. Schaffrin, and E. Grafarend (1987). GPS Design: Undifferenced Carrier Beat Phase Observations and the Fundamental Differencing Theorem. Department of Surveying Engineering Technical Report #116, University of New Brunswick, Fredericton, Canada.

Welsch, W. (1981). "Estimation of variances and covariances of geodetic observations." *Aust. J. Geod. Photo. Surv.,* No. 34, pp. 1–14.

Wempner, G. (1973). *Mechanics of Solids with Applications to Thin Bodies.* McGraw-Hill, New York, 623 pp.

Whitten, A. (1982). Monitoring Crustal Movement, Lecture Notes used in P.R. China.

Whittle, P. (1954). "On stationary processes in the plane." *Biometrika* 41 (1954), pp. 434–449.

Wilkins, F.J. (1989). Integration of a coordinating system with conventional metrology in the setting out of magnetic lenses of a nuclear accelerator, M.Sc.E. Thesis, Department of Surveying Engineering Technical Report No.

146, University of New Brunswick, Fredericton, New Brunswick, Canada, 163 pp.

Wimmer, H. (1981). Second order design of geodetic networks by an iterative approximation of a given criterion matrix. VI International Symposium on geodetic computations, Munich.

Wimmer, H. (1982a). "Ein Beitrag Zur Gewichtsoptimierung geodatischer Netze." Pul. *Deutsche Geodätische Kommission* c-269, Munchen.

Wimmer, H. (1982b). "Second Order Design of geodetic networks by an iterative approximation of a given criterion matrix." *Deutsche Geodätische Kommission,* Reihe B, Heft 258/III:112–127.

Wolf, H. (1961). "Zur Kritik von Schreibers Satz uber die Gewichtsverteilung in Basisnetzen." *Zeitschrift Für Vermessungswesen* 86 (1961), 177–179.

Appendix A

Computational Rules for Matrix Product and the "Vec" Operator

Rule (i): For any arbitrary k by j matrix **C** and n by m matrix **A** the *"Kronecker-Zehfuss product"* is defined as

$$\mathbf{C} \otimes \mathbf{A} := \left[c_{ij}\, \mathbf{A} \right] \tag{A.1}$$

yielding a matrix of size kn by jm.

Rule (ii): For any arbitrary k by m matrix $\mathbf{C} = [\mathbf{c}_1, ..., \mathbf{c}_m]$ and n by m matrix $\mathbf{A} = [\mathbf{a}_1, ..., \mathbf{a}_m]$ with the same number of columns the *"Khatri-Rao product"* is defined as

$$\mathbf{C}\, \Theta\, \mathbf{A} := \left[\mathbf{c}_1 \otimes \mathbf{a}_1, \; ..., \; \mathbf{c}_m \otimes \mathbf{a}_m \right] \tag{A.2}$$

yielding a matrix of size kn by m, with $\otimes$ being the *"Kronecker-Zehfuss product"* defined above.

Rule (iii): For any arbitrary k by m matrix $\mathbf{C} = [\mathbf{c}_1, ..., \mathbf{c}_m]$ "vec" denotes the operation which stacks one column of a matrix under the other, i.e.,

$$\text{vec}\,(\mathbf{C}) := \left[\mathbf{c}_1^T, \mathbf{c}_2^T, \; ..., \; \mathbf{c}_m^T \right]^T \tag{A.3}$$

yielding a vector of size km.

Appendix B
Problems of QP and LP

This appendix describes the problems of Quadratic Programming (QP), and Linear Programming (LP). For the details of the solution methods, some references are suggested.

B.1 The Problem QP

Generally, the standardized form of the problem of Quadratic Programming can be stated as:

Minimize

$$\frac{1}{2}\mathbf{x}^T \mathbf{H} \mathbf{x} + \mathbf{c}^T \mathbf{x} \tag{B.1}$$

Subject to

$$\mathbf{A} \mathbf{x} \leq \mathbf{b} \tag{B.2a}$$

$$\mathbf{x} \geq 0 \tag{B.2b}$$

where $\mathbf{x}$, $\mathbf{c}$ are n by 1 vectors, $\mathbf{H}$ is a symmetric n by n matrix, $\mathbf{b}$ a m by 1 vector; and $\mathbf{A}$ a m by n matrix.

When introducing a n+m by one vector of slack variables $\mathbf{v} = (v_1^2, v_2^2, \ldots, v_{m+n}^2) \geq 0$ as well as the vector of Lagrangian multipliers $\boldsymbol{\lambda}$, the Lagrangian function can be set up as:

$$L_1(\mathbf{x}, \mathbf{v}, \lambda) = \frac{1}{2}\mathbf{x}^T \mathbf{H} \mathbf{x} + \mathbf{c}^T \mathbf{x} + \lambda^T \left[\begin{pmatrix} \mathbf{A} \\ -\mathbf{I} \end{pmatrix} \mathbf{x} + \mathbf{v} - \begin{pmatrix} \mathbf{b} \\ 0 \end{pmatrix} \right] \tag{B.3}$$

with **I** the n by n identity matrix. The minimum value of L_1 (x, v, λ) can be obtained by setting the partial derivatives to zeros, i.e.,

$$\frac{\partial L_1}{\partial \mathbf{x}} = \mathbf{H}\,\mathbf{x} + \mathbf{c} + \left[\mathbf{A}^T \; -\mathbf{I}\right]\lambda = 0 \tag{B.4}$$

$$\frac{\partial L_1}{\partial \lambda} = \begin{bmatrix} \mathbf{A} \\ -\mathbf{I} \end{bmatrix} \mathbf{x} + \mathbf{v} - \begin{bmatrix} \mathbf{b} \\ 0 \end{bmatrix} = 0 \tag{B.5}$$

$$\frac{\partial L_1}{\partial v_i} = 2\,\lambda_i\,v_i \text{ for } i = 1, \ldots, m + n \tag{B.6}$$

Equation (B.6) may be reformulated as

$$\lambda^T\,\mathbf{v} = 0 \text{ with } \lambda \geq 0 \text{ and } \mathbf{v} \geq 0 \tag{B.7}$$

Equations (B.4)–(B.6) are the famous Kuhn-Tucker conditions for Quadratic Programming, which are necessary and efficient to get feasible solutions x for the problem QP. The problem QP has a unique minimizing solution when **H** is positive definite and the constraints (B.2) are feasible. Algorithm for the solution of a Quadratic Programming problem can be obtained, for instance, from Boot (1964).

B.2 The Problem LP

The standardized form of the problem of Linear Programming is written as Minimize

$$\mathbf{c}^T\,\mathbf{x} \tag{B.8}$$

Subject to

$$\mathbf{A}_1\,\mathbf{x} = \mathbf{b}_1 \tag{B.9a}$$

$$\mathbf{A}_2\,\mathbf{x} \leq \mathbf{b}_2 \tag{B.9b}$$

$$\mathbf{x} \geq 0 \tag{B.9c}$$

where **x** and **c** are the n by 1 vectors, $\mathbf{A}_1$ and $\mathbf{A}_2$ are m_1 by n and m_2 by n matrices, respectively, and $\mathbf{b}_1$ and $\mathbf{b}_2$ are the m_1 by 1 and m_2 by 1 vectors, respectively.

There are different ways to approach the solution of a Linear Programming problem, in which the Simplex Method due to G. Dantzig is first recommended. Details of the approach refer to G. Dantzig (1963).

Index